SECOND GEOLOGICAL SURVEY OF PENNSYLVANIA: 1875.

REPORT OF PROGRESS

IN THE

GREENE AND WASHINGTON DISTRICT

OF THE

BITUMINOUS COAL-FIELDS

OF

WESTERN PENNSYLVANIA,

BY

J. J. STEVENSON.

ILLUSTRATED

WITH 3 SECTIONS AND 2 COUNTY MAPS,

SHOWING THE CALCULATED LOCAL DEPTHS OF THE

PITTSBURG AND WAYNESBURG COAL BEDS

BENEATH THE SURFACE.

HARRISBURG:
PUBLISHED BY THE BOARD OF COMMISSIONERS
FOR THE SECOND GEOLOGICAL SURVEY.
1876

Stereotyped by
Singerly Printing and Publishing House,
Harrisburg, Pa.

Printed by
B. F. MEYERS, *State Printer*,
HARRISBURG, PA.

BOARD OF COMMISSIONERS.

His Excellency, JOHN F. HARTRANFT, *Governor*,
and *ex-officio* President of the Board, Harrisburg.

ARIO PARDEE, - - - - - -	Hazleton.
WILLIAM A. INGHAM, - - - -	Philadelphia.
HENRY S. ECKERT, - - - - -	Reading.
HENRY M'CORMICK, - - - - -	Harrisburg.
JAMES MACFARLANE, - - - - -	Towanda.
JOHN B. PEARSE, - - - - -	Philadelphia.
ROBERT V. WILSON, M. D., - - -	Clearfield.
Hon. DANIEL J. MORRELL, - - -	Johnstown.
HENRY W. OLIVER, - - - - -	Pittsburg.
SAMUEL Q. BROWN, - - - - -	Pleasantville.

SECRETARY OF THE BOARD

JOHN B. PEARSE, - - - - - - Philadelphia.

STATE GEOLOGIST

PETER LESLEY, - - - - - - Philadelphia.

UNIVERSITY OF THE CITY OF NEW YORK, N. Y.
May 1, 1876.

Prof. J. P. LESLEY,

State Geologist:

SIR:—Herewith I submit my report on work done in the Greene and Washington district during 1875.

Field-work was prosecuted from the beginning of May, until the close of November.

Throughout the season Mr. I. C. White was my aid. I cannot commend him too highly for the diligence and fidelity with which he labored. The frequent references to him in the report will attest, that he exhibited no small degree of skill and judgment in his work.

Very respectfully yours,

JOHN J. STEVENSON.

TABLE OF CONTENTS.

PART I.—INTRODUCTION.

	PAGE.
CHAPTER I. Physical features of the District, -	1
CHAPTER II. Surface geology, terraces, &c., -	11
PART II.—SYSTEMATIC GEOLOGY.	
CHAPTER III. The Stratified Rocks; Coal Measures,	23
CHAPTER IV. Folds in the Strata; anticlinals, -	25
CHAPTER V. Upper Barren Series, - - -	34
I. Greene County group, - - - - - -	35
II. Washington County group, - - -	44
CHAPTER VI. Upper Productive Coal Series, -	57
CHAPTER VII. Lower Barren Series, - - -	75
CHAPTER VIII. Lower Productive Coal Series, -	84
CHAPTER IX. Relations of the Strata, - - -	89
PART III.—GEOLOGY BY TOWNSHIPS.	
CHAPTER X. Greene County—	
1. Dunkard township, - - - - - -	90
2. Perry township, - - - - - - -	102
3. Wayne township, - - - -	106
4. Gilmore township, - - - - - -	109
5. Springhill township, - - - - - -	111
6. Monongahela township, - - - - -	115
7. Greene township, - - - - - -	122
8. Whitely township, - - - - - -	124
9. Cumberland township, - - - - -	125
10. Jefferson township, - - - - - -	135
11. Morgan township, - - - - - -	140
12. Washington township, - - - - -	145
13. Franklin township, - - - - - -	146
14. Centre township, - - - - - -	152
15. Morris township, - - - - - -	159

CHAPTER X. Greene County—*Continued.*
16. Jackson township, - - - - - - 161
17. Aleppo township, - - - - - - 163
18. Richhill township, - - - - - - 165
CHAPTER XI. Washington County—
19. E. Bethlehem, - - - - - - - 175
20. W. Bethlehem, - - - - - - - 181
21. Amwell, - - - - - - - - 185
22. Morris, - - - - - - - - 189
23. E. Finley, - - - - - - - - 192
24. W. Finley, - - - - - - - - 196
25. E. Pike Run, - - - - - - - 201
26. W. Pike Run, - - - - - - 205
27. Allen, - - - - - - - - 207
28. Carroll, - - - - - - - - 209
29. Fallowfield, - - - - - - - 213
30. Somerset, - - - - - - - - 216
31. Union, - - - - - - - - 220
32. Nottingham, - - - - - - - 223
33. Peters, - - - - - - - - 226
34. Cecil, - - - - - - - - - 229
35. Chartiers, - - - - - - - - 232
36. N. Strabane, - - - - - - - 235
37. S. Strabane, - - - - - - 239
38. Canton, - - - - - - - - 244
39. Franklin, - - - - - - - - 247
40. Buffalo, - - - - - - - - 253
41. Donegal, - - - - - - - - 261
42. Mount Pleasant, - - - - - - - 267
43. Robeson, - - - - - - - - 271
44. Hanover, - - - - - - - - 273
45. Smith, - - - - - - - - 275
46. Jefferson, - - - - - - - - 282
47. Cross Creek, - - - - - - - 284
(48.) Hopewell, - - - - - - - - 288
48. Independence, - - - - - - - 292
CHAPTER XII. Allegheny Co., (south)—
49. Jefferson, - - - - - - - - 296
50. Mifflin, - - - - - - - - 299

CHAPTER XII. Allegheny Co., (south)—*Continued.*
51. Snowden, - - - - - - - 303
52. Baldwin, - - - - - - - 305
53. Lower St. Clair, - - - - - - 307
54. Union, - - - - - - - - 310
55. Chartiers, - - - - - - - 311
56. Scott, - - - - - - - - 313
57. Upper St. Clair, - - - - - - 315
58. S. Fayette, - - - - - - - 317
59. N. Fayette, - - - - - - - 320
60. Robinson, - - - - - - - 323
61. Moon, - - - - - - - 329
62. Findlay, - - - - - - - 331
CHAPTER XIII. Beaver Co. (South)—
63. Hanover, - - - - - - - 335
64. Independence, - - - - - - 337
65. Hopewell, - - - - - - - 339
66. Moon, - - - - - - - - 341
67. Raccoon, - - - - - - - 344
68. (69.) Greene, - - - - - - 346

PART IV.—Tables of Depth.

CHAPTER XIV. Tables showing the *calculated* depth at which the Pittsburg and Waynesburg Coals lie at localities marked upon the maps, - - - - 350

In Washington County.

1. E. Bethlehem, - - - - - - 351
2. W. Bethlehem, - - - - - - 352
3. Amwell; 4. Morris, - - - - - 353
5. E. Finley, - - - - - - - 354
6. W. Finley, - - - - - - - 355
7. E. Pike Run; 8. W. Pike Run; 9. Allen, - 356
10. Fallowfield; 11. Carroll; 12. Somerset, - 357
13. S. Strabane, - - - - - - - 358
14. Franklin; 15. Buffalo, - - - - 359
16. Donegal; 17. Union, . - - - - 360
18. Nottingham; 19. Peters; 20. N. Strabane, - 361
21. Chartiers; 22. Canton. - - - - 362

23. Hopewell; 24. Independence, - - - - 363
25. Cecil; 26. Mt. Pleasant, - - - - - 364
27. Cross Creek; 28. Jefferson, - - - - - 365
28. Smith; 29. Hanover - - - - - - 366

In Greene County.

31. Dunkard; 32. Perry, - - - - - - 367
33. Wayne; 34. Gilmore, - - - - - - 368
34. Springhill; 35. Monongahela; 36. Greene, - 369
37. Whitely; 38. Cumberland, - - - - - 370
39. Jefferson; 40. Morgan, - - - - - - 371
41. Washington; 42. Morris; 43. Franklin, - 372
44. Center - - - - - - - - 373
45. Rich Hill; 46. Jackson, - - - - - 374
47. Aleppo - - - - - - - - - 375

PART V.—ECONOMIC GEOLOGY.

CHAPTER XV. Summary of resources—
Coal—Analyses of coal from different beds, - 376
Iron ore, - - - - - - - - - 883
Other metals, - - - - - - - 386
Limestone, - - - - - - - - - 387
Building stone, - - - - - - - - 388
Clays, - - - - - - - - - 389
Sand, - - - - - - - - 390
Petroleum, - - - - - - - - 390
INDEX, - - - - - - - - - - 399
ERRATA, - - - - - - - - - 420

REPORT OF PROGRESS, 1875.

GREENE AND WASHINGTON DISTRICT.

PART I.

SURFACE GEOLOGY.

CHAPTER I.

General Physical Features of the District.

The Greene and Washington district included Greene and Washington counties, and such portions of Allegheny and Beaver as lie south from the Ohio river. It occupies the extreme south-west corner of the State, and is bounded by the West Virginia line, and the Ohio and Monongahela rivers. The total area is not far from 1,800 square miles.

Hill and Valley Surface.

The surface is much diversified, and the character of the topography bears a close relation to the character of the rocks. In Greene county, along the river, the hills are usually abrupt; but above the bluffs, which seem so forbidding, there is a fine rolling country, whose slopes are gentle enough to admit of advantageous cultivation. Here the surface rocks for the most part belong to the Upper Coal series. As one recedes from the river, and comes into the region of the Upper Barrens, he finds a change. The broad valleys and glades so common near the river disappear, the valleys become narrower, while the hills are less frequently rounded, and are apt to assume the form of "hog-backs." Still in the northern and central portions of the county, where the upper barrens contain only limestones, shales and soft sandstones, there is a tendency to repeat the conditions observed along the river line; so that along Ruff's creek and Upper Ten-Mile, as well as on the main forks of the latter stream, the hills are often rounded, and the valleys are not quite so wide and fertile. But in the south and south-western portions of the county, where the prevailing rocks of the series are hard sandstones, the limestones and shales being almost wholly wanting, the character is altogether changed. Along Dunkard creek, above the line of Dunkard township, the valley is narrow, and the hills are steep; and this condition becomes more marked toward the head of the stream. Beyond that, on Fish creek, the valleys are almost mere gorges, while the enclosing hills are so abrupt that, in many places, to scale them is a matter of no slight difficulty. Yet here the general surface is by no means so harsh as one following the creek roads would be apt to suppose. Erosion has cut down many and deep valleys or gorges, and all this cutting seems like so much waste, for the valleys are rarely wide enough to admit of cultivation. But, above the walls of these ravines, there is a fine country with slopes not exceedingly steep, and with a general configuration resembling that of the best portions of the county. This is due clearly to the resistance of the upper sandstones, which have yielded little to the secondary erosion, so that although the valleys are narrow, and their walls cliff-like, yet this drawback is fully compensated by the larger amount of land available for farming purposes on the upland.

In Washington county the surface is rolling, and, except in the south-western portion, where the Upper Barrens become hard and sandy, while the limestones diminish, it is seldom rugged. The valleys made by the several streams are for the most part broad, and some of them are very beautiful. Those along Chartiers, the north fork of Ten-Mile, Raccoon and Buffalo creeks, are especially worthy of notice in this connection. Harshness of outline is almost wanting in this county, and in most of the townships the land can be cultivated without fear of much loss by washing of the soil.

In Allegheny county the surface is deeply cut, and the streams, as they approach the river, have worn their valleys so deeply, that the hills adjoining them appear quite as bluff, in many instances, as those bordering on the river itself. At the same time the valleys are very broad, and the hills, being composed of the yielding rocks of the Lower Barrens, have broken down so readily that, although they frequently rise from two to four hundred feet above the valleys, they show little harshness of outline. This condition changes somewhat in the north-western part near Beaver county, where the rocks of the Lower Barren become harder, and the slopes of the hills become steeper. Excepting in this portion, and that immediately facing the river, the hills are rarely too steep to admit of successful cultivation. In Beaver, the character of the surface is similar to that in Allegheny, except that, owing to the greater hardness of the lower barren rocks, the country is rather more abrupt.

Water Sheds.

The drainage system of this district is somewhat complex, owing to the peculiarity of shape. Topographically the narrow strip of West Virginia, known as the Ohio Panhandle, belongs here; for nearly all the streams flowing across it to the Ohio river take their rise in either Greene or Washington county. The area is, therefore, only a tongue, with the river on three sides. In Greene county there is little difficulty in tracing the divide between the streams flowing to the Monongahela and those flowing to the Ohio; but in Washington, where the streams flowing northward to the Ohio have their origin, the divides become quite complex, and the tributaries to the several streams dovetail in the most perplexing manner.

The principal streams entering the Monongahela river are Dunkard, in southern Greene; Ten-Mile, on the line between Greene and Washington; Pigeon, in Washington; and Peters, in Allegheny county. The first is a large creek rising in the highland of south-western Greene, and following an irregular course eastward, sometimes in Pennsylvania, at others in West Virginia. The fall of the bed is not great, as low dams are sufficient to throw the water back for miles. Ten-Mile is by far the most important stream entering the Monongahela from the west, and is formed by the union of the North and South forks at Clarksville, about three miles from the river. The South fork drains the greater part of Greene county, and is a bold rapid stream, beginning near Rogersville in Centre township, at the junction of Gray's and M'Courtney's forks. The latter is the more important, and rises in the highlands already referred to. In that region the surface is only partially cleared, so that water from rainfall does not evaporate or flow off as rapidly as in those sections where the clearing is more nearly complete; so that this stream always carries a very considerable body of water, and is the chief source of supply for South Ten-Mile. Gray's fork rises in the north-western part of the county, and carries much water, though much less than the other, as the country along its whole course is pretty well cleared. In Franklin township the creek is increased by the addition of Brown's fork above Waynesburg, which rises at the north near the Washington county line. Before reaching Clarksville it is farther increased by the accession of Ruff's creek and several other considerable streams from the north. The tributaries from the south, with the exception of Smith's creek entering near Waynesburg, are insignificant, as the drainage area on that side is very small.

The high dividing ridge between the North and South forks of Ten-Mile forms the boundary between Greene and Washington counties for a long distance, and the North fork, rising in Morris township, drains the most of the southern tier of townships. Though the area is much smaller than that of the other fork, yet this stream is large and carries little less water than the South fork. The other streams referred to, Pigeon and Peter's creeks, have comparatively small drainage areas, and are of slight importance, compared with those already described.

The chief streams entering the Ohio at the north side of the district are Chartiers and Raccoon creeks. The former rises in the south central portion of Washington county, and flows irregularly northward through Washington and Allegheny counties to the Ohio, which it reaches at a short distance below Allegheny City. The principal tributary in Washington is the North fork, which enters it about midway between Cannonsburg and the county line. In Allegheny it is increased by Robeson's run, along which the Panhandle railroad passes west from Mansfield. It is a strong and constant stream, draining a long but narrow area. Raccoon creek rises in the western part of Washington county and flows northward to the river through Washington and Beaver counties. In Washington county it has no important tributaries, and in Beaver the largest are Traverse and Service creeks. The drainage area is extensive and extremely irregular in outline. The creek carries much water and is rapid.

The streams entering the Ohio on the west side of the district are, for the most part, small. In northern Washington, Cross creek and Harman's creek (the latter being followed by the Panhandle railroad) are of some importance. Buffalo creek is larger than either of those, and drains the west central portion of Washington county. But the most important stream is Wheeling creek, which is formed just beyond the State line by the union of several large creeks rising within Greene and Washington counties. The Dunkard fork of Wheeling results from the union of two forks at Ryerson's station, in Greene county. Both of these rise in the highlands of south-western Greene, close to the head-waters of Ten-Mile and Dunkard. The drainage area includes Aleppo, and most of Jackson and Richhill townships. Hunter's fork, the other of the main tributaries, forms the western portion of the boundary line between Washington and Greene counties; the head-waters of the principal fork are close to those of the North fork of Ten-Mile, while the many tributary streams interlock with those forming Buffalo and Ten-Mile in Washington, and those leading into Gray's and Brown's forks of Ten-Mile in Greene. A small tributary to Wheeling creek rises near New Alexandria in Washington. Except near their sources, these streams have not a very rapid fall, though it is sufficient to make them useful as power when there is enough water. The last creek on this side is Fish creek, which rises in the south-west part of Greene county, near Wheeling, Ten-Mile and Dunkard, and flows across Springhill township into West Virginia, where it crosses the Baltimore and Ohio railroad at a short distance south from Belton. It is a very rapid stream, and in some places has a fall of nearly seventy-five feet per mile.

All the streams are liable to floods, but especially those which rise in the highlands of south-western Greene, as they fall rapidly for some distance from their sources. At such times they carry an enormous quantity of water, which, unfortunately is apt to disappear quite as suddenly as it appears.

Soil.

The soil varies in different portions of the district, according to the character of the rock from which it has been derived. Except in the immediate vicinity of the river, it is of local origin, so that even where poorest, it is very fair. In Greene county it becomes somewhat lean at the south-west, owing to the want of limestone in the upper rocks, which are the surface strata there, but in the rest of the county it is excellent, yielding good crops of grain and fine grass. The hills are usually steep enough to render loss of soil by washing common in the spring, when the ground has been freshly broken up by the plow. For this reason the chief agricultural employment is wool raising, which has proved very profitable. The nature of the soil, the remarkable frequency of excellent springs and the evenness of the climate fit this county admirably for the production of butter and cheese.

In Washington county the soil shows less variation than in Greene, owing to the remarkable development of the limestones of the upper barren series, as well as to the fact that the Great Limestone of the upper coal series is a surface rock over a large portion of the county. The greater regularity of the surface also aids to render this better adapted to general agricultural operations than is Greene. In both counties the sub-soil is thick and apparently inexhaustible, though in Washington it is clearly stronger. Grain is raised successfully in the latter county, but the chief business is that of raising sheep and cattle.

In Allegheny and Beaver counties the soil is hardly equal to that of Greene and Washington. The prevailing rocks are those of the Lower Barrens in which sandstones and shales predominate, while the limestones are very hard and do not yield readily to the weather, or form any considerable proportion of the soil. Still the soil is very far from being poor, and excellent crops are raised.

Productions.

Throughout this district the principal and most profitable employment is that of sheep-raising. The farmers claim that this is the only use to which their land can be put which will give a fair return for the capital invested. Yet, at the rate at which wool has ruled for two or three years back, this business has been carried on at a loss, and the outlook at present is not encouraging. I was informed by intelligent persons, well acquainted with the facts, that, during the past two years, the wool of Washington and Greene counties has been sold at an actual loss of from ten to fifteen cents per pound. The trouble is supposed to exist in the constant tinkering of the tariff on imported wools, and the consequent uncertainty in the market.

The wool from this region is by far the finest in the country, and it has long had a national reputation. Quite recently it was compared with selected samples of the finest Australian wools at the Agricultural Department, and proved to be vastly superior to them in quality. Within the past year a number of sheep were taken from Washington county to Australia for the purpose of improving the breed in that country. This wool, however, is not confined to Washington and Greene counties, for equally good wool is raised in the Ohio Panhandle of West Virginia, in Belmont, Harrison, Guernsey and Jefferson counties of Ohio, in Allegheny county of Pennsylvania, and in three or four counties of West Virginia, directly south from this district. It is worthy of notice that the limestones of the Upper Barrens, or the great limestones of the Upper Coals, occur in all these counties. Beyond the limits of these limestones the wool seems to be of inferior quality, though the breed of sheep is the same. Possibly this is only a coincidence, but if it be so it is a curious one in view of the fact that all wool-raisers regard the soil as having a marked influence on the character of the wool.

Agriculture.

The common method of farming in this district can hardly be called praiseworthy. Not enough is done toward keeping up the soil; deep ploughing is far from being the rule, and crops are extracted, year after year, from a thin layer on the surface which eventually is exhausted. The usual extremes of the wheat crop now are eight and twelve bushels per acre. In Washington the average yield is somewhat greater than in Greene. More attention is paid to keeping up the soil in the former than in the latter county, or perhaps a more accurate way of putting it would be to say, that there is less indifference respecting the matter, for even in Washington there seems to be little care about it. This method of farming must, in the end, prove ruinous, and its effects are seen even now in the smallness of the average crops. A farmer should understand that by putting a few dollars worth of fertilizers on each acre annually, he does not add so much to the cost of each acre. Increased crops repay the principal with usury. The unanimous testimony of judicious farmers is, that no investment made in their business yields so large and so prompt returns as that in fertilizers.

As alreauy intimated, the surface in this district is more or less steep and permits the water of rainfall to run off quickly, so that in cleared places the vegetation does not secure proper benefit. The rainfall is irregular, and farmers lose a considerable proportion of their crops by unseasonable dry weather. It seems to me that this might be avoided in great measure, and that too at no serious cost. Throughout the district the highest hills are usually provided with strong springs which flow constantly. Indeed, there are few regions more highly favored in this respect. Much of this water could be utilized for irrigation of the side hills; these springs could be turned into reservoirs, and the water could be applied to the fields at the proper season. The amount necessary for irrigation here is very small, as but one flooding would be required during any season. Most of the ditches need be only deep furrows made with the plow. The whole expense would be that of the first year, which, as already stated, would be very small. Many of the springs high up in the hills supply water sufficient for the irrigation of a hundred acres, even were there no rain during the whole season.

Avenues to Market.

The district is well provided with means of access to market. The Monongahela and Ohio together flow on all sides except the south, and afford an outlet to those portions bordering on them. Allegheny and northern Washington have excellent railroad facilities in the Panhandle road, which sends off a branch from Birmingham up the Monongahela river for nearly thirty miles, and another from Mansfield on the main stem, up Chartiers creek to Washington. The main line passes through Allegheny and north-western Washington. The Hempfield railroad, operated by the Baltimore and Ohio company, extends from Washington to Wheeling, West Va., and opens up the western part of Washington county. The extension of this road eastward from Washington to the river is partially graded, but the work of construction has been discontinued for the present. Greene county has been poorly provided with outlets. The Monongahela river on the east, sufficed for a large part of the region, but it is difficult of access for the interior townships. The Baltimore and Ohio railroad passes through West Virginia, very near the western boundary of Greene county. Both the river and the railroad are so far from the greater portion of the county, that they are of little service except for carrying such material as can be taken long distances in wagons without loss. During 1875, a narrow guage railroad from Waynesburg to Washington was begun, and the construction was pushed vigorously until winter put a stop to the work. If this road prove a pecuniary success, it will be extended east and west through Greene county, and will afford an outlet for the central portion. That part of Beaver county which is included within this district, has no outlet except the Ohio river.

Owing to the slack-water improvement, the Monongahela river is constantly navigable except for a short time during the winter, when it is choked with ice. In years of prolonged drouth, the upper pool, that nearest to the State line, occasionally becomes too low to permit the passage of steamboats, but the other pools are rarely impassable. The advantage of this improvement is almost incalculable. It has developed the country along the upper part of the river even more than could have been done by a railroad, as it affords cheap transportation for the coarser agricultural products, and it has given birth to the great coal enterprises which have been successfully carried on for many years. Without it the Monongahela would be no better than a broad creek, and of no value for transportation. This is its condition in West Virginia.

Unfortunately no such improvement exists on the Ohio, and

that great stream, which formerly was a constant outlet, is now useful only at time of flood, a period so uncertain, that no business calculations can be made respecting it. Vast quantities of coal accumulate at Pittsburg, having been sent down the Monongahela by the slack-water. At flood, this mass is hastily sent down the river. The effect of this condition is disaster to the coal trade. No control of the supply can be maintained, and owing to the quantity sent down at each flood, the market below is liable to be temporarily glutted, and prices often fall so as to be unremunerative to the producers. The iron trade is also injured, though not to the same extent, as iron can be shipped by rail. But were the river constantly navigable, ore could be brought much more cheaply from the west, and the iron could be shipped by water at much lower rates than by rail. Slacking the Ohio would be a costly work, but it will be necessary before many years in order to preserve the important interests along the upper portion of the river.

CHAPTER II.

Surface Geology.

Of the true or glacial drift, no direct traces occur within the district, except where, as along the Ohio river below Allegheny, and especially below the Beaver, fragments of it form part of the river terraces. The drift of this region belongs to a later time and is either local in its origin or it has been transported from the east and south by the rivers.

Terraces.

Terraces are usually well defined along the Ohio and Allegheny rivers and on several of their tributaries. Measurements were made at many localities to determine their relative heights above the streams, but sections showing their structure were obtained at only one locality. As this study was subsidiary to other matters, no connected investigation of the phenomena was made and the observations are sufficient only for a skeleton statement.

At the mouth of Dunkard creek, on the Monongahela, about two miles from the West Virginia line, the terraces are very indistinct; but at 180 or 190 feet above the stream, there is one which is quite well marked and can be traced for several miles up the creek, where at some localities it shows a curious conglomerate consisting of small pebbles and gravel, with some carbonaceous matter, the whole cemented by oxide of iron. At Greensboro', which is about six miles from the State line, the terraces are wholly destroyed in outline by erosion, but the rounded and polished fragments continue in sight to 275 feet above the river bed; thence up to 320 feet, the fragments are rare, but the soil is a river sand covered with red clay.

Below Greensboro', terraces are seen at 20, 180, 250 and 310 feet above the river bed. At the mouth of Whiteley creek, the second one is very distinct; and it can be traced up the creek to Mapletown, where the dark conglomerate is seen. Rolled and polished stones are of large size and frequent occurrence on this stream. At the mouth of Muddy creek, there is a fine terrace at 260 feet above the river, which is continuous along the creek to beyond Carmichaels, where it forms a handsome plain on which that village stands. The soil there, at somewhat more than three miles from the river, is sand mixed with some clay, and contains vast numbers of rolled and polished stones, which are exposed by the plow in the fields, and by cuttings in the roads leading to the village. The deposit of sand and gravel is from fifteen to thirty feet deep, the thickness having been ascertained from the statements of well diggers. A precisely similar condition prevails on Pumpkin run, which enters the river at Rice's landing, three or four miles below the mouth of Muddy creek. There the upper terrace on the river hills is at 300 feet above low water mark, and extends back for about two miles and a half, being crossed near its edge by the road leading from Carmichaels to Millsborough. The fragments are numerous, polished and vary in diameter from one inch to two feet; most of them are oval, and they had all been subjected to much wear before they were finally deposited.

At the mouth of Ten-Mile creek, the terraces are 20, 185 and 310 feet above the river. No pebbles of large size occur on the highest shelf, which is marked only by the sandy soil and the distinctly level surface. A still higher one is indicated by some knolls rising about 400 feet above low water, and covered with sand mixed with heavy ferruginous clay. The Carmichaels terrace is missing at the mouth of the creek; but it certainly did exist there, for at Clarksville on this stream, three miles from the river, rolled fragments mark the line at 250 or 260 feet above the stream. On the South fork of Ten-Mile this terrace is very distinct. At Jefferson there is a plain precisely similar to that at Carmichaels and on the same level. At Waynesburg is seen the peculiar blackened conglomerate already mentioned as occurring on Whitely and Dunkard. At Clinton, on the same stream, the conglomerate was found, but the terrace could not be traced from Waynesburg to that locality. The thickness of the detrital deposit varies from five to twenty-five feet.

On the North fork of Ten-Mile, the terraces cannot be traced so as to determine their relations to those on the river; but the evidence of successive halts is sufficiently clear as far up as three miles above the village of Ten-Mile in Washington county. The soil on these terraces is altogether local in its origin, and in no way resembles the material distributed by the river.

Below the mouth of Ten-Mile, the terraces seen at that locality remain in sight for several miles and on a little stream, entering just below Frederick, a fourth one is reached at two miles from the river, having an elevation of 410 feet. This is covered by sand and ferruginous clay, 15 feet thick and stratified, which contains a few rolled fragments. On both sides of the river at Belvernon, there is a handsome terrace whose top is now 180 feet above low water mark; but it has suffered by erosion to the extent of not less than ten feet. The sand in this deposit is of such excellent quality, that it is taken out extensively for use in the manufacture of glass. Opposite Belvernon two excavations exhibit the structure. In the upper one the detailed section is as follows:

1. Alternations of fire-clay, gravel and coarse sand, with fragments of varying size; the whole containing much carbonaceous matter in streaks, mostly broken coal; very ferruginous towards the base, - - - - - 12—16 feet.
2. Sand, fine and angular, excellent for manufacture of fine window and mirror glass; containing thin, irregular layers of blue plastic clay, with occasional layers of conglomerate cemented by oxide of iron; contains also numerous rounded fragments of rock, some of them being of large size; the pebbles are limestone, sandstone and conglomerate. The latter being much like the Great Conglomerate, 16—22 "
3. Coarse sand and gravel, with many small rounded fragments; much carbonaceous matter, coal and imperfect lignite; occasionally yields large fragments of trees. This is often a ferruginous conglomerate, - - - 2 "
4. Ferruginous sand, frequently conglomerate; contains some transported fragments of considerable size, - - - - - - 2—4 "

5. Blue plastic clay, - - - - - - -	0—4 feet.
6. Blue laminated shale of the Lower Barren series, - - - - - - - - -	—— "

The plastic clay, No. 5, is evidently derived from the underlying shale. The lower excavation was visited by Mr. White, who gives the following, as the section there:

1. Clay, containing rounded fragments, lumps of coal, etc., - - - - - - - -	10 feet.
2. Dark sand, used for moulding, - - - -	7 "
3. White sand, used for making glass, - - -	5 "
4. Stratum containing rounded fragments and many nodules of iron ore, - - - - - -	1 "
5. Dark sand, - - - - - - - -	4 "

The excavations are barely half a mile apart, so that it is clear that the deposit is subject to sudden variations. At the upper pit it is from 40 to 45 feet thick, while in the lower one it rarely exceeds 35 feet. The same terrace is on the opposite side of the river and is the plain on which Belvernon is built. The sand from that side is said not to be equal to that obtained from these excavations.

From this point the terraces are easily followed down the river, and at Monongahela City Mr. White made the following measurements:

Seventh terrace, - - -	480	feet above	river.
Sixth terrace, - - -	400	"	"
Fifth terrace, - - -	340	"	"
Fourth terrace, - - -	290	"	"
Third terrace, - - -	190	"	"
Second terrace, - - -	120	"	"
First terrace, - - -	40	"	"

Below this locality, to Pittsburg, the constant terraces are the third and fourth. The latter is a marked feature on both sides of the river, being seen at Peter's creek, Thompson's run, and opposite the Allegheny county poor-house. This seems to be the upper limit of large waterworn fragments along the lower portion of the river, and all the shelves above are covered only by sand and ferruginous clay. These higher terraces are fragmentary, as they have suffered severely from erosion.

At the mouth of Chartiers creek, below Pittsburg, there are six terraces, the highest being 390 feet above the river. A similar series occurs on that creek at the mouth of Miller's creek. A handsome terrace can be followed on Chartiers from Cook's station to Bridgeville. At the former place it is 25 feet above the creek, but at the latter it is 70 feet, a difference in fall of 45 feet; the total fall of the creek between the two points being about 180 feet. This terrace, or the one immediately above it, forms the broad shelf on which the Panhandle railroad runs near Grafton station The general structure of the country along the upper part of this stream shows that the terraces reached to near the headwaters, and ample evidence to this effect is afforded by the discovery of fragments of the Middle Washington Limestone near Cannonsburg, in Washington county, at about 200 feet above the present level of the creek. These came from the south, and have been rounded by friction, and not by simple weathering. It is quite true that fragments of limestone are not very satisfactory proof, for in weathering they frequently assume the form of water-worn fragments; but here there can be no doubt, since the hills in the vicinity of Cannonsburg, and northward, are all far too low to catch this limestone.

On Montour's run a similar terrace is seen, which is even more marked than that on Chartiers creek. It is traceable along the stream for several miles between Findlay and North Fayette townships of Allegheny county, and falls less rapidly than the bed of the run, the difference being about 35 feet in four miles. It shows many rolled and water-worn fragments, even so far from the river as Findlay township. Between Robinson and Moon townships of the same county it is much interrupted, and the attempt to determine its relation to the river series was not altogether successful. It seems to be connected with the one occupying the third place in the list already given.

Back of Middletown, on the Ohio, a little way below the mouth of Montour's run, the terraces are 30, 200, 350 and 410 feet above low water. The first is known as the "Second Bottom," and is very rarely reached by the river at high water. On it is the village of Middletown. The second has lost its outline almost wholly, and is recognized only by the rolled fragments which lead up to the rocky escarpment of the third. The top of the latter is reached at the church on the hill directly back of the village. It has been sadly cut up by erosion, but the remaining patches, though small, are numerous. A deep ravine separates it from the fourth, which is on the summit of the river hills. Standing on this, one sees a broad plain several miles wide, north and south, and extending back from the river, on both sides, to low hills which separate it from the higher land in the interior. On this highest terrace there are no polished or transported fragments, but the sands and clays of the soil bear ample testimony to its origin. On the third, rolled stones are of occasional occurrence, but are not numerous. The second is rich in such fragments, and as might naturally be expected, the lowest terrace is in some places almost paved with them.

Mr. White reports that in Beaver county, along the river line, there are three persistent terraces at 30, 60 to 80, and 100 to 120 feet above the low water mark. On the third of these Beaver and Georgetown are built, and Phillipstown is on the second. The highest line of water-worn fragments observed by him is 120 feet. Along Raccoon creek, which flows from Washington through Beaver to the Ohio river, the terraces are well-marked at many places. At New Sheffield there is one at 165, and another at 180 feet above the creek. The latter may be continuous with one that I saw on this stream in Washington county, which is very distinct where the Pittsburg and Steubenville pike crosses the creek. The second and third terraces of Beaver county were not observed along the Ohio river front of Allegheny county, though they may be and probably are present there. That line was not closely examined with a view to working out these relations.

Satisfactory exposures of the escarpments are by no means so numerous as one hastily passing down the Monongahela might suppose, but the few that were seen are sufficient to give a full conception of their nature. At Belvernon the detrital deposit is found from 30 to 40 feet thick, and resting on the shales of the Lower Barrens, which form the wall down to the lowest shelf, which is only 20 feet above the river, the intermediate terraces being absent there. A precisely similar condition exists opposite Braddock's Fields. At Middletown, on the Ohio, the escarpment of the third terrace shows a thin deposit of detritus resting on the Lower Barren rocks, which form the slope to the second. At every locality where an exposure occurred, it was seen that *the terrace is merely a shelf worn out of the stratified rocks on which the river has spread a thin layer of detritus.* From the bottom of this layer to the next shelf below there is a rocky escarpment, never one of detritus.

That the principal terraces do not indicate all the lines of halting during the elevation of the land, or the deepening of the water-ways, is quite evident, for the total number of benches is nine, of which only five can be traced. As recognized along the Monongahela, the series may be grouped as follows:

Fifth terrace, -	- 400 to 450	feet above river.
Fourth terrace, -	- 310 " 320	" "
Third terrace, -	- 250 " 280	" "
Second terrace, -	- 180 " 200	" "
First terrace, -	- 20 " 30	" "

The highest terrace rarely shows any water-worn fragments of considerable size, and such as do occur have been transported for only short distances. The fourth terrace shows a similar condition, except that the transported fragments are larger and more numerous. Yet at many places this horizon seems to be entirely destitute of these fragments. At each of these levels the river sand is found in abundance, and is usually covered by a stiff ferruginous clay, containing more or less of the sand. The effect of this on the soil has long been noticed by many of the more observing farmers; and I have frequently been asked to explain how it occurs that, in Washington county the river townships have along their eastern border a rather poor sandy soil, while in the interior the soil is a rich loam; and this too, when the great limestones which render the interior townships so fertile are equally well exposed along the river. The explanation is quite simple. The soil along the river border is derived from the coarse sandy material of the mountain region, whereas that of the interior is of local origin, having been produced by disintegration of the rocks found there in place.

On the third terrace rounded fragments become numerous. They have been referred to already as occurring at Greensboro', as well as along Muddy creek and Ten-Mile. They are found at many other localities farther down the river. This was evidently one of the most marked of all the halts; for connected with it are the terraces extending to Carmichael's on Muddy creek, and to Jefferson and Waynesburg on Ten-Mile. It is persistent down the river, being the one seen opposite Braddock's Fields and on Thompson's run. And it is evidently associated with one at the mouth of Chartier, on the Ohio.

The second terrace marks an equally important horizon. It can be traced down the river without difficulty from the mouth of Dunkard creek to below the mouth of the Allegheny. With it are connected the main terraces of Dunkard, Peters' creek, Chartiers', and most probably those of Montour's run and Raccoon creek. The sections showing its character, as exposed at Belvernon, exhibit the variable structure of the detrital deposit. The transported fragments found at this level are much larger than those seen at any either above or below it. At Belvernon one mass of conglomerate is 4 by 2 by 1 foot. And there are many others two-thirds as large. At the mouth of Ten-Mile, as well as at many numerous other localities along the upper part of the river, fragments weighing from 150 to 300 pounds are of very frequent occurrence. Farther down, as might be expected, the fragments become smaller, and below Pittsburg there are more of them, for the material from the Allegheny is there added. Along the Monongahela the fragments are, in great measure, derived from the Mahoning, the Great Conglomerate and the sandstone, which those of us, who have worked in the southern portion of the coal-field, have been accustomed to call Tionesta.* It is probable that the greater part of these came down Cheat river from the mountains, though some of the Coal Measure rocks may easily have come down the Tygart's Valley fork of the Monongahela, or have been brought by the streams of southern Fayette county in this State. There is a total absence of metamorphic rocks, though some pieces of the Tionesta are much like quartzite. Below Pittsburg the rocks represented in the terraces are very different,

* This is the Piedmont Sandstone of Maryland.—[J. P. L.]

for there one finds not only Upper and Lower Carboniferous, but also the drift materials, granites, gneiss and schists, together with jaspery conglomerates. The latter are usually small, and seldom exceed six inches in diameter. This commingling extends, at Middletown, to more than 200 feet above the river. At the mouth of the Beaver the composition of the terrace material changes very greatly. Above that the finer gravel predominates, but at the mouth of that river the coarser material at once preponderates, and the drift fragments are more numerous than those of rocks found in place along the rivers. This condition continues along the Ohio at least as far as Wheeling, the material meanwhile growing coarser. Whether the great gravel terrace at and above Wheeling is the same with that at the mouth of the Beaver, I am unable to say, as the line has never been studied out.

Ancient River Beds.

In Ohio, it has been ascertained by actual borings, that the Ohio and other rivers, once flowed at a much lower level than now, and that the old valleys are filled to a considerable extent with debris. At Steubenville the Ohio flows on a bed of gravel not far from 250 feet thick; and at Belleair, below Wheeling, the bottom of the gravel was not reached in sinking for the piers of the Great Bridge at that place. Information of this kind is utterly wanting along the Upper Monongahela, for no borings have been made on the lower terrace. Petty changes in the course of the river have occurred at different times, such as the one at Belvernon, which took place immediately after the date of the second terrace there. But these are utterly insignificant. There is no evidence in the structure of the whole country that the river ever flowed along a line lying west from its present course, for the old tributary channels of Dunkard, now occupied by Dunkard, Whitely, Muddy, Ten-Mile and other important streams are separated by low ridges of hills which always did separate them. What the condition on the east side of the river may be, I am not prepared to say, but a glance over the whole region shows no evidence of any depression through which the stream might have flowed at any previous time. At all events it is quite clear that the river never flowed along its present line above the mouth of the Youghiogheny river, at a lower level than it does now. Along the river from Fairmount, in West Virginia, to Elizabeth, the stratified rocks are frequently seen in the bed of the stream at low water, though the exposures are most common above the mouth of Ten-Mile creek.

Date of Erosion.

The relations of the terraces on the tributaries to those along the river afford some means of determining the time when the erosion was begun, which has given so varied a character to the topography of this district. The whole of this erosion was produced by running water. Its extent may be imagined when one sees the high points like Turkey knob and the cluster surrounding it in Greene county, Krepp's knob near Brownsville, and several almost equally high hills south-east from Washington, all of which rise from 100 to 250 feet above the neighboring hills, and contain outliers of strata, which originally overspread the whole region, and therefore reached to still greater elevations on the crests of the anticlinals; for these knobs are in no case near their crests or axes. Looking at these high points, projecting so far above the surrounding country, it may seem to the casual observer, almost ridiculous to assert that in all probability the whole region was covreed with water at a comparatively recent period. Yet the thick coating of debris, the regular arrangement of what can be no other than shingle, which is constantly seen on even the highest places, lead directly to the conclusion that the highest terrace marks only a stage in the withdrawal of the water which had previously covered the whole surface.

During this first stage, no doubt, much of the erosion to which is due the absence of the higher rocks over a large portion of the country was performed; but by far the greater part of the earlier erosion dates to the time when the river rested at the second level—that between 180 and 200 feet. When the river had passed this, the valleys of Dunkard, Ten-Mile, Chartier's Montour's and Raccoon had all been outlined, and the hills which now border them and are crowned by rounded summits had somewhat of their present shape. Since that time the erosive process has been confined to widening the valleys or rounding off the summits of the hills, all of which had been formed or shaped before.

If these terraces belong to the terrace epoch, as they undoubtedly do, then we must regard all this erosion as subsequent to the Champlain period.* Each terrace is a shelf with a thin layer of detritus resting on the stratified rocks, which form by far the greater portion of the escarpment. This certainly appears to be the case as far as my present information goes; but it is quite possible that the work to be done during the coming season in Fayette, Westmoreland and the northern part of Allegheny, may disclose facts leading to a different conclusion. The evidence of the filled channel of the Ohio seems to conflict with the conclusion just reached, for it appears hardly possible that the Ohio could have flowed at 200 feet below its present level, on the west side of the district, without inducing extensive erosion, on that side, anterior to the Champlain. Respecting the phenomena there I have no information, as the Ohio is at some distance west from the Pennsylvania line, and time could not be afforded during the past season for making any investigations outside of the district aside from those which had a direct bearing on the economic geology. At the same time, all the evidence along the Monongahela clearly points to the conclusion that the valleys there are of comparatively recent origin. The terraces of the tributary streams show a regular slope from the head-waters to the corresponding terraces

* Unless we prefer to consider the erosion of the present terraces to have been a rehandling of the side-walls of valleys which had been previously excavated to nearly their present size. The subject is surrounded with difficulties of both a mechanical and a geological kind. It is granted on all hands that since the close of the coal era no Middle Secondary or Tertiary deposits have been made upon our Coal Fields; for even the most comprehensive erosion could hardly fail to spare some scattered relics of such formations had any existed. The Coal Measures must therefore have remained through Middle Secondary and Tertiary ages exposed to surface erosion; and we can point to no consequence of such erosion except our valleys. But valleys eroded through horizontal deposits of alternate sandstone, shale, &c. must necessarily be terraced at all stages of their excavation, and irrespective of any changes in the amount of flowing water or general rate of erosion. To conclude that our valleys, as valleys, are of recent creation we must premise that no rain fell from the beginning of the Permian to the close of the Pleiocene age.—[J. P. L.]

along the river, which is precisely analogous to the slope in the present beds of those streams, though as would be expected, the fall in the old channels is less abrupt than in those now used.*

Quarternary Fossils.

Animal remains belonging to this epoch are very rare. In Mt. Pleasant township, Washington county, the tooth of a mastodon was found, which is now in the collection of Washington and Jefferson college. It was obtained on the high dividing ridge between Raccoon and Chartier's creeks. No other remains are known to have been found within the district; but within the past year, numerous fragments of teeth and bones of mastodon were discovered in the river bank, at the junction of the Allegheny and Monongahela rivers, which are now in the collection of the Pittsburg high school. Bits of wood, usually more or less carbonized, occur occasionally in the sands at Belvernon and other localities, where the terrace sands have been worked, but no good specimens have come under my observation.

* There is a confessed difficulty in explaining the slope of terrace lines.—On the supposition of a submerged continent, emerging by stages, terraces should have no slope, but be absolutely horizontal. Also on the supposition of glacial lakes, terraces should be horizontal, like those of Glen Roy in Scotland. On the supposition of periodical eras of extra flood-waters in a diminishing series, no conceivable volume of water would suffice for an operation of more than a few days at any of the higher levels; and the power of descending floods is so great that vast cubical blocks of the lower Coal Measure sandstones and conglomerates would have been stranded along all the valley sides at high flood mark. On the supposition that the terraces represent side relics of the valley bed in successive ages, a date must be assigned for the uppermost terrace far back in Middle Secondary times, and dates for the other terraces less and less remote. Any supposition that all the terraces belong to one geological age, and that one the most recent, involves the idea of a re-filling and re-excavating of the valleys in connection with glacial phenomena, to discuss which intelligently we have yet too little information.—[J. P. L.]

PART II.

SYSTEMATIC GEOLOGY OF THE DISTRICT.

CHAPTER III.

The Stratified Rocks.

The stratified rocks found in this district belong, in all cases, to the Coal Measures. In oil-borings the Conglomerate and the Lower Carboniferous have been reached, but these in no instance come to the surface.

Classification.

In the description of the Coal Measures, the most convenient classification is that adopted by the Professors Rogers and their assistants on the surveys of Pennsylvania and Virginia. It is applicable throughout the whole bituminous trough west from the Allegheny mountains.

In this district the *Lower Productive Coal Series,* from the Conglomerate up to the Mahoning Sandstone, occurs only in the north-western portion along the Ohio river and its tributaries in Beaver county. As exposed here, it contains the *Upper* and *Lower Freeport* and *Kittanning Coals;* lower beds not coming to the surface. Resting on this is the *Lower Barren Series* which, reaching to the base of the *Pittsburg Coal*, is partially exposed along the Monongahela and Upper Ohio rivers, as well as at various localities in Washington and Allegheny counties, but is seen in its full extent only in Beaver. It contains several thin coals and some uncertain deposits of lean iron ore, but its economical interest is very small. The *Upper Coal Productive Series* from the *Pittsburg Coal* to the *Waynesburg Sandstone* inclusive, is finely exposed in a large part of the district. It barely crosses into Beaver county, where only small outliers of it are found on top of the highest hills. In this series are five very persistent beds of coal, and an enormous mass of limestone. Each of the coals is workable over a greater or less area. The last division is the *Upper Barren Series*, which includes all the rocks above the *Waynesburg Sandstone*. It contains a number of variable coals, only one of which ever attains to any considerable importance. The limestones are numerous, and in some localities are of extraordinary thickness. This series covers the greater portion of Washington and Greene counties, and occasional outliers of its lower members are seen in Allegheny county. In its greatest development it is found only in Greene.

The total thickness of strata in Greene county, including all from the highest in the *Upper Barren Series*, to the lowest pierced in the oil-borings on Dunkard creek, is not far from 2,800 feet. This fact, ascertained only with great labor, and the exercise of the utmost patience in tying together fragmentary sections, induces me to entertain much respect for the ingenuity and skill of M'Kinley, who, forty years ago, announced that the pile of rocks in Greene county is nearly 3,000 feet thick. Those who are aware of the difficulty encountered in the attempt to work out the Upper Barrens even now, when the country is comparatively speaking thickly settled, and roads form a close network over the surface, cannot fail to feel a certain degree of astonishment at the accuracy of M'Kinley's statement. That was made at a time when the region was in great part unimproved, when roads were few, and when geologists were without the convenient instruments now regarded as indispensable. The result was the more remarkable, as all attempts to obtain detailed sections to a distance of more than 500 feet above the *Waynesburg Coal* had proved unsuccessful.

CHAPTER IV.

FOLDS IN THE STRATA.

Five Principal Anticlinal Axes.

The plication of the rocks within the district is much greater than was suspected, and five strong axes have been traced. With one exception these become more abrupt northward, so that in that direction there is an apparent rise of the whole series. Their general trend is north north-east; but they are not altogether parallel, nor do they follow straight lines. Followed southward nearly all of them are found to shift their axes occasionally, the sidethrow being toward the east. In this report provisional names have been applied to these for cenvenience of description, but they are only provisional, as in all probability these axes will be found identical with others previously described from the more northern districts of the State. Owing to the want of accurately levelled lines much difficulty was met in the endeavor to trace out these curves. The only dependence was the barometer, and this frequently indicated a change in the direction of the dip when no change had occurred; the rate of dip being usually so small that a very slight alteration in atmospheric conditions is often sufficient to destroy the value of the results.

Fayette County Anticlinal.

The anticlinal axis which makes so strong a mark across Fayette county, and crosses the Kiskiminetas below East Liberty, barely escapes the south-eastern corner of the district, and is probably the same with that which passes through Morgantown in West Virginia, at a few miles south from the State line.

Waynesburg Anticlinal.

The first well defined fold at the south-east is the Waynesburg, so called because it passes near that borough, in Greene county. It enters Washington, on the Monongahela river, about a mile and a half below Belvernon, passes through Allen township half a mile west from Speer's mill, and through West Bethlehem township near Centreville, where it brings up the *Pittsburg Coal.* It crosses the North fork of Ten-Mile creek near Hawkins' mill, and enters Morgan township in Greene county. In this township it crosses Ruff's creek about a mile below Martinsburg, and enters Franklin township, where it crosses the South fork of Ten-Mile below Dodysburg, and Smith creek near the brick school house. If this course were maintained the axis should cross Dunkard creek near the western limit of Wayne township, but no axis is found there. In the interval exposures are so rare and unsatisfactory that tracing is altogether out of the question. It is probable that near the southern boundary of Franklin township it is thrown off to the east, for a well marked anticlinal was found at Blacksville on Dunkard, near the eastern border of Wayne township. This is the place where the synclinal east from this axis should have crossed the creek. It is quite evident that the fold disappears soon after entering West Virginia.

Lisbon Synclinal.

The synclinal east from this axis enters Greene county near the mouth of Muddy creek in Cumberland township. It passes nearly a mile south-east from Carmichaels; then, through Greene township, and crosses Whiteley creek near the eastern border of Whiteley township; there it is evidently thrown eastward, for it reaches Dunkard creek at a little way east from the border of Wayne township, not far below Blacksville. The precise line of this axis is ascertained with some difficulty, as along it there occur some subordinate anticlinals, very narrow but quite sharp, one of which is distinctly marked on Glade run near Carmichaels.

The extremes of the anticlinal axes already mentioned are well exhibited on the Monongahela river by means of the *Pittsburg Coal.* Thus, at the mouth of Cheat river, near the crest of the Fayette county axis, that coal is 370 feet above low water mark; ten miles below Greensboro', at the bottom of the synclinal, it is 185 feet below that mark; while below Belvernon on the crest of the Waynesburg axis, it is 180 feet above the river. The flattening of the Waynesburg axis is nicely shown by comparison of the levels at Belvernon and Waynesburg, each place being very near the crest of the anticlinal. Below Belvernon the *Pittsburg* is seen at an elevation of about 900 feet above tide-water; whereas, at a short distance below Waynesburg, on Ten-Mile creek, the *Waynesburg* is barely 940 feet above tide level. The interval between these coals in this neighborhood is not far from 360 feet, so that the flattening southward from Belvernon to Waynesburg is not far from 300 feet. From the summit of the Fayette axis to the bottom of the trough there is an average dip of 110 feet per mile, while from the bottom of the trough to the crest of the Waynesburg axis the average is barely 70 feet, in an east south-east direction.

Waynesburg Synclinal.

The west slope of the Waynesburg axis is short and not very abrupt, being barely four miles long, and having a dip of little more than fifty feet per mile. The synclinal, like that east from the same axis, occasionally exhibits a subordinate anticlinal. Thus, at Monongahela City there is quite a sharp fold, and the coal above the city is at the same elevation as at Houston's station about two miles below. The synclinal may be regarded as entering the district along the line of the subordinate axis at Monongahela City; whence, crossing through Carroll, Fallowfield and Somerset townships, where it is closely followed by the channel of Pigeon creek, it passes a little east from Bentleysville and enters West Bethlehem, where it crosses the North fork of Ten-Mile creek near Zollarsville. Beyond that locality it becomes very shallow and is traced with difficulty. In Greene county it crosses Ruff's creek near the eastern line of Washington township, Ten-Mile near the mouth of Brown's Fork, and Dunkard not far from Dent post office in Wayne township. It is so shallow at each of these places that its existence was ascertained only by means of leveled lines.

Pin-hook Anticlinal.

The course of the second anticlinal in Washington and the interior of Allegheny county, has been compiled chiefly from Mr. White's observations, and I apply to it the name by which it is designated in his notes. He terms it the Pin-hook Axis, from a locality in Amwell township of the former county. It enters this district just below Braddock's station in Mifflin township of Allegheny county, and follows a straight line passing near the head of Pine run into Jefferson township; thence, through the north-western corner of that township it was traced into Snowden, where it crosses both Lick run and Piney fork at about two miles above the junction of those streams. It enters Washington county at the north-west corner of Union township, and passes through the south-east corner of Peters' township near Thomas' saw-mill; thence it describes almost a straight line to the southern portion of Nottingham township, where it crosses Mingo creek near Mr. Leyda's residence. As it enters Somerset township it seems to be suddenly thrown eastward, for it is seen crossing the North fork of Pigeon creek near school house No. 7, fully one mile out of its course. No other deflection occurs in this county. The crest crosses the South fork at Pigeon creek near Vanceville and the National road in West Bethlehem, about four and one-half miles west from Hillsborough. In Amwell township it passes north from Pin-hook, or Pleasant Valley, and directly through the village of Amity, reaching Ten-Mile creek on the line of Morris township. As it enters Greene county the course is again thrown eastward, and the axis appears in Washington township of that county as an obscure anticlinal, crossing Ruff's creek above the store; in Franklin, crossing Brown's fork near Rees' mill; in Centre, Gray's fork at Clinton; and it is probably the insignificant fold observed just west from Jolleytown on Dunkard creek.

Want of absolute levels prevents me from giving any definite statement respecting the rate at which this anticlinal increases in height northward. And for the same reason it has been found impossible to make fully satisfactory comparisons of the river sections with those obtained in the interior. But an approximation can be made at several localities. The flattening southward, taking the coals as the basis, is very great. At Pin-hook the *Waynesburg* is 20 feet above the stream, while at the crossing of Ten-Mile, the *Washington* is at the same distance above the creek. The latter locality is about 60 feet lower than the former, and the interval between the coals is about 140 feet, so that the flattening, in a distance of less than seven miles, is not far from 200 feet. Taking the coals as a base, the height of the wave in Greene county is about 25 feet on Dunkard, and 40 feet on Gray's fork; in Washington county it is 80 feet on Ten-Mile, 260 feet at Pin-hook, 300 feet at Peters' creek; and in Allegheny county 340 feet at Braddock's station.

Nineveh Synclinal.

The synclinal north-west from this axis is traceable only with extreme difficulty in its northern extension. It crosses the Monongahela river near Six-mile ferry, but thence into Washington county its line cannot be made out with any degree of certainty. It enters North Strabane township, Washington county, at its north-eastern corner, and passes out of it near Clokeysville into South Strabane township. It crosses the National road a little east from the village of Martinsburg, and enters Morris township near the point where Amwell, Morris and Franklin join. In Morris it passes almost directly through Prosperity. Here it is thrown eastward to coincide with the deflection of the Pin-hook anticlinal, and enters Greene county about two miles east from its true line. In that county it crosses Brown's fork of Ten-Mile about half a mile north from Ninevah; Gray's fork about a mile east from Graysville; the Dunkard fork of South Wheeling at the south-east corner of Richhill township; the Aleppo fork of the same stream near the line of Aleppo and Springhill township; and passes into West Virginia at the extreme south-west corner of the State. From North Strabane township, Washington county, this trough is very distinct and is traceable without difficulty.

Peter's Creek Anticlinal..

South-east from the Pin-hook axis there is another of equal height, which barely enters the district. Measurements along the river show that at the mouth of Peter's creek the *Pittsburg Coal* is 350 feet above low water mark, and that to the north and south the rocks are dipping away from this place. This axis can be traced from the mouth of Peter's creek, through Jefferson township, quite to the line of Washington county, but no farther. In that county, the Pin-hook axis overrides it. The trough between this and the Pin-hook, crosses the river near the northern boundary of Jefferson township, Allegheny county and reaches the Washington county line near the mouth of Lick run.

Washington Anticlinal.

The third important anticlinal is the Washington, so named because it passes within a short distance west from that borough in Washington county. It is exceedingly important and is well marked in Allegheny, Washington and Greene counties. It enters the district near Temperanceville on the Ohio river; passes a little south-east from Bowerhill station in Scott township, Allegheny county; crosses the Chartiers railroad near Hasting's station in South Fayette township of the same county, and touches the same railroad near Ewingsville and Cannonsburg in Washington county. It crosses the Hempfield railroad at Chartiers creek, about a mile west from Washington; thence southwardly it is found passing through the south-east corner of Buffalo township, and through East Finley near the village of that name. It enters Greene county a little above the mouth of Owens' run in Richhill township, and crosses South Wheeling creek barely half a mile below the mouth of Crabapple. Beyond that it soon passes into West Virginia. At the Ohio river this axis lifts the *Pittsburg Coal* to about 400 feet above low water mark. At the extreme south-western point where it was seen the axis is still strong, and the estimated height of the wave is nearly 200 feet. It brings up the *Pittsburg Coal* on Chartiers creek in Washington county, and the *Waynesburg* on South Wheeling in Greene, as well as on Hunter's fork between the two counties. Its economical importance therefore is very great.

Nineveh Synclinal.

The chief source of difficulty in tracing the synclinal between the Washington and Pin-hook anticlinals is its boat shape. Followed northward it rapidly grows shallow, until in Allegheny county the barometer is of little service in the attempt to trace it. Indeed, if the barometer can be trusted, there are numerous slight folds in this portion of the trough, and I am much inclined to believe that at no great distance north from the Ohio river, in Allegheny county, these two axes will be found united, the resulting fold occupying the line of the trough in this district.

Mansfield Synclinal.

The north-western slope of the Washington axis is short, barely three miles. The synclinal crosses the Pan-handle railroad at Mansfield in Allegheny county, and becomes very indistinct where it reaches the river near Nimick's station. In Washington county it crosses Miller's run in Cecil township below Venice; the West fork of Chartiers creek in Chartiers township at Connellsville; the Hempfield railroad about one mile west from the eastern line of Buffalo township; Robinson's fork in West Finley about three miles below the village of Good Intent; and thence soon passes into West Virginia. It is exceedingly shallow everywhere, and along the West fork of Chartiers is not sufficient to carry the *Pittsburg Coal* under.

Claysville Anticlinal.

At a short distance farther west there is another anticlinal, which I have named the Claysville. It cannot be traced through Allegheny county, but seems to be the same with an obscure axis observed on the Ohio river at Wood's run below Allegheny City. In Washington county, it crosses the West fork of Chartiers creek just above the line of Mt. Pleasant township; Brush run at a short distance west from the line of Canton township; Buffalo creek about half a mile below Taylorstown, and the Hempfield railroad near Claysville. Beyond the railroad it could not be identified. The trough west from it crosses Buffalo creek about a mile from the West Virginia line, and can be traced northward to the northern line of Hopewell township, where it is seen passing near the village of West Middleton; beyond that it becomes obscure.

I have said that this anticlinal and its trough cannot be traced through Allegheny county. A more accurate statement would have been that they become exceedingly indistinct, owing to the fact that they lose their individuality and are merged into the Bulger anticlinal, at the west. The crest of the Claysville axis crosses the Pan-handle railroad near Hays' station, and continues northward, passing near the village of Remington on the Pittsburg and Steubenville road, and thence to the river which it reaches in the vicinity of Wood's run. But the synclinal has practically disappeared, and in its stead there is a flattening which continues for several miles.

Bulger Anticlinal.

At Bulger station on the Panhandle railroad there is a well-marked anticlinal, the same with that already referred to as overriding the Claysville axis in Allegheny county. Southward from the railroad it rapidly disappears, and at a distance of ten miles it is wanting. Were this axis persistent southward, it should be cut by Cross creek near Patterson's mill, but there, and for the whole distance to the West Virginia line, the rocks are dipping to the south-east; and they continue to do so until the point is reached where the West Middletown synclinal crosses the creek. Northward, in Washington county, it is difficult to make out this axis, and in Allegheny it does not exist.

It is however sufficiently evident from the observations of Mr. White in Beaver county that this, like the other axes, has been shifted eastward in its southern course, for he finds there a gentle fold extending from the Ohio river to the Washington county line. The shifting is about three miles, which is rather more than in the case of any of the other axes. In Beaver, the Bulger anticlinal enters the county near the mouth of Logstown run on the Ohio river; crosses Raccoon creek near Independence; Big Traverse creek not far from its mouth, and reaches Washington county about three miles east from Frankfort. Throughout it is very obscure.

Burgettstown Synclinal.

The trough west from this is extremely shallow, and is said by Mr. White to enter Beaver near Hog Island on the Ohio, to cross Raccoon below the mouth of Tramp run, Big Traverse near Keifer's mill, and to reach the Washington county at Frankfort. It is very obscure in Beaver; and in Washington cannot be traced near the Beaver line. Like the axis, it is evidently shifted to the east, and is well-marked at the railroad, which it crosses at Burgettstown, four miles west from Bulger. This trough disappears southward, and cannot be found in the southern part of Smith township, nor in Cross Creek township.

Five Main Synclinals.

There are then in this district five troughs:

The first, between the Fayette and Waynesburg axes.

The second, between the Waynesburg and Pin-hook axes.

The third, between the Pin-hook and Washington axes.

The fourth, between the Washington and Claysville axes, and

The fifth, lying north-west from the Claysville axis.

The fifth includes the Bulger axis which lies but little away from the line of the synclinal. It is not unlikely that the north-western boundary of this trough is the anticlinal which crosses the Ohio at Brown's island; but it has not been recognized within the district. Of these troughs the first and fifth are wide but the others are quite narrow.

CHAPTER V.

The Upper Barren Series.

Though it is by no means improbable that, in the high ridge forming part of Monongalia, Marion and Wetzel counties in West Virginia, there may be found rocks belonging to a higher horizon in this series than any thus far observed in Pennsylvania, yet it is quite certain that the series as a whole is much better exhibited in this district than in any portion of West Virginia, since we find here a greater number of limestones; whereas in West Virginia the prevailing rocks are sandstones; limestones being almost wanting, except in the southern portion of the Ohio Panhandle.

As has been already stated, the surface rocks in the larger part of Washington and Greene counties belong to the Upper Barren series. Of these the sandstones and shales, and frequently even the limestones, yield so readily to atmospheric agents that the unaffected portions become deeply buried under debris, and the geologist is not merely often, but almost constantly perplexed to determine his horizon. In Greene county, where the series shows its greatest development, the extreme poverty of exposures is not the only obstacle. It was soon ascertained that, owing to excessive variation in the character of the rocks, the section obtained in northern Greene was of little service in the southern portion of the county; while in Washington the difference from the Greene county section was so remarkable that at first it seemed impossible to reconcile the evident contradictions. Careful tracing, however, showed that these contradictions were only apparent and not real, and that the difficulty arose in part from the fact that the strata pay no heed to parallelism, and in part from great changes in structure of the rocks themselves.

Two Groups.

The mass of rock in this series is so great that, for convenience, I have thought well to divide it into two groups, of which the line of separation naturally falls at the great limestone termed in this report the "Upper Washington." The lower portion, which may be designated the *Washington County Group*, extends from the top of the Waynesburg Sandstone to that limestone, and varies in thickness from about 450 feet in the southern portion of the district, to little more than 150 feet at the north. Though thicker in Greene, it is vastly more important in Washington. The upper portion of the series, which may be known as the *Greene County Group*, includes all the rocks above the Upper Washington Limestone, and has an extreme thickness of not far from 800 feet. It is satisfactorily exposed only in Greene; for though it has a considerable thickness in Washington, yet the horizons in the two counties cannot always be identified with certainty.

I. Greene County Group.

The section which I have used as typical of this group, is that obtained in Centre township of Greene county. In it the limestones are better shown than in any of the other sections, and their relations are well exhibited. It is as follows:

	Rocks.		Thickness in feet.
1.	Concealed		80′
2.	Limestone	(XIV)	fragments.
3.	Reddish shale		80′
4.	Limestone	(XIII)	4′
5.	Sandstone		50′
6.	Limestone	(XII)	10′
7.	Sandstone and shale		80′
8.	Limestone	(XI)	2′ 6″
9.	Argillaceous shale		12′
10.	Sandstone		30′
11.	*Nineveh Coal Bed*		1′ 8″
12.	Sandstone		36′
13.	Bituminous shale		1′
14.	Limestone	(X)	2′ 6″
15.	Sandstone, shaly, massive, (Fish Creek)		100′
16.	*Dunkard Coal Bed*		1′ 6″
17.	Limestone	(IX *b*.)	3′
18.	Sandstone and shale		30′
19.	Limestone	(IX *a*.)	6′—15′
20.	Sandy shale		70′
21.	Limestone	(VIII)	2′—5′
22.	*Coal, Local Bed*		1′ 8″
23.	Sandstone		19′—30′
24.	Limestone	(VII)	2′ 6″
25.	Sandstone		31′
26.	Shale and iron ore		10′
27.	Upper Washington limestone	(VI)	

In the extreme south-western portion of Greene county, the difference from the typical section is so great that the succession must be given in detail. It is well shown on Fish creek, as follows:

Rocks.		Thickness in feet.
1. Concealed		80′
2. Limestone	(XIV)	4′
3. Shale		25′
4. Bituminous shale		2′
5. Shale		30′
6. Sandstone (Gilmore)		30′
7. Shale		20′
8. Limestone	(XIII?)	1′
9. Sandstone and shale		255′
10. *Nineveh coal bed*		2′ 6″
11. Sandstone and shale		30′
12. Limestone	(X)	6′
13. Sandstone (Fish Creek)		40′
14. *Coal bed*		blossom.
15. Shale or sandstone		45′
16. Limestone		1′
17. Sandstone		20′
18. Ferruginous shale		10′
19. *Dunkard coal bed*		10″
20. Shale		3′
21. Sandstone		10′
22. Shale, mostly argillaceous		50′
23. *Coal bed*		1′
24. Shale and sandstone		45′
25. *Coal bed*		8″
26. Upper Washington limestone?		—

The exposures in Pennsylvania end with No. 22, and the rest of the section was obtained in West Virginia. The top of the section is found in the high dividing ridge between the waters of Fish, Dunkard, Ten-Mile and South Wheeling creeks, which forms part of Gilmore, Springhill, Jackson and Aleppo townships. On the Dunkard side of the divide, the section is the same down to the *Dunkard Coal*, but below that the difference is quite marked, as may be seen by the following section obtained at a little west from Jolleytown:

Rocks.		Thickness in feet.
1. *Dunkard Coal Bed*		1′ 3″
2. Limestone	(IX *b.*)	2′
3. Shale		25′
4. Limestone	(IX *a.*)	1′ 6″
5. Shale and sandstone		35′
6. *Coal Bed*		1′
7. Upper Washington limestone		—

Descending from the highlands north-eastwardly along M'-Courtney's fork of Ten-Mile the typical section is obtained; north-westwardly along Dunkard fork of South Wheeling, a section nearly allied to it is seen; whereas on the Aleppo fork of that stream heading in the same vicinity, a wholly different section appears. These two sections will be found in the descriptions of Richhill and Aleppo townships of Greene county.

The lower strata of this group are found in all the south-western townships of Washington county, but the higher rocks occur only in East and West Finley. There the sections are to be obtained on the ridges, but the localities are so separated and the intervals vary so widely, that it is almost impossible to make out the section. By dovetailing the several fragmentary sections secured in these townships, the following has been made out as probably quite near the condition:

	Rocks.		Thickness in feet.
1.	Limestone	(—)	6′
2.	Red shale		50′
3.	Limestone	(—)	6′
4.	Sandstone		50′
5.	Limestone	(X)	6′
6.	Concealed		50′
7.	Limestone	(—)	5′
8.	Sandstone and shale		40′
9.	*Coal Bed*		1′ 6″
10.	Shale		5′
11.	Limestone	(VII?)	2′
12.	Sandstone		20′
13.	Limestone	(VI)	—

Diminution in Thickness Northward.

In this series, the intervals between the more important strata, show a constant diminution in thickness northward from the southern boundary of the district. For this reason, although a much greater thickness of rocks is exposed in south-western Greene and along Dunkard creek, yet I feel assured that the type section embraces strata belonging to as high a horizon in the series as any represented in the sections obtained on either Dunkard or Fish creek. For the same reason it is quite possible that, although we may find a much thicker mass of rock in this series in West Virginia, the Pennsylvania section may still reach a higher horizon.

It has been found impossible to reach the top of the group anywhere; the nearest approach being that made by Mr. White, at the head of Aleppo fork of South Wheeling creek, where the concealed portion is 60 feet. Elsewhere the concealed interval varies from 80 to 300 feet.

Limestone XIV.

Limestone XIV is seen only on the ridge in northern Centre township, separating Gray's and Brown's forks of Ten-Mile, and on the dividing ridge in the south-western part of Greene county. It cannot be identified in Washington county, on the ridge in Centre township. A line of limestone fragments occurs a little way east from Hopewell church, which is so persistent that I have no hesitation in placing it in the section, though its thickness cannot be ascertained, nor can the position of the stratum be defined within several feet. It is there 80 feet above limestone XIII. On the Dunkard and Fish creek sides of the dividing ridge in the south-west this limestone is concealed, but it is satisfactorily exposed at the head of the Aleppo fork of South Wheeling. It is a dark blue somewhat earthy rock, and contains minute crystals of blende with occasional fish scales. It was traced along this dividing ridge to the head of M'Courtney's fork, where its relations to the Nineveh coal leave no room to doubt that it is the same with the limestone XIV of the typical section.

Gilmore Sandstone.

In the south-west corner of the district one finds on the crests of the higher hills in Springhill, Gilmore, Aleppo and Jackson townships a coarse massive sandstone, the No. 6 of the Fish Creek section. In northern Greene, on the Hopewell Church ridge, it is represented by dull red shales; and very possibly some similar shales, seen in East and West Finley township of Washington county, but not given in the section, may belong to this horizon. This no doubt is the massive sandstone referred to in the reports of the former survey as capping the whole series in the south-west; but in those reports its thickness is greatly overestimated, since at no place has it been found more than 30 feet. As exposed there it is a massive rock, forming cliffs, which are best seen at the head of Dunkard creek in Gilmore township. It exhibits much cross-bedding and is of very uneven texture, so that it weathers into great cavities. One of these observed by Mr. White in Aleppo township is almost rectangular in form, and is 40 by 20 by 15 feet in dimensions. It is regarded by the inhabitants as the work of the Aborigines. In Gilmore township these cavities are numerous, frequently large and, in some instances, apparently communicating. They are the infallible refuge of foxes from the hunters.

Limestone XIII.

Limestone XIII occurs on the ridge in northern Centre, and can be followed from nearly a mile east from Hopewell church to almost within sight of Graysville in Richhill township. It is about four feet thick, and consists almost equally of shale and limestone. On the divide in south-western Greene county it was found at the head of Dunkard creek in digging a well, and at the head of M'Courtney's fork; it is not far from 20 feet below the great sandstone of that region, and seems to be about one foot thick. No exposure of it was seen anywhere in that portion of the county, and nothing is known of its character farther than that it is a dull, blue and earthy rock. Near it is a *plant-bearing shale* which was struck in the same well. Specimens of this were seen, but were not preserved for us, though the person having them agreed to do so. This shale, like the limestone, was not seen at any exposure.

Limestone XII.

In Morris and Centre townships of Greene county a heavy sandstone is found under limestone XIII. It is well exposed near the blacksmith's shop, east from Hopewell church, and at the head-waters of Brown's fork of Ten-Mile creek. Immediately below it comes limestone XII, which is well shown on the dividing ridge between Gray's and Brown's forks, as well as on Gray's, Brown's and Brushy forks of Ten-Mile. Near Gray's fork, it is seen in the road to Hopewell church, which leaves the creek at the Baptist church, about two miles below Graysville, and is again seen on this road about two miles east from Hopewell church. On Brown's fork, it comes down to the stream about half a mile above Nineveh and remains in sight for nearly two miles. At this distance a full section of it is exposed in a bluff, showing it to be a mass of limestone and shale fifteen feet thick. With this I identify the limestone at the head of Robinson's fork, in East Finley township, Washington county, which is eight feet thick and 230 feet above the upper Washington. The interval here is about 300 feet. This rock is also found on the high ridge in West Finley, west from Robinson's fork, on South Wheeling creek, in Jackson and Richhill townships, Greene county.

Local Limestone XIb.—Limestone XIa.

A limestone occurs in the interval between limestone XII, and the one which there I hesitatingly identify with XI*a*. It occurs at thirty feet above the latter and seems to be entirely local, as it has not been observed elsewhere. Above it at fifty feet is a thin coal, but beyond that nothing could be ascertained respecting its relations to the overlying rocks, as all the higher strata are utterly concealed. In Greene county limestone XI*a* was seen in Centre, Morris and Richhill townships. It is exposed imperfectly on the Ridge road between Gray's and Brown's forks, at nearly two miles below Graysville, but is best shown in the bluff on Brown's fork opposite Nineveh. I have identified it with the seventh limestone of the section obtained on South Wheeling creek in Richhill township. But I am by no means certain that the identification is correct. This stratum may be the same with that at the head of Robinson's fork in Washington county, which is No. 3 of the Washington county section.

In south-eastern Greene, on Dunkard, Fish creek, M'Courtney's fork and the Aleppo fork of South Wheeling, a coal is found about 360 feet below limestone XIV. On M'Courtney's fork the connection is easily traced to the type section of northern Greene and the coal is found to be the *Nineveh Coal*. In the interval from limestone XIII to the coal at the south-west, the exposures are not continuous at any locality, but it is believed that the whole section has been seen and that it contains nothing except sandstones and sandy shales except on South Wheeling creek. The *Nineveh Coal* is very persistent, naving been observed in Gilmore, Springhill, Jackson, Aleppo, Centre and Morris townships of Greene county; but it was not found in Washington county. In Gilmore it is mined by John Taylor, at the head of Dunkard and is fourteen inches thick. In Springhill, it is seen at several localities above New Freeport, where it is about one foot. In Jackson it is not more than six inches, but was once mined by stripping at a little distance above White Cottage. On Gray's fork of Ten-Mile, it is imperfectly exposed above Rutan post-office and at a mile below Graysville. Below Nineveh, on Brown's fork, it is from eighteen to twenty inches thick and is mined by stripping. Though very thin it is of no slight importance in these localities, where coal mines are from ten to fifteen miles away.

Limestone X.

Limestone X is seen at many localities in western Greene, and is no doubt the limestone found at the top of the section in Jefferson township in that county. On Fish creek it is exposed at several places, and is always brecciated and associated with a black shale above it. In Gilmore, it occurs near the head of Dunkard, where it is about four feet thick, and its overlylng shale contains many remains of plants and occasional scales of fish. In Aleppo, it is five feet thick, dark blue and earthy, and underlies carbonaceous shale which yields leaves and fish remains. In Richhill, it is seen near the township line on South Wheeling, and shows the characters just mentioned. It occurs also on Gray's fork, at the Baptist chuch below Graysville, and at several places below Nineveh, on Brown's fork of Ten-Mile. At forty feet below it, a *thin coal* was seen in Aleppo and Springhill townships, but this was not found elsewhere. In Washington county, this limestone has been identified at the head of Hunter's fork of South Wheeling creek, and at several other localities in East and West Finley townships. It shows the same characteristics as in Greene county.

Dunkard Coal Bed.

The *Dunkard Coal*, though very thin, is of much importance in south-west Greene county where coal is almost wanting. Northward it thins out and disappears before reaching the Washington county line. It is usually double and, except in

Springhill township, rests almost directly upon limestone IX*b*. In Richhill township it is concealed or wanting, since it could not be found there. In Gilmore township, it is mined by Mr. Garrison on the North fork of Dunkard creek, where it shows 11 inches of coal, separated nearly midway by three inches of shale. Immediately above it are three feet of clay shale, containing many impressions of plants, of which specimens were collected by Mr. White. In Jackson township the bed is thicker, and shows—

Coal, 8–12 inches; Clay, 1 inch; Coal, 5–8 inches; total, 14–21 inches.

It is mined by stripping on several of the little tributaries entering M'Courtney's fork below White Cottage in Jackson township, and is in high repute as an excellent coal. In Springhill township it is worked at several places, and is the bed from which White's mill is supplied. It is very thin, seldom more than ten inches Limestone IX*b* is absent in this township. In Centre township it was formerly mined along Gray's fork at a number of places below Rutan post office, and varies from 18 to 20 inches. Northward from the line of this stream it does not occur, being absent on Brown's fork of Ten-Mile, and on Hunter's fork of South Wheeling creek.

Fish Creek Sandstone.

The interval between Limestone X and the coal is occupied by sandstone. At varying distances from the latter there is a massive layer ten to forty feet thick, which forms fine bluffs along Gray's and Brown's forks of Ten-Mile. On Fish creek this sandstone is a striking feature for miles. Other portions of the mass occasionally show a tendency to become compact, but this rarely occurs to any considerable extent except in Springhill township where the sandstones throughout tend to be massive, so that the Fish creek region is famed for its *fine building stone.* In Washington county a very compact sandstone is found on top of the high knob at Hillsborough, which may belong somewhere in this interval.

Limestone IXb—IXa.

Limestone IX*a*, like Limestone IX*b* and the *Dunkard Coal* is of limited extent, and seems to be confined to the southern portion of the trough between the Washington and Pin-hook axes. It has not been positively identified east from the latter axis and, except in Gilmore township, disappears before reaching it. If we follow the line of the synclinal between these axes we find the interval between Limestones X and VI diminishing, so that before reaching the Washington county line it is little greater than the interval between X and IX on M'-Courtney's fork in Jackson township, and meanwhile Limestones IX*a* and IX*b*, with the *Dunkard Coal*, have disappeared. Not one of these can be found in Washington county. Limestone IX*a* is seen on Dunkard creek above Jolleytown, eighteen inches thick; on M'Courtney's fork, below White Cottage, 6 to 8 feet; on Gray's fork, at the mouth of Scott's run, and on Clover run, 6 to 10 feet; but on Brown's fork and Ruff's creek it is absent. On Dunkard fork of South Wheeling it is seen near Slatt's mill. The rock is somewhat variable in character. On Gray's fork it is slightly ferruginous, but contains only a small proportion of earthy matter; whereas followed southward it becomes more earthy and lighter in color, until on Dunkard creek it is extremely impure, and on Fish creek it has entirely disappeared. The relations of this stratum will be discussed in another connection. The rocks below the horizon of this limestone are well exposed in northern Greene and in a considerable part of Washington county.

Limestones VIII and VII.

Limestone VIII seems not to be persistent, having been found only on Ten-Mile creek in Centre township, Greene connty. A limestone holding this relative position occurs in Richhill township, but I am not sure that it is the same stratum. The coal associated with it was seen only in eastern Centre township, and is purely local. Limestone VII is much more persistent, having been seen in Franklin, Morris, Centre and other townships of Greene county, while in Washington, a thick limestone is almost constantly found at from fifteen to twenty feet above the Upper Washington Limestone. In Greene county, this is usually blue and earthy, becoming more and more earthy southward, until it disappears along the southern line of the district; but in Washington it is quite pure, and at times is of very considerable thickness. In that county it has a thin coal above it in East and West Finley townships.

II. Washington County Group.

This group, as already stated, finds its chief expansion in Washington county, though it attains to greater thickness in Greene. It is exposed to a greater or less extent in every western township of the latter county, and in more than two-thirds of the former; only petty outliers of the lower portion are found in Allegheny county. The group is more than 400 feet thick on Dunkard creek, and thins out northward to not more than 140 or 150 feet in north-western Washington. It is difficult therefore to give an exact section, but the following may be taken as approximately representing the *extreme development* of the group in Washington county:

General Washington County Section.

Rocks.	Thickness in ft.
1. Coal or shale	1′
2. Upper Washington Limestone........(VI)	30′
3. Sandstone	40′
4. *Jolleytown Coal Bed*	1
5. Sandstone	40′
6. Middle Washington limestone........(IV)	15
7. Sandstones and shales	60′
8. Limestone........(III)	8′
9. Sandstone and shale	20′
10. Bituminous shale or *coal Bed*	1′
11. Lower Washington limestone........(II)	20′
12. *Washington Coal Bed*	10′
13. Laminated sandstone	12′
14. *Little Washington Coal Bed*	1′
15. Shale	6′
16. Limestone........(I*b*)	20′
17. *Waynesburg "B" Coal Bed*	1′
18. Sandstone	30′
19. Limestone........(I*a*)	8′
20. *Waynesburg "A" Coal Bed*	2′
21. Waynesburg sandstone	——

General Greene County Section.

In Greene county the section is very different. The limestones are much thinner and some of those given in the section are wanting. The series there is approximately as follows:

	Rocks.	Thickness in ft.
1.	Dark shale	3′
2.	Upper Washington limestone........(VI)	3
3.	Shaly sandstone	20′
4.	Limestone........(V)	5′
5.	Shale and sandstone	30′—45′
6.	*Jolleytown Coal Bed*	1′
7.	Shale and sandstone	180′
8.	Limestone........(III)	3
9.	Sandstone	18′
10.	Black shale	3′
11.	Lower Washington Limestone........(II)	3′ 6″
12.	Shale	6′
13.	*Washington Coal Bed*	2′ 4″
14.	Clay and laminated sandstone	18′
15.	*Little Washington Coal Bed*, with shale	7′
16.	Sandstone	20′
17.	*Waynesburg "B" Coal Bed*	1′
18.	Sandstone and shale	18′
19.	Limestone and shale........(I)	13′
20.	*Waynesburg "A" Coal Bed*	1′ 6″
21.	Variegated shale	10′
22.	Waynesburg sandstone	——

General Section in West Virginia.

Followed southward beyond the limits of the district into West Virginia, this group changes in character; the limestones wholly disappear, and on the Parkersburg branch of the Baltimore and Ohio railroad, it is represented only by shaly sandstones which extend upwards for nearly 400 feet above the Washington coal. In northern Ohio, that is north from the Central Ohio railroad, the greater part of this group is not seen, and the highest rock exposed is probably the Middle Washington Limestone, which occurs on the railroad summit about twenty-two miles west from the Ohio river.

White Upper Washington Limestone VI.

Within this district the Upper Washington Limestone is a fine stratum of such marked characteristics, that it cannot be mistaken for any other in the series. In all portions it weathers to an *almost snowy whiteness* with the slighest tinge of blue. The upper part is quite slaty and is blue on the freshly exposed surface. The middle layers are dark, almost black, and frequently mottled with drab. They are exceedingly brittle, ring sharply when struck, and yield a limestone of superior quality.

The brittleness of this portion and its ability to withstand the weather, fit it admirably for use on roads, and it is extensively employed in macadamizing the National Road. The lower part is ordinarily of a light flesh color, and in point of purity is scarcely inferior to the middle portions. It is less brittle and yields more readily to the weather. The top and bottom divisions are persistent, but the *middle or dark portion* disappears soon after entering Greene county.

The greatest thickness of this rock is seen in Washington county on Cemetery Hill, near the borough of Washington, where it is a mass of shale and limestone, 30 feet 30 inches thick, sub-divided as follows:

Limestone laminated argillaceous	2 feet.	
Dark shale	5 "	
Calcareous shale	6 "	
Shale with vegetable markings	2 "	
Limestone		10 inches.
Bituminous shale		10 "
Limestone	2 "	
Calcareous shale	1 "	3 "
Limestone	1 "	6 "
Shale		10 "
Limestone	3 "	
Shale	2 "	
Limestone	3 "	

At the tunnel on the Hempfield extension, about one mile east from Washington, this bed shows nearly 20 feet of solid limestone.

In Washington county it is well exposed in Smith, Mt. Pleasant, Cross Creek, Donegal, Buffalo, Canton, Cecil, Franklin (North and South), Strabane, Somerset, Amwell, Bethlehem, Morris, and East and West Finley townships, and is occasionsionally seen in Nottingham and Peters. Its thickness varies from six to thirty feet, being greatest in the central portions of the county, and the interval between it and the *Washington Coal* below, varies from 100 feet on the State road in southern Smith, to 180 feet on Ten-Mile creek in Amwell township.

In Greene county it is readily identifiable everywhere except in the Fish Creek region. On Dunkard it is quite impure and bears little resemblance to itself as it ordinarily appears. It has been recognized in Morgan, Washington, Morris, Richhill, Aleppo, Centre, Franklin, Greene, Jefferson, Cumberland, Perry,

Wayne and Gilmore townships. Its thickness is from four to eight feet, and the interval between it and the *Washington Coal* varies from 325 feet on the Dunkard, to 260 on Ten Mile near Waynesburg, 190 on Ruff's creek, in eastern Washington, and 135 on Hunter's fork of South Wheeling creek in Richhill. No difficulty is encountered in following this rock from its northwestern exposure in Smith township, Washington county, to Whitely creek in Greene township, Greene county.

Fish Bed.

Directly *overlying* this limestone there is a dark shale, more or less bituminous, often sufficiently so to show a cannel-like fracture. Under such circumstances it contains vast numbers of bivalve crustaceans and fish scales, all well preserved, and in some instances it is crowded with long slender leaves, which seem to belong to *Sigillaria Menardi.* On Dunkard creek, about a mile above Jolleytown in Gilmore township, the bituminous matter is collected so as to form a *thin bed of coal.* In Washington county, there is frequently a little coal at a few feet above the limestone, and the shales are without much carbonaceous matter. When such is the case, the shales are wonderfully rich in impressions of leaves and stems, all of which seem to belong to the same *Sigillaria.* The best locality of this kind is at the tunnel, east from Washington. It is worthy of note, that while *Calamites* and *Sigillaria Menardi* are so plentiful here, frequent examinations have been rewarded by the discovery of but a single fern leaflet, and that an imperfect specimen of *Neuropteris.* In Greene county, ferruginous shale occurs above the carbonaceous shale, but except within a very small area in Centre township, the ore is so distributed as to be utterly unavailable.

Crustacean Bed.

The middle or dark portion of Limestone VI contains great numbers of little bivalve crustaceans, which are well preserved. In most cases the tests are destroyed in breaking the rock, and on the weathered surface the specimens are rarely in satisfactory condition. These crustaceans are common to nearly all of the limestones in the Upper Barrens. The lower layer of the middle portion of Limestone VI shows comminuted fragments of molluscan shells. This layer is exceedingly fetid when struck.

Limestone V.

In Greene county, at from twenty to thirty feet below Limestone VI, there occurs another, V, which is exceedingly persistent throughout the county, and is doubtfully recognized in the south-western townships of Washington county. It is hard and coarsely brecciated, weathering to dull-gray, slightly tinged with yellow. Owing to its compactness, its very peculiar appearance, and its ability to withstand the weather, it is an exceedingly important guide. It can be well seen at many places along Ten-Mile creek, above Waynesburg. It comes down to the creek near the line of Centre township, and at Rogersville is only a few feet above the bridge-level. For the most part it seems to be absent in Washington county, and the identifications in West Finley are at best very doubtful.

Jolleytown Coal Bed.

At from twenty to seventy-five feet below the Upper Washington Limestone there is a very persistent little coal, which Mr. White, in his notes on Gilmore township of Greene county, has named the *Jolleytown Coal.* It is of some value along Dunkard creek, in that vicinity where coal is not within easy reach. In that region it is about twenty inches thick, and is of very fair quality. References to it will be found in the descriptions of Perry, Wayne, Gilmore, Whitely, Franklin, Centre and Richhill townships of Greene county. In Washington county it is exceedingly variable though very persistent, and was observed at nearly every locality where its horizon was exposed.

Middle Washington Limestone IV.

The Middle Washington Limestone, IV, is a massive rock, which occurs in all townships of Washington county where Limestone VI is seen. In Jefferson township it is shown on the high land near the eastern border of the township, along the Burgettstown and Eldersville road. This is its most northwestern exposure. No satisfactory identification of it has been made in Allegheny county. If it occur there, it can be only in Snowden and Upper St. Clair townships, and there its horizon

SECOND GEOLOGICAL SURVEY OF PA.

GREEN COUNTY GROUP.

UPPER BARREN MEASURES, B.

SEE PAGE 35. K.

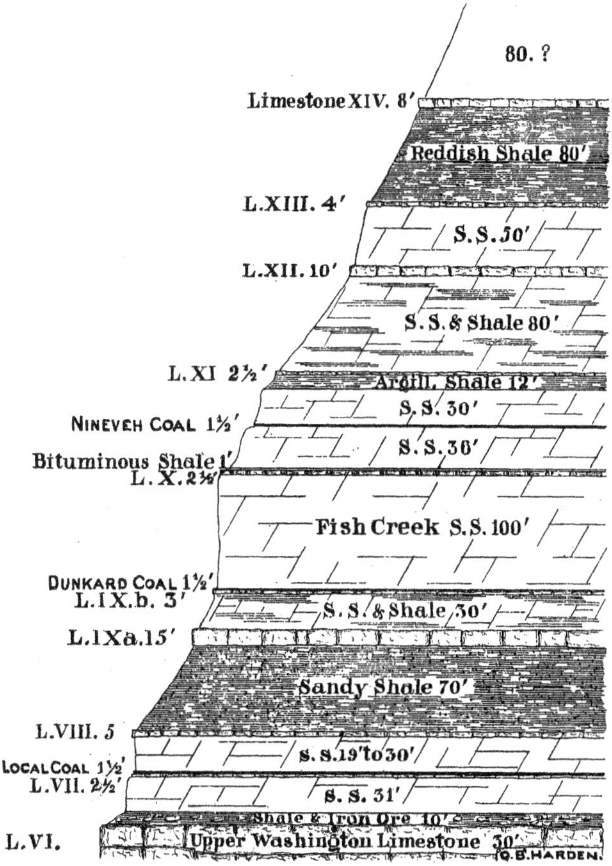

80. ?

Limestone XIV. 8′

Reddish Shale 80′

L.XIII. 4′

S.S. 50′

L.XII. 10′

S.S. & Shale 80′

L.XI 2½′

Argill. Shale 12′

S.S. 30′

NINEVEH COAL 1½′

S.S. 36′

Bituminous Shale 1′

L.X. 2½′

Fish Creek S.S. 100′

DUNKARD COAL 1½′

L.IX.b. 3′

S.S. & Shale 30′

L.IXa. 15′

Sandy Shale 70′

L.VIII. 5

S.S. 19′ to 30′

LOCAL COAL 1½′

L.VII. 2½′

S.S. 31′

Shale & Iron Ore 10′

L.VI.

Upper Washington Limestone 30′

O.B. HARDEN.

PHOTO-ZINCOGRAPH, by F. A. WENDEROTH & CO., 1328 Chestnut Street, Philada.

SECOND GEOLOGICAL SURVEY OF PA.

GREEN COUNTY GROUP

UPPER BARREN MEASURES B.

Section on Fish Creek

SEE PAGE 36. K.

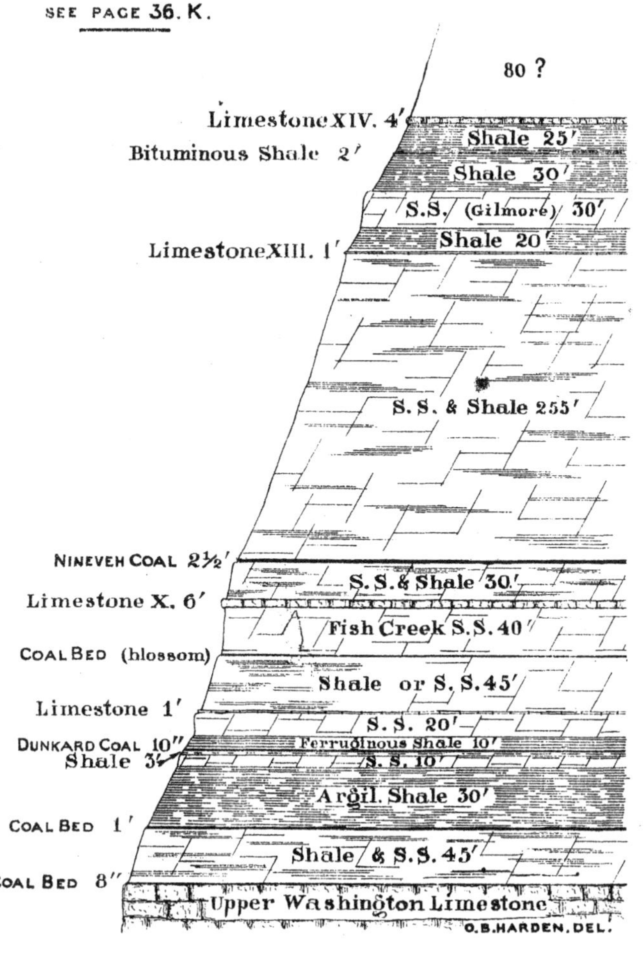

PHOTO-ZINCOGRAPH, by F. A. WENDEROTH & CO. 1328 Chestnut Street, Philada.

SECOND GEOLOGICAL SURVEY OF PA.

WASHINGTON COUNTY GROUP

UPPER BARREN MEASURES. A.

GENERALIZED SECTION IN GREENE CO.

SEE PAGE 45. K.

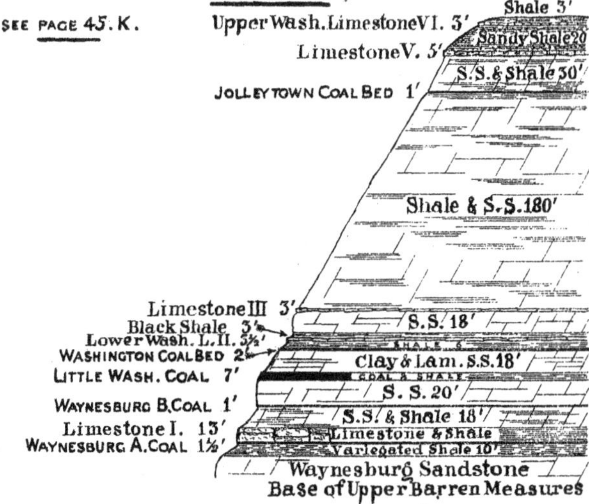

WASHINGTON COUNTY GROUP

UPPER BARREN MEASURES. A.

GENERALIZED SECTION IN WASHINGTON CO.

SEE PAGE 44. K.

COAL BED 1'
Upper Washington L. VI 30'
S. S. 40'
JOLLEYTOWN COAL BED 1'
S. S. 40'
Middle Washington L. IV. 15'
S. S. & Shale 60'
Limestone III. 8'
S. S. & Shale 20'
Bitu. Shale & COAL 1'
Lower Washington L. II 20'
WASHINGTON COAL BED 10'
Laminated S. S. 12'
LITTLE WASH. COAL BED 1'
Shale 6'
Limestone Ib. 20'
WAYNESBURG B COAL 1'
S. S. 30'
Limestone Ia. 8'
WAYNESBURG A COAL 2'
Waynesburg Sandstone
Base of the Upper Barren Measures
O. B. HARDEN. DEL.

PHOTO-ZINCOGRAPH, by F. A. WENDEROTH & CO., 1328 Chestnut Street, Philada.

is concealed. As found in Washington county, it is a massive buff limestone, from six to fifteen feet thick, which, when freshly broken, is of dull flesh color and glistens with innumerable bits of calcspar. So great is the quantity of iron present that the rock after long exposure weathers rusty yellow to a depth of several inches, and finally, after exfoliation, crumbles. Toward the base there are some thin earthy layers which have a slaty fracture and, containing less of iron, do not yield so readily to the weather as does the upper portion.

The upper or ferruginous part of the bed is *richly fossiliferous;* but as the fossils are not silicified, it is rarely possible to obtain specimens in identifiable condition. They are replaced by calcspar, so that upon the freshly fractured surface they cannot be recognized, while on the weathered surface all specific, and frequently all generic characters have been destroyed. In the cut on the Hempfield extension, immediately east from Washington, minute univalves were observed belonging to *Bellerophon* and *Euomphalus;* and on the surface of many blocks there are readily traceable the forms of branching *bryozoans.* Prof. Adney, of Washington and Jefferson college, has a fragment of a *diplodus*-like tooth in a piece of rock evidently from the middle portion of this bed. The most industrious searching failed to secure any other specimens. In the cut leading to the first tunnel east from Claysville on the Hempfield railroad, one of the lower layers yielded what is evidently a *fish spine.* Prof. Jones, of Washington and Jefferson college, has a very fine spine of *Ctenacanthus Marshii*, which he obtained near Cannonsburg from a transported fragment of the lower layer. This spine is nearly five inches long, and somewhat more than one wide at the insertion.

This Limestone was followed into Greene county by way of Hunter's fork of South Wheeling creek. It is also found on Ruff's creek in Washington township of that county, but cannot be identified on Ten-Mile, unless, indeed, it is represented by a line of nodular limestone, seventy feet below Limestone V, about a mile above Waynesburg. On Ruff's creek and Hunter's fork it is quite thin, and its disappearance at no great distance southward would naturally be expected. On Dunkard creek a very coarse and impure limestone is found not far from this horizon, but I hesitate to regard it as the same. In Washington county a little *coal* is usually found *underlying* this rock. On Dunkard creek a little coal often underlies the impure limestone just referred to. It is possible that they may be the same, but it has been found impossible to connect the points, and one is not justified in regarding them as equivalent.

Limestone III.

Limestone III is found throughout Greene county at from forty to seventy feet above the *Washington Coal*, and varies from light-blue to buff in color. In Washington county it cannot always be recognized, as the intervals have thinned considerably, and the Lower Washington Limestone has become so thick that one is liable to mistake this for an upper layer of that stratum. The more so, as it is usually thin, being from ten inches to four feet. But about two miles north from the Williamsport pike, and near the western border of South Strabane township, it is eight feet thick, with a bituminous shale resting upon it, and is twenty feet above the black shale associated with Limestone II. In the shale above it there are many lamelli branch shells, along with scales of fish, some of which are quite large and belong to *Rhizodus*.

Lower Washington Limestone II.

The Lower Washington Limestone, II, everywhere accompanies the *Washington Coal*. In Greene county it is commonly found at from six to eight feet above the coal, though at one locality on Dunkard creek the interval is twenty feet. It is dark blue, with some flesh-colored layers, weathers bluish-white and has a slaty fracture. Its thickness varies from six inches on Hunter's fork to three or five feet at most of the other localities in the county. In general appearance it is quite as well marked as is the Upper Washington, so that one is frequently enabled to determine the horizon of the *Washington Coal*, when everything is concealed except fragments of this limestone.

Fish Bed.

Over it there is invariably a black carbonaceous shale, which occasionally holds a thin coal. This shale is quite fossiliferous, and at an exposure on Purman's run near Waynesburg, fine specimens of bivalve crustaceans and fish scales may be found. Accompanying these there is sometimes a leaf impression, which clearly shows that it had undergone long maceration, for only the skeleton was buried.

In Washington county this shale is quite different. As seen at many localities, it no longer shows the neat slaty structure or cannel-like fracture, but is a rudely bedded shale with barely enough carbonaceous matter to give it color. In that county, the limestone shows great variations in thickness. At the extreme north-west exposure, near Eldersville in Jefferson township, it is barely five feet, but followed southwardly, it is found increasing with a certain degree of regularity, until in the vicinity of Washington, it is nearly twenty feet, not including the layers of shale. Along the Hempfield railroad, it is very thin at Claysville, but just beyond the line in West Virginia, it is almost twenty feet. The interval between it and the *Washing Coal* varies from three to twenty feet, and the color of the rock from dark blue to brown.

Washington (Brownsville) Coal Bed.

This bed is the same with that called "Brownsville," by Mr. White, in his paper published in the Annals of the Lyceum of Natural History, and under that name I have described it in my Notes on the Geology of West Virginia and elsewhere. The name originally applied to it was taken from an obscure hamlet in West Virginia, and I have thought best, with Mr. White's consent, to change the title to *Washington*, because in a large portion of Washington county it becomes of much economical importance, which it rarely does elsewhere. Though quite variable in thickness it never disappears, and is persistent equally with either the *Pittsburg* or *Waynesburg*. In the Ohio Report I have identified with this Coal XII of that State; but in so doing I was in error, as its equivalent there is the one which I have numbered XIII. The most northerly exposure in this district is at Eldersville in Jefferson township, Washington county, and at the north-east it is probably the coal seen in Snowden and St. Clair townships, Allegheny county, at somewhat more than fifty feet above the *Waynesburg*. Its thickness varies from six inches to eleven feet.

This bed is mined in Amwell, Morris, Donegal, Buffalo, Canton, Franklin and South Strabane, Washington county, and is of workable thickness at several localities in Jefferson, Hopewell, Independence and Mt. Pleasant townships of the same county. In this region it is apparently as important as the *Waynesburg* is in Greene county, and there are many points of similarity in the structure of the two beds. Each is badly broken up by thick clay partings, which render mining inconvenient and somewhat expensive.

The greatest complexity of structure is found in Franklin, Canton and Buffalo townships, along the Hempfield railroad. In a cut just west from Washington, the bed shows a total thickness of seven feet one inch, sub-divided as follows:

Coal		3 inches.
Clay		8 "
Bituminous shale		10 "
Clay	1 foot	3 "
Coal		5 "
Clay		1 "
Coal		2 "
Clay		2 "
Coal		3 "
Clay		3 "
Coal	2 feet	9 "

As exposed in another cut, about eight miles farther west, it is much more broken up, so much as to be practically worthless. It is there ten feet two inches thick, sub-divided thus:

Coal		4 inches.
Clay		2 "
Coal		2 "
Clay		2 "
Coal		3 "
Clay		6 "
Coal		5 "
Shale, with streaks of *coaly* matter	4 feet.	
Sandstone	1 "	
Coal	1 "	4 "

The ordinary structure is much simpler, and resembles the following section, obtained in Amwell township, on Ten-Mile creek; total, three feet five inches:

Coal	1 foot.	
Clay	3 "	7 inches.
Coal		11 "
Clay	4 "	6 "
Coal		8 "

In Greene county, this bed diminishes greatly in importance, and is seldom sufficiently thick to admit of working: On Hunter's fork of South Wheeling, it is barely six inches, and usually throughout the county it is little more than eighteen inches. It is always a double bed with a thick clay parting somewhat above the middle. Immediately south from the Virginia line, it becomes quite complex, showing five divisions at Brown's mills on Dunkard creek; and at Farmington, on the Baltimore and Ohio railroad, it has, as measured by Mr. White, fourteen divisions, making a total thickness of ten feet seven inches. Followed farther southward it becomes of little importance, and near Long run on the Parkersburg branch of the Baltimore and Ohio railroad it is in all five feet three inches thick, subdivided thus:

Coal		9 inches.
Shale	2 feet.	
Bituminous shale	1 "	
Clay		3 "
Coal	1 "	3 "

Followed still farther southward, it becomes only six inches thick in all, and apparently disappears about twenty miles south from the railroad. It is traceable across the northern Pan-handle of West Virginia into Ohio. In the latter State it is occasionally seen north from the Central railroad, on the tops of the highest hills in Belmont county, but, excepting near Martinsville, it is always thin and single. Everywhere the quality of coal from the *Washington* is quite inferior, and the sudden variations in thickness, together with the frequent occurrence of "clay veins" and "horsebacks," render mining an uncertain business.

Washington Sandstone.

Separated from the coal by only a thin layer of fire-clay, is a thinly laminated micaceous sandstone, varying in color from dark gray to bluish gray, and in thickness from eight to eighteen feet. It is always quite micaceous and contains minute fragments of carbonized vegetable matter as well as imperfect leaves. On Dunkard creek, near Brown's mills, and on the North fork of Ten-Mile, near Ten-Mile village, some thin clayey layers toward the top show many well preserved and beautiful impressions of leaves; but in each case the rock is badly broken by weathering, and good specimens can be obtained only after much excavation. In Greene county the Upper Barrens show no other stratum at all resembling this; but in Washington county two others occur above, one almost directly on limestone IV, and the other at a short distance above limestone VI. These are altogether local, each having been seen at only one place. One familiar with the lower rock is not likely to confound either of these with it, as they are less distinctly laminate and are not so crowded with fragments of carbonized vegetable matter. *The lower sandstone contains vertical gashes*, extending from top to bottom, from one inch to one foot wide, and filled with vertically laminated shale. They have no definite strike. No similar gashes occur in either of the upper sandstones referred to.

Little Washington Coal Bed.

Below this sandstone in Washington county, is a thin but persistent coal which I have called the *Little Washington Coal.* Throughout Greene county this is a mass of dark, slightly bituminous shale, sometimes containing two or three inches of coal. The carbonaceous matter is not well aggregated at any place in the county except on Mill run, in the extreme north-west corner, where it forms a coal bed ten inches thick, and the rest of the shale is not black, but drab. The dark shale usually contains some *blackband ore* but of poor quality. In Morris, East and West Finley townships of Washington county, the black shale is quite persistent; but as one descends Ten-Mile creek from Morris into Amwell township, he sees the gradual aggregation of the carbonaceous matter and the change of the color of the shales from black to gray.

This coal was seen in West Bethlehem, Somerset, North and South Strabane, Franklin, Amwell, Buffalo, Donegal, and the northern portion of West Finley, in Washington county, and is from six to fourteen inches thick. It is not even approximately parallel to the *Washington Coal*, but invariably describes short waves, two to three feet deep and six to thirty feet long. In these irregularities the upper bed does not participate, and all variations are at the expense of the intervening rocks. In a cut on the Hempfield railroad, about one mile west from Washington, the interval between the beds varies within 100 yards from eight to twelve, and then to nine and seven feet; and at the west end of the cut, the lower coal is tending upward at 35°. If this continue for ten yards farther, the interval between the coals would disappear. Whether it does or not cannot be told, as the hill is cut away at this point.

Limestone Ib.

In Greene county there occurs below this coal, a sandstone with some sandy shale, the whole about twenty feet thick and resting on the *Waynesburg* (*B*) *Coal*. This is the case at all the eastern exposures, but at the north-west, where the *Washington Coal* again comes to the surface, there is as on Hunter's fork, an important limestone in this interval, from two to ten feet below the *Little Washington Coal*. This is the one marked *Ib* in the section. On Hunter's fork it is ten feet thick, brecciated on top and bottom, with a ferruginous layer in the middle. From this line, northward to the Hemfield railroad, it increases in thickness, becoming twenty feet at some localities. The upper portion, northward, is very ferruginous for six or eight feet, while the lower layers are light colored and somewhat brecciated. In the western part of Washington county, this rock is persistent to the extreme north-west outcrop of the *Washington Coal*, but it does not appear on North Ten-mile or Pigeon creek, in the eastern part of that county. It is a massive compact rock, but is too impure for the manufacture of lime. On exposure to the weather it exfoliates so readily as to be useless for building purposes.

Waynesburg (*B*) *Coal Bed.*

The *Waynesburg* (*B*) *Coal*, though thin, is persistent. It was seen at the greater number of localities in Greene and southern Washington, where its horizon is exposed. Occasionally it degenerates into a bituminous shale, but it is usually a coal of fair quality. Its thickness varies from six inches to eighteen inches.

Limestone Ia.

Limestone I*a* is found in the greater part of the district where its place is reached. At Waynesburg in Greene county it is double, and is exposed just below the borough. At many localities in Washington it becomes very thin, or is represented only by a calcareous shale. It is frequently pure enough to be burned for lime.

Waynesburg (a) Coal Bed.

The *Waynesburg* (*a*) *Coal* is as persistent as any member of the whole series, having been observed in West Virginia and Ohio, as well as at every place in Greene and Washington counties where its horizon is exposed. At the extreme northwest, in Jefferson township, Washington county, it is wanting, as the interval between it and the Waynesburg has thinned out. The same is probably the case in Snowden and St. Clair townships, Allegheny county. It is the *Coal XII* of the Ohio series, the connection between the two States having been made along the Hempfield railroad. At many places it overlies a thin limestone which is seen also in Ohio, on the Central Railroad. The coal is rarely more than two feet thick, and the extremes of variation are one to three feet. The bed is seldom of any economical importance, and the only openings which have come to my knowledge are those on Muddy creek, above Carmichaels, in Cumberland township, Greene county. Between this coal and the Waynesburg Sandstone, there is usually some argillaceous shale.

CHAPTER VI.

The Upper Productive Coal Series.

This important series is well exposed on the Monongahela river from the West Virginia line to Pittsburg. In that distance the top of the series never passes below the surface of the river, and the base never runs out in the hills, except where those bordering on the stream are very low. It embraces five seams of coal, all of them persistent but, excepting the *Pittsburg*, exceedingly variable in thickness and quality. The section is so variable, that one finds much difficulty in endeavoring to present any general scheme of the succession, but an approximate section showing the extreme development of the series as a whole is as follows:

			Thickness.	Total.
1.	Waynesburg Sandstone		70′	82′
2.	Shale		0 to 12′	
3.	*Waynesburg Main Coal Bed*		6′	
4.	Fire-clay		3′	
5.	Sandstone		20′	88′
6.	Limestone		5′	
7.	Sandstone and shale		60′	
8.	*Uniontown Coal Bed*		1′ to 3′	
11.	Limestone and shale,	Great Limestone	18′	
10.	Sandstone and shale....		60′	173
9.	Limestone and shale,		55′	
12.	Sandy shale		40′	
13.	*Sewickley Coal Bed*		1′ to 6′	
14.	Sandstone		10′	
15.	Limestone		18′	53′
16.	Sandstone or sandy shale		25′	
17.	*Redstone Coal Bed.*		1′ to 4′	
18.	Limestone		10′	
19.	Pittsburg (Upper) Sandstone		40′	60′
20.	Shale		0 to 10′	
21.	*Pittsburg Coal Bed*		5′ to 12′	458′ to 487′

The Waynesburg Sandstone.

The Waynesburg Sandstone in its greatest development in Greene county, is not far from seventy feet thick, and is in two nearly equal divisions, separated by a sandy shale, which occasionally becomes a flaggy sandstone. One is at a loss to discover any special characteristics, by which to distinguish this from the almost equally imposing Pittsburg Sandstone below. So closely do the two strata resemble each other, that the upper has frequently been mistaken for the lower.

The inferior division of the Waynesburg Sandstone is often massive, and as below Waynesburg, shows fine cliffs nearly forty feet high, but it is liable to change into a flaggy rock. When massive it is soft, and in its mode of weathering greatly simulates the Pittsburg. Erosion has produced many large cavities, whose surface is either honey-combed or ornamented with handsome scroll work. On Crabapple creek, in Greene county, large rooms have been excavated, some of which seem to communicate with others behind, by narrow passages, and quite a cave is said to exist on the property of Mr. J. Huston, in Cumberland township of the same county. A stone so soft can hardly be regarded as of much value for building purposes, yet as it dresses easily and is of an agreeable color, it is much used. The weather attacks it promptly and the material is far from being durable.

The upper division of the stratum is always a cross-bedded flaggy rock. There is no regular dip in the cross-bedding.

Northward and north-westward, this rock diminishes in thickness, until in Washington county it wholly disappears. At from ten to fifteen miles west from the Monongahela river, it loses the massive character and finally degenerates into a mere shale. Yet along the West Virginia line, in north-western Greene and south-western Washington, on the forks of South Wheeling creek, it exhibits its massive form and stands out in bluffs. In Greene county and southward, it contains layers of conglomerate, and for the most part shows much feldspar in small grains; it is a persistent stratum in West Virginia southward to the centre of the State. In an oil boring at Rogersville, in Greene county, it proved to be *saliferous*.

2. *Plant-bearing Bed.*

Usually there intervenes between the sandstone and the coal, a variable shale, from zero to twelve feet thick, somewhat sandy and commonly breaking down readily upon exposure to the weather. Special interest attaches itself to this shale, in that it is the repository of the *finest leaf impressions obtained in the upper series.* This is characteristic of it, not in this district only, but also in West Virginia, where throughout an area of several thousand square miles, it is always plant-bearing. The species are not numerous, and appear to be the same at all exposures, whether in West Virginia or in Pennsylvania.

The best locality for making collections is on Muddy creek, near Carmichaels, in Greene county, where, for almost half a mile, the shale and coal are finely exposed along the stream, and the very numerous openings give ready access to the roof. Mr. White, who has given some attention to the study of fossil plants, has made out the following partial list of genera occurring at this locality and its vicinity:—*Neuropteris*, *Pecopteris*, *Alethopteris*, *Sphenopteris*, *Anotopteris ?*, *Goniopteris ?*, *Sphenophyllum*, *Annularia*, *Pinnularia*, and *Hymenophyllites.*

Where this stratum is absent, the sandstone rests directly on the coal.

3. *The Waynesburg Main Coal Bed.*

The *Waynesburg Coal* shows great variation in thickness, and at times becomes exceedingly thin, though it never wholly disappears within this district. Where not eroded by the overlying sandstone it is double, triple, or even farther divided. Northward, in Washington county, it is quite insignificant; and at the exposures in Allegheny county is rarely of any value whatever.

Southward, in Greene county and the adjoining portion of West Virginia, it attains its greatest importance; but followed still farther southward, it is seen dwindling as at the north, until at the farthest point where I have examined it in central West Virginia, it is barely six inches thick.

In Ohio, it is the *Coal XI* of that State, and is extremely variable.

In Greene county it is the chief source of supply. In eastern Washington county it is thick and is mined; but in the western portion of that county it is very thin and is inferior in value to the *Washington Coal.*

In Washington county the extremes of thickness are six inches and seven feet, and the variations are quite abrupt, these extremes being sometimes found within the limits of a single township. In the eastern portion, the bed is usually not very complex in structure, and the following section obtained at the village of Pin-hook, in Amwell township, may be regarded as a typical form:—

Bed 5½ feet,	*Coal*	5″ to 16″
	Clay	10″ to 18″
	Coal	27″ to 34″
	Clay	4″ to 20″
	Coal	5″ to 7″

But an opening on Brush run in Buffalo township, shows extreme sub-division, almost equal to that exhibited at many localities by the *Washington Coal.* There one finds:—

Bituminous shale	6″ to 8″
Coal, with thin partings of clay	10″ to 14″
Coal	10″ to 12″
Clay	1″ to 3″
Coal	8″ to 10″
Coal	30″ to 36″
Clay	2″ to 8″
Coal	2″ to 4″

In Greene county, openings in this coal were seen in Dunkard, Greene, Monongahela, Cumberland, Jefferson, Morgan, Franklin, Perry and Richhill townships. In all of these it is thick and seldom shows such sub-division as that given in the last section. In Franklin, the bed is usually double, being divided by a variable layer of clay, though occasionally it is triple. The ordinary structure in a total of six feet is:—

Coal	12″ to 18″
Clay	12″ to 48″
Coal	12″ to 42″

In Dunkard it is commonly triple, the lower division consisting of from twenty-seven to forty-five inches of coal, and a similar structure prevails in Jefferson. At one locality near Carmichaels, in Cumberland, the division is more complex than at any other in the county, being as follows:

Coal	12″
Clay	1″ to 2″
Coal, bony	6″
Coal	15″
Clay	2″ to 10″
Coal	26″
Clay	36″
Coal	3″

Clay	4″
Coal	5″
hale	10 feet.
Coal	8″

This lowest bench was seen only in Cumberland and Monongahela townships, and the shale separating it from the body of the bed varies from two to twelve feet in thickness. In Richhill township, on South Wheeling and Crabapple creeks, the coal is only double, and at some openings, owing to erosion of the upper bench during the deposition of the overlying sandstone, it is single. The thickness of the lower or main bench is somewhat greater here than at the eastern exposures.

Followed into West Virginia from Washington county, the bed is found diminished in thickness, but retaining its characteristic tendency to abrupt change. At Roney's Point mills on the Hempfield railroad, the benches are two and twenty-six inches respectively, separated by eighteen inches of clay. The upper bench soon disappears and in the Pan-handle to the Ohio river the bed is single. The upper bench re-appears in Ohio, where the bed displays such extraordinary changes in thickness, that it is known over a considerable extent of country as the "Jumping Six-foot vein."

In West Virginia, on Scott's run, about five miles south from the line of Greene county, the coal shows variations unequalled by any in this district. There, within a distance of barely four miles, the following three separate measurements were obtained:—

I.		II.		III.	
Coal	1′ 9″	*Coal*	9′ 0″	*Coal*	1′ 3″
Bitumin's shale	8″			Clay	3′ 6″
Coal	4′ 8″			*Coal*	2″
				Clay	2″
Total	7′ 1″			*Coal*	7″
				Clay	1′ 1″
				Coal	4′ 2″
				Total	7′ 9″

From this line it is easily traced in West Virginia to the Parkersburg branch of the Baltimore and Ohio railroad, where it shows the same mode of sub-division and is accompanied by *plant-bearing shale*, similar to that of Pennsylvania. Southward from the railroad, it is traceable only with difficulty, as it becomes very thin, and probably disappears not far south from the Staunton pike.

6. *A local limestone.*

In a large portion of Washington and all of Allegheny county, the interval between the *Waynesburg* and *Uniontown* coals is occupied by sandstone or sandy shale, but in southwestern Washington and throughout Greene, a *limestone* is invariably found at from 20 to 40 feet below the former coal, the interval between the coal and limestone increasing southwardly, but decreasing northwardly until it finally becomes zero. This limestone is from 4 to 8 feet thick, and is ordinarily pure enough to yield excellent lime. It is found in West Virginia south from our district, but the interval between it and the Waynesburg coal diminishes, and the rock itself becomes argillaceous, so that on the Parkersburg branch of the Baltimore railroad the interval is barely ten feet, and the limestone is little better than a compact calcareous shale. Beyond this line southward it disappears, there being no evidence of its existence at twenty miles south from the railroad. In West Virginia, west from Washington county, it is quite irregular. At Roney's Point mills it is eight feet below the coal and three feet thick. Beyond this it disappears and is not found in Ohio.

8. *The Uniontown Coal Bed.*

The Uniontown Coal is persistent either as coal or as bituminous shale. At several localities in Greene and Washington counties it has been mined on a small scale, but with the exception of one opening in Cumberland township of Greene county all work on it has been abandoned. At the opening referred to the section is as follows:—

Coal	1′	6″
Sandstone	10′	0
Coal		6″

The little coal, or the lower bench, is frequently seen, although commonly in much closer proximity to the upper bench. When the sandstone parting is absent, the bed occasionally becomes three feet thick. It is true that the two benches have never been seen united and then diverging, but there is no room to doubt that they are parts of the same bed. Wherever the exposure is good the lower bench can be recognized, and it is always separated from the upper bench by a well-defined parting. The case is no more remarkable than the instances where the *Washington Coal* doubles its total thickness within a short distance, the amount of coal remaining approximately the same. When the *Uniontown Coal* is a bituminous shale it has a cannel-like fracture, and *contains great numbers of fish teeth and scales*, which are usually quite small. I have been unable to trace this coal to any distance in West Virginia, and am inclined therefore to regard it as confined to the eastern portion of the trough.

It is not the same with *Coal X* of the Ohio section.

The Great Limestone.

The *Uniontown Coal* rests directly upon the upper division of the Great Limestone of the old Pennsylvania reports.

I propose to limit this term, "The Great Limestone," to the double mass occurring between the *Uniontown* and *Sewickley Coals.* Limestones occur below the *Sewickley* as well as below the *Redstone*, and in some localities are so important that it might appear perfectly proper to include them under this name. But they are by no means persistent, the intervals below those coals being frequently occupied by sandstones or shale. It seems preferable therefore to confine the name to the persistent portion—that between the *Uniontown* and *Sewickley.*

Ordinarily this mass is in two divisions, separated by sandstone or shale. At one place in Dunkard township of Greene county, and along the lower portion of Ten-Mile creek, it shows three divisions. But these are the only instances of such subdivision, and the mass will be described as simply double.

Upper Division of the Great Limestone.

The upper division varies from six to eighteen feet in thickness. It lies directly under the *Uniontown Coal*, and is easily distinguished by its bright buff color. It disappears in the northern portion of the district, and is absent in Allegheny county, at least it could not be identified there. In north-west Washington county it was doubtfully recognized in Jefferson township. In Ohio it is absent, and *Coal IX* of that State, which is not identified in Pennsylvania or West Virginia, rests directly on the lower division.

Lower Division of the Great Limestone.

The lower division is found throughout the district wherever its horizon is exposed, except in north-eastern Allegheny county. It varies in thickness from 50 to 90 feet. The character of the rock is variable. The lower portion is commonly more or less *magnesian*, breaks with a smooth surface, which is sometimes lustreless, while at others it is quite bright. This part is employed for the manufacture of *cement* at many places, and is available throughout eastern Washington county. It is the more persistent part of the mass, having been identified in Allegheny county, on the Pittsburg and Steubenville pike. When exposed to the weather it eventually breaks up into small angular fragments. Its thickness is sometimes fully 50 feet, nearly all of which is limestone. The relative proportion of shale increases northward, and in Allegheny county the shale seems to predominate. This portion is wholly *non-fossiliferous* except in a thin layer at the base. The upper portion contains every variety of limestone, from that pure enough for the manufacture of lime, to that which is utterly worthless for any purpose whatever. Some of the layers are *fossiliferous*, but the fossils are rarely silicified, and specimens cannot be obtained in identifiable condition. For the most part the species are *Univalves*, being in this respect unlike those found at the very base of this division, which are almost all *Lamelli-branchs.*

At Eldersville in Jefferson township, Washington county, this Great Limestone has thinned away so that it is represented by only three thin layers, aggregating but fourteen feet. In Allegheny county, near Pittsburg, it seems to be altogether wanting.

In Ohio the lower division varies in thickness from zero to seventy feet north from the Central Ohio railroad. It rests directly upon the *Sewickley Coal,* and the layer in contact with the coal is fossiliferous. This condition exists also at Wheeling. The greatest thickness is in the vicinity of the railroad to a distance of probably eighteen miles west from the river. Northward from that line it diminishes until it disappears, per-

SECOND GEOLOGICAL SURVEY OF PA.

UPPER PRODUCTIVE COAL MEASURES

GENERALIZED SECTION ALONG THE MONONGAHELA RIVER

SEE PAGE 57.K.

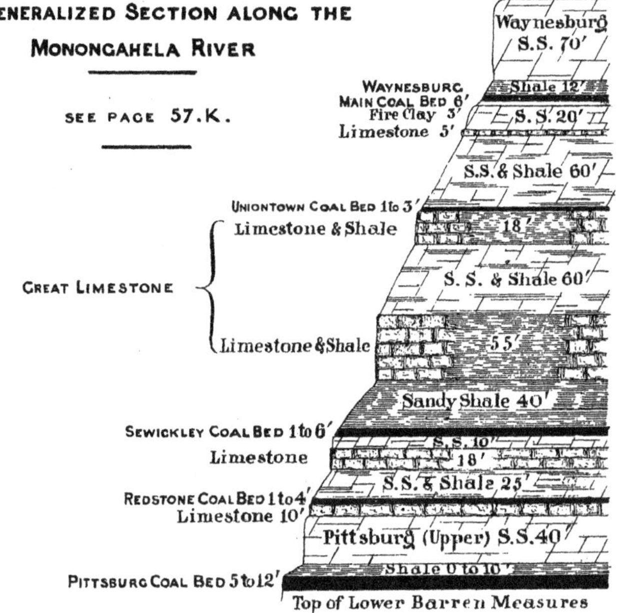

LOWER BARREN MEASURES

SECTION S. E. OF GREENE CO.

SEE PAGE 76.K.

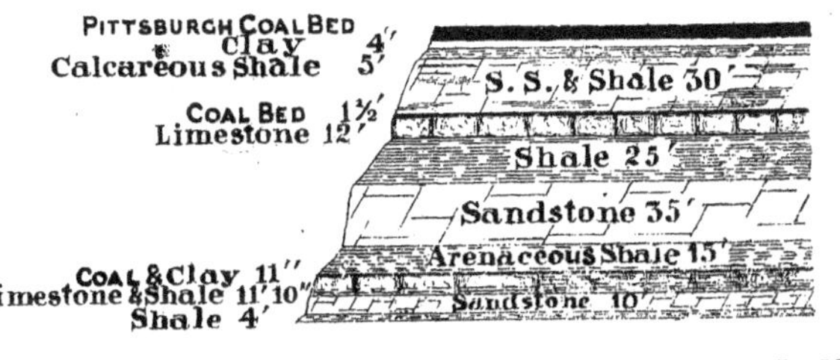

O.B. HARDEN. DEL.

PHOTO-ZINCOGRAPH, by F. A. WENDEROTH & CO., 1329 Chestnut Street, Philada.

mitting *Coals VIII* and *IX* to come together. In general character it does not differ from its equivalent in Pennsylvania.

In West Virginia, south from this district, the limestone thins out quite rapidly, and on the Parkersburg branch of the Baltimore and Ohio railroad is represented by only some thin argillaceous limestones, while at twenty miles farther south no trace of it exists.

In the vicinity of the Great Kanawha river, in that State, the whole mass of limestones belonging to the Upper Coal Series is represented only by a few calcareous nodules.

The Sewickley Coal Bed.

This coal attains its chief importance in the southern part of the district and is available only in the south-eastern portion of Greene county, being too thin elsewhere to be of any value. It has been traced along the Monongahela river to the Washington county line, and has been seen at many places in that county. In the northern part of the district it is frequently a bituminous shale, and at numerous localities it seems to be wholly absent. In Allegheny county it cannot be identified.

A laminated sandstone overlies this bed in south-eastern Greene county, as is also the case in the adjoining portion of West Virginia. This rock appears along the river to where the coal passes under the stream. Under the coal is a similar sandstone, which is quite thin along Dunkard creek. But where the coal again appears above the river, the upper sandstone has almost disappeared, while the lower one has greatly thickened, so that the coal seems to lie immediately below the Great Limestone. This condition maintains itself throughout the rest of the district.

On Dunkard and Whiteley creeks this bed is mined. The variations in structure are not great, and the following section on Dunkard is typical for the township:—

Coal	2 feet	1 inch.
Clay		1 "
Coal	3 "	4 "

On Whiteley creek near Mapleton, the openings are quite numerous and show a structure as follows, the total thickness being 4 feet 7 inches:—

Coal..........		9 inches.
Parting..		¼ "
Coal..	1 foot	7 "
Parting..		1 "
Coal..		10 "
Parting..		2 "
Coal..	1 "	2 "

This is an extreme case of division, as one rarely finds more than three distinct benches in the bed.

Northward from this line the coal becomes very irregular, and not unfrequently is represented only by a bituminous shale. Along Ten-Mile creek it is a coal from 12 to 22 inches thick. Thence along the river to the mouth of Fishpot creek it is with difficulty recognizable as a bituminous shale, but at lock No. 5 it is three feet thick, and at a mile above Brownsville it is of considerable importance, as appears from the following section :—

Coal..	3 feet	6 inches.
Clay.		2 "
Coal..		8 "

Thence it thins out northward, and cannot be recognized until near Monongahela City, where it occurs as a bituminous shale. Beyond this, along the river, it was not found.

In the interior its horizon is reached in Washington county, on Chartiers and Cross creeks, as well as on the Panhandle railroad. On Chartiers it is a bituminous shale directly underlying the Great Limestone. Possibly the black shale seen at several localities along the Pittsburg and Steubenville pike, in Allegheny county, may be the representative of this bed, but in the absence of all connected sections one would hardly be justified in pronouncing it the equivalent.

In West Virginia, south from Greene county, this bed is quite as variable as it is within the limits of the district. Near the Pennsylvania line it is from four to six feet thick, but southward it diminishes in importance, being traceable with difficulty along the Baltimore and Ohio railroad, and becoming only three inches thick on the Parkersburg branch of that road. In Ohio and the Panhandle of West Virginia it is divided, and is represented by the Ohio *Coals VIII* "*B*" and *VIII* "*C*." The upper division, or *VIII* "*C*," is of workable thickness in eastern Belmont county of Ohio, but disappears within twenty miles west from the Ohio river; and, along the river, it has disappeared long before Steubenville is reached. In Monongahela township of Greene county, in this State, the *Sewickley* shows a sub-division quite as great as that in Ohio and the West Virginia Panhandle.

The Fishpot Limestone.

The interval between the *Sewickley* and *Redstone Coals* is occupied by two beds of sandstone, separated by a bed of limestone. In the northern portion of the district, the limestone is not persistent. Along the river line in Washington and Greene counties, this limestone shows some interesting variations. Beginning at the West Virginia line, it is quite thin; thence it increases to the mouth of Fishpot run in southern Washington, a little beyond the trough between the Fayette and Waynesburg axes. There it is thirty feet thick.

From this locality it diminishes and at length disappears before reaching the summit of the Waynesburg axis.

From this line northward along the river, the interval between the two coals is occupied by sandstone or sandy shale. Between the Pin-hook and Washington axes, limestone occurs in this interval as far north as the southern portion of Allegheny county, but beyond Upper St. Clair township of that county, it seems to be filled with sandstone. In the trough west from the line of the Claysville axis, the limestone is persistent as far north as the Steubenville pike in Allegheny county. Whether it reaches farther north cannot be determined, as there are no exposures. In the southern portion of this trough, in Ohio and West Virginia, the limestone is quite thick in the immediate vicinity of the Ohio river. In West Virginia, it stretches south from the State line for not more than fifteen or twenty miles. No traces of it occur on the Parkersburg branch of the Baltimore and Ohio railroad.

The Redstone Coal Bed.

The Redstone coal, though exceedingly variable, is persistent throughout the greater portion of this district. Its horizon is reached at nearly all points along the Monongahela river, while in the interior it is exposed in the north-western part of Washington county, and in central Allegheny county along the Pan-handle road, as well as in the valleys of nearly all the streams.

Beginning at our southern boundary on the Monongahela river, the *Redstone* is found near Dunkard creek as a bituminous shale, barely one foot thick, which occasionally contains three or four inches of coal. Near Greensboro' it is a *coal*, 18 inches embedded in a mass of richly bituminous shale 13 feet thick. This condition continues (the mass becoming thinner meanwhile) until, at a little above Whiteley creek, the shale disappears and the coal has the same thickness as at Greensboro'.

Near Hatfield's ferry, the Redstone bed goes under the river. When it comes up again, as it does below the mouth of Ten-Mile creek, it is represented by 18 inches of bituminous shale. Near Fredericktown, the bed is much thicker and consists of bituminous shale, three feet, and coal six inches. From this locality to the vicinity of Monongahela city, it is only a bituminous shale, varying from one foot to five feet in thickness; but at that place it suddenly acquires considerable importance, and becomes three feet six inches thick.

It retains the coal-phase along the river to beyond the Allegheny county line, everywhere yielding a good coal.

Following the river down, in Allegheny county, the strata are seen rising rapidly, and the coal is soon so near the hill-tops that it is rarely exposed; and even what exposures there are are so imperfect as to give no satisfactory information.

Throughout this county, except in the extreme south-eastern corner, the coal is much degraded in character, and varies from two inches to two feet. The localities at which it has been identified are very few. In Washington county, it has been recognized only in the vicinity of the Washington axis and along the river border. For the most part it is quite thin but shows some astonishing changes. These may be seen by comparing the shaft-records from the vicinity of Washington.

In Ohio and the Pan-handle of West Virginia, this bed is represented by *Coal VIII* "*A*" of the Ohio section, which is a variable bed, of economical importance at no locality.

In West Virginia, south from our area, it is very persistent, and near the Pennsylvania line is of much importance. Followed southward, it is found diminishing in thickness, until on the Parkersburg branch of the Baltimore and Ohio railroad, it varies from two inches to four feet and is of no value whatever. Twenty miles south from the railroad it has disappeared. On the east side of Laurel Hill, in the eastern central part of West Virginia, there is a coal at from forty to eighty feet above the *Pittsburg*, which may possibly be this, though I am inclined to regard it as being the *Sewickley*.

The Upper Pittsburg Sandstone.

The interval between the *Redstone* and *Pittsburg Coals* is extremely variable, both in respect of thickness and of composition. The variations in thickness will be considered in another connection. Near the West Virginia line there is in the upper portion a limestone, but this soon disappears, and is absent where the bed passes under the river. No traces of it occur north-west from the trough lying east from the Waynesburg axis. Everywhere, except in the extreme south-east portion of the district, this interval is occupied by sandstone or sandy shale.

The Pittsburg Coal Bed.

The Pittsburg Coal Bed alone of all the beds in the Upper Productive Coal series has such narrow limits of variation, as to be an available bed throughout the district, wherever it is within reach. It nowhere becomes too thin, or too impure, to repay working; and ordinarily its quality is superior to that of any other of the upper coals. It is accessible nearly all the way along the Monongahela from the State line to Pittsburg, being carried under water level, for a few miles only, in the trough east from the Waynesburg axis.

At the mouth of Cheat river, near the State line, the *Pittsburg Bed* is 370 feet above the Monongahela river. Thence it descends until, at a little way above Gray's landing, it goes under the river bed, and is 185 feet beneath it at about sixteen miles from the State line. From this point it rises northwestward, until it crosses the Waynesburg axis below Belvernon; but, owing to changes in the course of the river, its height above the stream is variable. At Brownsville it is at low water line. Below Belvernon it is constantly above the river, as the trough at Huston's station is not deep enough to carry it under. At Braddock's Fields it is 360′, and below Pittsburg fully 400 feet above the low water mark.

With rare exceptions this bed is double, consisting of a roof and a lower division, separated by a clay parting, which varies in thickness from one-fourth of an inch to nearly three feet, and frequently contains thin strings of coal which are connected with the roof division.

Where the structure is completely exposed, there is seen resting on the roof a bituminous shale, 8″ to 12″ thick, always laminated, and having a fracture like that of the poorer cannels; it is not persistent in the southern portion of the district.

The Roof Division shows extreme variations. Its thickness is from two inches to eight feet, but there is a distinct increase in thickness northward. Occasionally it is a single bench; but commonly it contains two or more benches of coal, separated by clay; and at one locality it is broken into twenty divisions. The coal is invariably poor, owing to the large proportion of ash. The clay partings are subject to abrupt variations, for on the Panhandle railroad the roof shows twenty divisions at a little distance east from Raccoon station, while at the station it shows 5′ of coal broken only by partings so thin that they can hardly be distinguished on the weathered surface. The changes in thickness of the whole division are equally abrupt, several instances having been observed where within a short distance it varied from a single 2″ bench of coal to a mass of coal and shale 3′ thick.

I have said that this roof-division thickens northward. This statement is the result of many comparisons, for if one were permitted to select examples he could without difficulty find many cases in Allegheny and northern Washington where the roof is as thin as at any place in Greene or south-eastern Washington. But taking all the measurements in the south-eastern portion, and comparing them with all those made in the northern portion, it becomes apparent that the roof is thicker northward, and that in north-western Washington and Allegheny the thickness is suddenly and greatly increased.

The *lower division* of the Pittsburg coal is from 3′ 6″ to 9′ thick and contains three persistent partings, usually thin, which divide it into four benches known as the "Upper," the "Bearing-in," the "Brick," and the "Lower Bottom."

In the first or *Upper Bench*, there is occasionally a parting which is rarely seen except at the extreme north-west, where it seems to be a common feature. This is the thick bench and usually yields the best coal.

The "*Bearing-in*" *Bench* varies from 2″ to 4″, and is invariably distinct, except where the bed is a block coal, and all the partings are missing. The name is applied because on this bench the miner works in to gain a face against which to bring out the other portions of the bed. This is generally a good coal, but in removal it is reduced to slack.

The "*Brick*" *Bench* is characterized by cleavage planes which break the coal into blocks in size and shape like a common brick, whence the name. It yields a good coal, hardly inferior to that from the Upper Bench.

The "*Lower Bottom*" *Bench* is the lowest of all, always of inferior quality, and for the most part utterly worthless. It is broken by numerous thin layers of clay, as well as by cleavage planes, so that it is brittle, and full of ash.

The Upper Bench contains thin partings or binders of pyrites, one of which, at from ten to fifteen inches from the top is quite persistent. This impurity sometimes occurs in the Brick, and is always present in the Lower Bottom.

The thickness of the whole Lower Division of the Pittsburg Bed diminishes northward, as the Roof Division seems to increase in that direction; but, with the exception already noted, the various benches are persistent throughout. In the south-eastern part of the district the total thickness is from seven to nine feet; greatest at Brownsville, where the roof is 4″ and the lower division is 9′. In the vicinity of Pittsburg and the adjoining portions of Allegheny county, it varies little from 5′ 6″; while in north-western Washington, it varies from 3′ 6″ to 5′; the former (3′ 6″) being found at midway on the Panhandle railroad, where the coal is a block.

The coal from the Lower Division of the Pittsburg coal is somewhat brittle, caking, rich in volatile combustible matter, and containing a variable percentage of sulphur. In some portions of the district, it exhibits layers of cannel near the top and occasionally, as along the Pan-handle railroad in Washington county, it becomes a very superior *block coal.*

The *Pittsburg Coal Bed* is exposed in Dunkard, Monongahela and Jefferson townships, Greene county, in East Bethlehem, East and West Pike Run, Allen, Fallowfield, Carrol, Union, Peters, Hanover, Robeson, Smith, Jefferson, Chartiers and North Strabane townships, Washington county, and in all the townships of Allegheny county within the district. It barely crosses from Washington into Beaver, there being in the latter county only a few small outliers.

In West Virginia, south from this district, the Pittsburg Coal Bed is accessible throughout an extensive area, of which I determined the eastern boundary as far south as Braxton connty in 1874. In Monongahela county, which adjoins Greene county of this State, the roof and lower divisions are distinct and the latter at one locality on Scott's run, attains the usual thickness of ten feet. Southward, however the roof apparently disappears, and at Fairmount, twenty miles from the State line, the bed is single and nine feet thick. So also at Clarksburg and other places along the Parkersburg branch of the Baltimore and Ohio railroad, but the singleness is only apparent, for the structure at Clarksburg, is as follows:—

Coal, Roof Division	3′	0″
Main clay parting		$\frac{1}{8}$″
Coal, Lower Division	5′	4″

In the lower division, here, there are three subordinate partings at 12″, 15″ and 18″ from the bottom, giving the "Bearing-in" and "Brick" benches of the Pennsylvania mines.

The main clay parting occasionally becomes more marked than in the section, as near the tunnel east from Clarksburg and near Shinnston, a few miles north from that town.

The distinction between the divisions is shown also in the character of the coal. That from the roof having a tendency to become bony, while the lower division contains little ash. In the latter, also, is found the *persistent pyrites band* which occurs in Pennsylvania in the Upper bench. It is evident then, that the clay partings of the roof division have disappeared in northern West Virginia, and that that division becomes solid coal towards the south, just as it does at some localities in the northern part of this district.

Followed southward and south-eastward the bed is found growing thinner, so that, where it passes under the Little Kanawha at Glenville, the total thickness of both divisions is only five feet; and at the extreme eastern exposure in Upshur county, it is somewhat less than four feet. The lower division being only two feet.

A very interesting feature of the Pittsburg Coal Bed throughout that region, is the growing resemblance of its Upper and Lower Divisions; for, going eastward, one finds more and more difficulty in distinguishing the two divisions; until at last, along the eastern outcrop in Barbour county, there are several openings in which no difference exists, the Roof division being precisely like the other, and the parting between them has disappeared.

To the westward of this district, the *Pittsburg Coal* is found on Wheeling creek and other streams in the Ohio Panhandle of West Virginia, but no sufficient examination of it has been made north from Ohio county. In the vicinity of Wheeling, the bed is always double, and the roof has the same character as in Pennsylvania, always yielding a bony coal and varying from one to two feet in thickness; the Lower Division being from five to six feet thick.

Crossing into Ohio, one finds the coal bed double along Wheeling creek, the lower division varying from 5′ to 6′, while the roof itself is usually double. In this vicinity its section is sometimes badly broken up into thin layers, but the amount of coal is quite constant. Northward and westward the thickness of the *lower division* decreases, but the four benches are always present, and show the same peculiarities as in Pennsylvania. The *roof* shows an extraordinary increase in Jefferson county, becoming at one locality five feet thick, and frequently showing three feet of coal. Near Unionport, in Jefferson county, on the Panhandle railroad, the *lower division* has an extreme thickness of four feet nine inches, and at the village of Jefferson, north from the railroad in the adjoining county of Harrison, it is four feet eight inches. In the western portion of Harrison, on the extreme point of the outcrop line, at Hanover, the following structure is shown:—

Roof division, 1′ 6″;—Clay, ¼″;—*Lower division*, 3′ 11″.

South-west from this on the same outcrop line, the *lower division* shows a thickness of four feet at Deersville. Between this point and the Central Ohio railroad the *roof division* is wanting, having been removed by erosion during the formation of the overlying sandstone, and it is not certain that the lower division is fully represented. In the second geological district of that State the bed becomes, according to Prof. Andrews, only one foot thick at its western outcrop.

CHAPTER VII.

The Lower Barren Series.

This series consists chiefly of sandstones and shales with some variable limestones and coals.

Along the Monongahela river, the upper portion is exposed near the West Virginia line, as well as along the whole river front of Washington and Allegheny counties.

Along the Ohio river it is constantly in sight as far as Rochester, beyond which it rises rapidly, and at the State line is barely caught by the hills on the north-western side of the river.

The extreme thickness *exposed* along the Monongahela is 375 feet, this being reached only near the West Virginia line and at Pittsburg. In Allegheny county away from the river, and in Washington county north from the Panhandle railroad, this thickness is occasionally shown. Throughout Greene, away from the river, and in the greater part of Washington the Lower Barrens are deeply buried. The whole series is seen in detail only in Beaver county, where owing to the predominance of sandstones the exposures are good and clear.

In the south-eastern portion of the district, as shown by oil-borings on Dunkard creek, the interval between the *Pittsburg Coal* and the Mahoning Sandstone is not far from 425 feet; but this interval increases northward and north-westward, until in Beaver county, according to Mr. White, it is from 530 to 540 feet.

The section as obtained in Beaver county by Mr. White is as follows:—

Barren Measures in Beaver County.

1. *Pittsburg coal bed*	——
2. Shale	8′
3. Limestone	5′
4. Concealed	100′
5. Shale	10′
6. *Coal bed*	1′ 6″
7. Sandy shale	35′

8.	Limestone	4′
9.	Calcareous shale, fossiliferous	3′ 6″
10.	*Coal bed*, slaty	1′
11.	Shale	10′
12.	MORGANTOWN SANDSTONE	60′ to 70′
13.	*Coal Bed*	0′ to 3′
14.	Flaggy sandstone	25′ to 35′
15.	Limestone, crinoidal	2′ to 5′
16.	*Coal Bed*	0′ to 1′ 4″
17.	Variegated shale	25′ to 30′
18.	Blue sandy shale	50′ to 60′
19.	Red shale, argillaceous	0′ to 20′
20.	*Coal Bed*, cannel	0′ to 6′
21.	Sandstone, laminated	90′ to 100′
22.	Limestone, black, fossiliferous	0′ to 5′
23.	Dark shale	9′ to 15′
24.	*Coal Bed*	0′ to 2′ 6″
25.	Sandy shale	25′ to 35′
26.	MAHONING SANDSTONE	——

The concealed interval, No. 4, was not found exposed at any locality in Beaver county, but was seen more or less in detail at several places in Greene county, and portions are occasionally exposed along the Panhandle railroad in Allegheny county. It is strange, however, that this interval should be so thickly covered with debris, that even the numerous excavations along the railroad for coal tramways, in nearly every instance, failed to reach the rocks.

In the extreme south, south-east of Greene county, the following section was obtained:—

1.	*Pittsburg Coal Bed*	—
2.	Clay	4′
3.	Calcareous shale	5′
4.	Sandstone and shale	30′
5.	*Coal Bed*	1′ 6′
6.	Limestone	12′
7.	Shale	25′
8.	Sandstone	35′
9.	Arenaceous shale	15′
10.	*Coal* and clay	11″
11.	Limestone and shale	11′ 10″
12.	Sandstone	10
13.	Shale	4′

At Fort Pitt station, on the Panhandle railroad, the interval for 110 feet below the *Pittsburg Coal* contains nothing but sandstone, aside from two thin limestones, one immediately below the coal, and the other at fifteen feet lower.

So great and so frequent are the changes in the shales and sandstones of this series, that a detailed discussion of the whole would be intelligible only by a comparison of a large number of sections; but for the most part the rocks are of so little interest or importance that such a comparison would be only a waste of time and space, edifying to neither the author nor the reader. A few of the rocks, however, are sufficiently persistent to enable us to discuss their characteristics intelligently.

3. Limestones under the Pittsburg Bed.

At a distance varying from a few inches to eight feet below the *Pittsburg Coal* is a limestone, which is rarely wanting, though in some localities it is represented only by calcareous shale or nodules. It is coarse, light blue to gray, sometimes brecciated, and always contains vast numbers of minute univalve mollusks.

At from ten to thirty feet below this is another and almost equally persistent limestone, which in the south-eastern portion of the district sometimes becomes twelve or fourteen feet thick. Northward it diminishes in thickness, and in north-western Washington, as well as in western Allegheny, it is usually absent, though the upper one is present. In central Washington, as well as in southern Allegheny, it is present, and sometimes becomes six feet thick. From the conditions at Mansfield in Allegheny county, I am inclined to suppose that the two limestones unite, and that the single limestone, so persistent northward and which occupies the place of the upper one, represents them both. The lower limestone resembles the upper in general characteristics, but is not so constantly fossiliferous.

Little Pittsburg Coal Bed.

Resting on this lower limestone is a little coal, which is quite persistent in the eastern part of the district, being rarely absent where its horizon is exposed. It is present also on Chartiers creek in Washington county, where the Washington axis brings up the *Pittsburg Coal,* as well as on the same creek in Allegheny county near Mansfield, where the interval between the two limestones is only ten feet as against thirty feet in south-eastern Greene. No evidence of the existence of this coal was found northward or westward from this locality, and in north-western Washington and western Allegheny it certainly is wanting. This bed I have identified with the *Little Pittsburg Coal* of the old reports. It seldom exceeds eighteen inches in thickness, and is of no economical value.

Local Mansfield Limestones.

Toward the base of the concealed interval of the Beaver section there are two limestones which were seen on the Panhandle railroad west from Mansfield, and on the Chartiers Valley railroad south from that place. Evidently these are wanting along the Monongahela river, and they are not mentioned by Mr. White in his notes on Beaver county. They are quite ferruginous, and each is fossiliferous. Above the upper is a flaggy sandstone, whose variations will be referred to in another connection.

6, 7, 8, Coal and Limestone.

Below this comes an argillaceous shale, under which is a little coal, from six inches to eighteen inches thick, which seems to be wide spread but very irregular in its distribution. It occurs at a number of localities along the river in south-east Greene, but appears to be absent opposite Pittsburg, and no evidence of it was seen in Washington along the river. The exposures there are very poor, and this coal being thin, might easily be overlooked. It is found on the Panhandle railroad, at and beyond the great tunnel near Nimick's station, where it is four to five inches thick, but no other exposure was observed in Allegheny county. In Greene county, it rests directly *on a variable limestone* which there sometimes becomes ten feet thick; but on the Panhandle railroad, ten feet of variegated clay, and in Beaver thirty-five feet of sandy shale separate it from the limestone which has become quite thin. The little coal (No. 10) mentioned in Mr. White's section as *underlying this limestone* is altogether local.

12. Morgantown Sandstone.

At from 140 to 160 feet below the *Pittsburg Coal*, there is a widely persistent sandstone which prevails throughout the whole district, though it shows great variations in thickness and character. It is undoubtedly the same with that which

I have denominated the Morgantown Sandstone, from the fact that it is so extensively employed for building purposes at the village of that name in West Virginia. The thickness in Beaver county, as shown in the section, is extreme, and there the upper portion is massive, while the lower is flaggy. In the vicinity of Nimick's station, on the Panhandle railroad, the whole mass is shaly and contains not a little argillaceous shale; but on the Monongahela river the massive character is regained, so that opposite Pittsburg it is fifty-two feet thick, and farther up forms cliffs along the stream. In south-east Greene, it is from thirty to seventy feet thick, and is of interest as it is the *first oil rock of the Dunkard oil region* in that county. The rock is blue to gray, somewhat coarse grained and ferruginous and is an excellent building stone.

13. Local Coal Bed.

Underlying the Morgantown sandstone, Mr. White finds a coal in Beaver county, which is confined to that county and north-western Allegheny, including the corner of Washington adjoining. It is irregular, being often absent. The extreme thickness is three feet, which occurs on Bigger's run in Washington county, on the border of Beaver. It was seen also on Moon run, in Allegheny county.

Birmingham Shale.

Below the Morgantown Sandstone at Pittsburg, there is a dark thinly laminated shale, which is nearly fifty feet thick, and forms one of the marked features of the bluff. It is jointed, the joints passing through the mass, so that the face of the hill is insecure, and between Birmingham station and Temperanceville, is covered with extensive slides. Before reaching Chartiers creek this rock changes, and in the railroad cuts is represented by a poor sandy shale.

15. Crinoidal Limestone.

The Crinoidal Limestone is the Black Fossiliferous Limestone of the old reports. I have rejected the latter name as improper, because it conveys an altogether inaccurate conception of the rock which is a black calcareous shale at only one locality. Instead, I employ the name by which I have designated it in the Ohio Report and in other publications, as that describes the peculiar feature of the stratum, by which it can be recognized at nearly all localities, except that opposite Pittsburg.

This limestone is dark blueish or greenish gray, tough and breaks with a granular surface, much resembling that of a coarse sandstone. It weathers into blocks, which are sharply defined, but the weathering proceeds very slowly and *the rock is invaluable as a horizon in the Lower Barrens* where everything else is variable. It is unlike any other rock in the series and seldom changes in character. In all cases it is fossiliferous and contains immense numbers of crinoidal stems and spines or plates.

Mr. White states that it occurs throughout southern Beaver, and that although only three feet thick, it is the only persistent limestone in the series. Along the Ohio river, from Nimick's station to Birmingham, this stratum is constantly in sight, but there it differs strangely from its common structure. The limestone is of irregular thickness, sometimes embedded in a black calcareous shale, extremely rich in fossils, while at others it forms a rude shapeless mass of limestone, six to eight feet thick. At a short distance up the Monongahela river, the usual character is resumed, so that at Six-Mile ferry, Braddock's Station, Peter's Creek and other localities, it is as described above. It passes under the river near Peter's creek and is not exposed again within the district, though its horizon is reached below the State line.

In the limestone I observed *Productus Nebrascensis*, *Productus Prattenianus*, *Productus Longispinus*, *Productus semi-reticulatus*, *Hemipronites crassus*, *Spirifer cameratus*, *Spirifer plano-convexus*, *Athyris subtilita*, *Lophophyllum proliferum*, *Zeacrinus mucrospinus*, together with many indeterminate plates and stems of crinoids. In addition to these, the black shale at Pittsburg contains *Chonetes granulifera* and *Nautilus occidentalis*.

16. Ohio Coal "VII*b*."

Directly underlying the crinoidal limestone is a very persistent though variable little coal. Mr. White states that it is occasionally absent in Beaver county. It is rarely absent in Allegheny county, where it varies from three inches to one foot

SECOND GEOLOGICAL SURVEY OF PA.

SOUTH BEAVER COUNTY

GENERALIZED SECTION

BY I. C. WHITE

SEE PAGE 75. K.

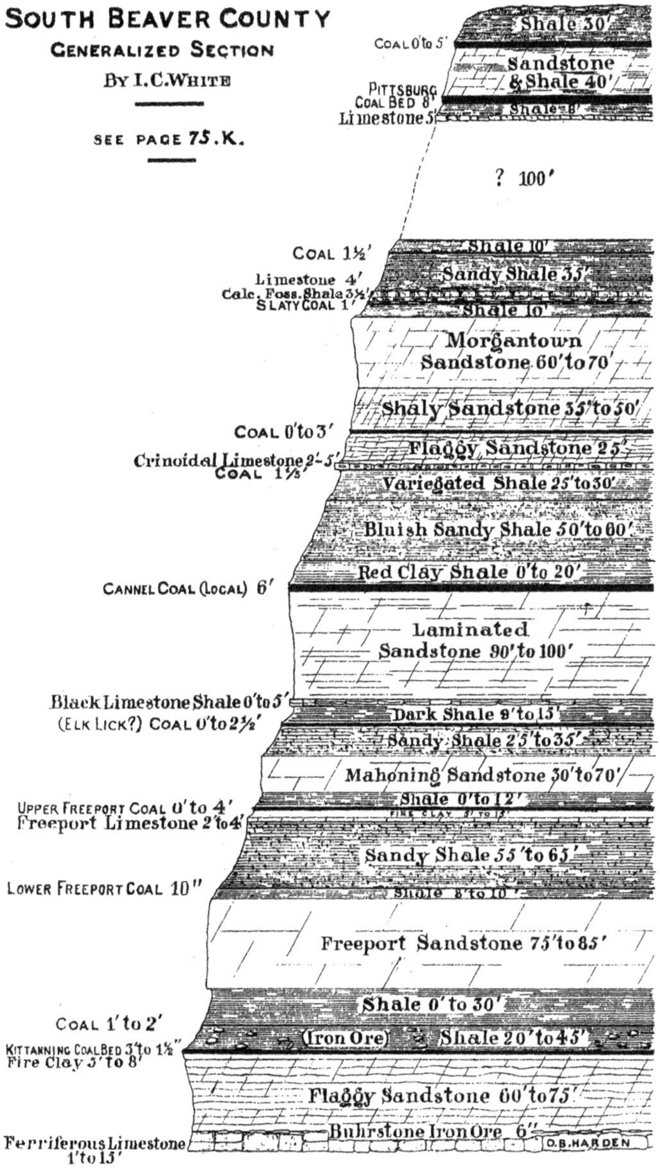

Shale 30'
Coal 0' to 5'
Sandstone & Shale 40'
Pittsburg Coal Bed 8'
Shale 8'
Limestone 5'
? 100'
Shale 10'
Coal 1½'
Sandy Shale 35'
Limestone 4'
Calc. Foss. Shale 3½'
Slaty Coal 1'
Shale 10'
Morgantown Sandstone 60' to 70'
Shaly Sandstone 35' to 50'
Coal 0' to 3'
Flaggy Sandstone 25'
Crinoidal Limestone 2'-5'
Coal 1½'
Variegated Shale 25' to 30'
Bluish Sandy Shale 50' to 60'
Red Clay Shale 0' to 20'
Cannel Coal (Local) 6'
Laminated Sandstone 90' to 100'
Black Limestone Shale 0' to 5'
Dark Shale 9' to 15'
(Elk Lick?) Coal 0' to 2½'
Sandy Shale 25' to 35'
Mahoning Sandstone 30' to 70'
Shale 0' to 12'
Upper Freeport Coal 0' to 4'
Fire Clay 3' to 13'
Freeport Limestone 2' to 4'
Sandy Shale 55' to 65'
Shale 8' to 10'
Lower Freeport Coal 10"
Freeport Sandstone 75' to 85'
Shale 0' to 30'
Coal 1' to 2'
(Iron Ore) Shale 20' to 45'
Kittanning Coal Bed 3' to 1½"
Fire Clay 5' to 8'
Flaggy Sandstone 60' to 75'
Buhrstone Iron Ore 6"
Ferriferous Limestone 1' to 15'

O. B. HARDEN

PHOTO-ZINCOGRAPH, by F. A. WENDEROTH & CO., 1328 Chestnut Street, Philada.

in thickness. It is of no economical importance. This is *coal VIIb* of the Ohio section.

Nos. 17, 18, 19, 20.

Below this is a mass of variegated clay which seems to be persistent. It rests on a mass of sandstone and shale, which is well exposed on the north side of the Ohio, at Wood's Run station on the Pittsburg, Fort Wayne and Chicago railroad, where it has a thickness of about 130 feet, showing no break except occasional changes from massive to flaggy, or *vice versa.* It is the sandstone at M'Kee's rocks, near the mouth of Chartiers creek. The same rock is partially exposed near Nimick's station, below the track level. In Beaver county, Mr. White found at somewhat more than one-third of the way from the top, a mass of argillaceous shale and a bed of impure cannel coal, six feet thick, both of which appear to be purely local, as they were found at only one place. These are Nos. 19 and 20, of the section. As seen at Wood's run and M'Kee's rocks, the lower portion of this sandstone is *massive*, blueish and micaceous, containing some thin layers of *conglomerate*, and is well adapted for use as building stone.

At Wood's run, the following section is seen below this stratum:

Dark shale	7'	
Coal		2''
Dark fissile shale	12'	

No lower strata than these are exposed in Allegheny county, for westward along the Fort Wayne railroad the sandstone forms great cliffs for several miles. In Beaver county, Mr. White finds at this horizon, Nos. 22, 23 and 24 of the section, which however are quite local, as they cannot be traced. No. 22 is a fossiliferous calcareous shale, containing many specimens of the more common Coal Measures species. The coal underlying it may be the *Elk Lick*, but it is difficult to determine exactly where that name belongs.*

Below this to the Mahoning Sandstone there is a sandy shale, which passes gradually into the sandstone below.

* [The name was given by me in 1840 to a four foot bed of coal, apparently overlying the Mahoning Sandstone, at the falls of the Elk Lick creek in southern Somerset county, and it has been productive of nothing but confusion ever since. J. P. L.]

The Lower Barrens in West Virginia.

In West Virginia, south from this district, the Lower Barren Series is well exposed on both flanks of Laurel Hill, but owing to the flattening of that axis southward, the upper rocks cross the fold south from the Baltimore and Ohio railroad, and the Lower Barrens are exposed only in part, where streams cut deeply into the axis. The *Little Pittsburg* and the *Coal No.* 6 of the Beaver section, seem to be persistent, having been found at some distance south from the railroad. The limestones are quite numerous in Monongalia county, but southward they become earthy and disappear, so that at the railroad there are only three of them left, of which the *Crinoidal* is one; even these disappear at an inconsiderable distance further south, so that on the east flank of Rich mountain the series is about 400 feet thick, and consists wholly of strange red shales with here and there an irregular sandstone.

In the disturbed region embracing the oil-break of West Virginia, the lower barren series is well exposed on both sides of the break, beginning at Ellenborough, on the Parkersburg branch of the Baltimore and Ohio railroad.

From the line of fault near that place the rocks are horizontal, or nearly so, to the immediate vicinity of the break. They are shales and sandstones, the former dull brick-red, the latter gray and apt to change into the prevailing shale. A thin limestone occurs near the top of the series and two, perhaps three variable coals are found, one far up, which may be equivalent to the *Little Pittsburg*, while the others are near the base, and have no representatives in the Greene and Washington district of Pennsylvania.

The *Crinoidal Limestone* is readily recognized in northern West Virginia. It still retains the peculiarity of color already noted, but disintegrates more easily upon exposure. Some portions weather into nodules, while others break down into a blue mud. The fossils are numerous, but the species are fewer than in Pennsylvania. Immediately underlying it is a mass of variegated shale, which is fossiliferous throughout, and rests on a thin coal. This group disappears altogether at a short distance from the Baltimore and Ohio railroad.

The Lower Barrens in Ohio.

In Ohio this series has its greatest thickness on the Ohio river not far from Steubenville, where the whole interval from the Pittsburg to the *Upper Freeport Coal* is 505 feet, the *Crinoidal Limestone* being 225 feet below the former. It is difficult here, as is frequently the case in north-eastern Ohio, to determine precisely the upper line of the Mahoning Sandstone, so that in calculating the thickness of the series, it is sufficient for our purpose to take the interval between the two coals referred to. At the western outcrop of the *Pittsburg Coal*, in the first geological district of the State, this interval is only 420 feet.

In the first district of Ohio there is a limestone under the *Pittsburg* which contains the minute univalves characterizing it in Pennsylvania. Below this there is no other persistent limestone until the *Crinoidal* is reached which, in that district, is from 140 to 225 feet below the *Pittsburg*. Its features are the same as in Pennsylvania, but occasionally it loses its flinty, granular structure and becomes a compact rock with a semi-conchoidal fracture. The number of species of fossils is larger than in Pennsylvania, and includes some which I have never found at any locality east of the Ohio river. Two localities in Ohio have yielded *teeth of fish.* Dr. Newberry states that he obtained a tooth from it opposite Pittsburg. This is the *fossiliferous limestone* which is given in Dr. Hildreth's section, as occurring at from 80 to 100 feet below the *Pomeroy* or *Pittsburg Coal*, and it is the *Ames limestone* of Prof. Andrew's in the Second Geological District of Ohio. The little coal underlying this limestone is persistent in Ohio, and in my reports on portions of that State, I have designated it as *Coal VII* "*B.*" It is sometimes of economical importance.

In Ohio there occurs at from 70 to 100 feet above the *Upper Freeport*, a wide spread but extremely variable coal, which Dr. Newberry has numbered *Coal VII.* It occupies the same relative position with the little coal at Wood's run below Pittsburg, and that observed in Beaver county by Mr. White. It has been hesitatingly identified with the *Elk Lick Coal.*

CHAPTER VIII.

The Lower Productive Coal Series.

The Lower Coal Series occurs only in Beaver county. It is not shown in full, as the section reaches no lower than to the *Ferriferous Limestone.* Beaver county was examined by Mr. White and the following summary, in so far as it relates to that county, is made up wholly from his notes. The section as obtained is approximately as follows:

Lower Productive Coal Measure in South Beaver.

1.	MAHONING SANDSTONE	30′ to 70′
2.	Shale	0′ to 12′
3.	*Upper Freeport Coal bed*	0′ to 4′
4.	Fire-clay and shale	5′ to 15′
5.	Freeport Limestone	2′ to 4′
6.	Sandy shale	55′ to 65′
7.	*Lower Freeport Coal bed*	0′ 0″ to 0′ 10″
8.	Shale	8′ to 10′
9.	Freeport Sandstone	75′ to 85′
10.	Shale	0′ to 30′
11.	*Coal bed* [*Strip Vein?*]	1′ to 2′
12.	Shale containing iron ore	20′ to 45
13.	*Kittanning Coal bed* [Creek Vein]	3′ to 1· 6″
14.	Fire-clay	5′ to 8′
15.	Flaggy sandstones	60′ to 75′
16.	Iron ore	0′ 6″
17	Ferriferous Limestone	1′ to 15′

1. MAHONING SANDSTONE.

The Mahoning sandstone is a variable rock, sometimes occurring in massive cliffs seventy feet high, while at others it is a poor flaggy sandstone or even sandy shale barely thirty feet thick. It first comes up from the bed of the Ohio at a short distance above the Beaver county line, whence it rises rapidly, so that at the mouth of the Beaver it is 250 feet above the river. Thence to the State line, along the Ohio, it maintains about the same elevation, as the stream flows irregularly with the strike. On this line it is for the most part a compact rock, which is a most excellent building stone. It consists chiefly of quartz grains and is speckled with slight stains of oxide of iron, which give it a peculiar appearance, so that the quarry-men usually speak of it as the "pepper and salt rock." Near the base it shows a layer of conglomerate containing pebbles as large as a hazel-nut. It passes under Raccoon creek in Beaver county near Independence, and form cliffs along the stream below that village. Mr. White makes no reference to any thin coal in this stratum and as he had opportunity to examine the whole at many localities it is quite certain that no such coal exists in southern Beaver. But in West Virginia at many localities, a thin coal does occur about midway, and a similar condition is found over a large portion of Ohio.

2. Shale.

Between the Sandstone and the *Upper Freeport Coal* there is a shale which varies in thickness from a few inches to twelve feet, and at some places is altogether wanting.

3. Upper Freeport Coal Bed.

The Upper Freeport Coal here shows as strange, though by no means such extreme variations, as those which I have described in my "Notes on the Geology of West Virginia." In some localities it is absent, in others only six inches, while in its extreme development it is a mass of coal and shale eleven feet six inches thick. Between these extremes every gradation occurs. Near Moffatt's Mill, on Raccoon creek, it is two feet five inches thick and separated from the Mahoning Sandstone above by two feet of shale. At Swearingen's coal works, near Hookstown, in Greene township, it shows, in five feet seven inches of measures—

Coal, Slaty		4″
Bituminous shale	1′	
Coal	3′	6″
Clay		3″
Coal		6″

and on Raccoon creek, near the mouth of Service creek, a total section of eleven feet six inches is thus sub-divided—

Coal		8″
Shale	6′	6″
Bituminous shale	1′	
Shale	2′	
Bituminous shale		4″
Coal	1′	

and always shows the clay parting near the base, as in the Swearingen section. The bench below this parting is from six to ten inches thick, and in quality its coal is rather inferior to that from the bench above.

5. Freeport Limestone.

The Freeport Limestone is quite persistent, varying from light dove to buff in color, and from two to four feet in thickness. Occasionally it is absent, as at Phillipsburg, where its place is occupied by a sandy shale containing little calcareous matter. It sometimes shows minute univalves, but for the most part it is non-fossiliferous.

7. Lower Freeport Coal Bed.

The Lower Freeport Coal is so insignificant in Beaver county that Mr. White hesitated to make the identification. Some inconsistencies in the old report, due doubtless to clerical errors or to oversight in correcting the proof, would lead to uncertainty respecting the true place of this coal, as in a few sections it is placed under instead of over the Freeport Sandstone. The typical section, however, leaves no room for doubt, and *places the coal over the sandstone*, so that this little coal must represent the one referred to by Prof. Rogers as the Lower Freeport. In that portion of Beaver included within the district, this bed never is more than ten inches thick and is traceable only with difficulty.

9. Freeport Sandstone.

The Freeport Sandstone is micaceous and constantly compact. It is very hard, never less than seventy-five feet thick, and forms vertical cliffs along several of the streams. Near the middle is a layer four feet thick, which is excessively hard and in structure resembles quartzite. Water-worn specimens of this portion have a peculiar polished surface, which at once distinguishes them from fragments of any other rock. It is evidently persistent, having been observed at numerous localities.

11. Strip Vein Coal Bed.

Below the sandstone and separated from it by a shale, varying from zero to thirty feet in thickness, is another coal. Respecting its relations to the coals farther east no definite information has been obtained, as these lower rocks are in great measure concealed on this side of the river by the terraces. The numerous sections obtained on the opposite side during the progress of the former survey afford no assistance, as they are too far apart. The difficulty arises from the fact that the interval between the *Lower Freeport* and the *Kittanning* is abruptly diminished in that direction from one hundred and fifty feet to less than forty feet. The whole section, as displayed in this portion of Beaver, is found at Smith's ferry on the Ohio State line, so that a series of detailed sections from that point to the mouth of the Beaver would doubtless remove all the difficulty. In all probability such a series would show the presence of an old anticlinal not far from the line of the Bulger axis. In Ohio such folds occur between the *Kittanning** and the *Upper Freeport.* This coal is persistent along the Ohio below the State line, and is identical with the "*strip vein*" of Yellow Creek, Ohio, which has been regarded as the equivalent of the Ohio *Coal IV* or the *Kittanning.* In Beaver county it is from one to two feet thick and yields a good coal. Though so thin, it has been opened in Moon, Greene and Hopewell townships to supply local demands. The shale underlying this coal contains much low grade kidney iron ore, but it is so scattered throughout the mass as to be of no economical value.

13. Kittanning Coal Bed.

The Kittanning Coal rises from the river bed about a mile and a half above Freedom, and soon is 75 feet above the low water mark. Along this line it is usually concealed by the heavy terraces lining the stream. Few openings were seen, and those are all in the vicinity of the Ohio river. Southward from the river the coal becomes so bad that it is familiarly known as the "Sulphur vein." It is three feet thick at the mouth of Raccoon creek, two feet one inch near Phillipsburg, and the same near Georgetown. At both of the last two localities it is very bad. This is the "creek vein" of Yellow creek, Ohio, and is easily traced down the river from the State line.

* Or rather with the coal, which is there regarded as equivalent to the *Kittanning.*

14. Fire-brick Clay.

Underlying this bed is an important deposit of fire-clay, which is manufactured into fire-brick at Phillipsburg. It seems to be persistent on the Ohio below the State line, and is employed extensively both as fire and potter's clay at numerous localities along the river in Ohio. It is not persistent eastward and seems to disappear before reaching the mouth of the Beaver. The same is true of the similar clay beneath the coal underlying the Freeport sandstone.

17. Ferriferous Limestone.

The Ferriferous Limestone is for the most part concealed by the terraces, but it is well exposed at two and a half miles below the mouth of Raccoon creek, where it is twelve feet thick, on the land of Mr. Allen. Followed down the river it is found to become thin and impure, so that at half a mile below the last locality it consists of only one foot of impure limestone, which has a cone-in-cone structure. On Mr. Allen's property it is a compact bluish rock, wonderfully rich in fossils, and yielding excellent lime. It answers well as a flux for iron ore. Resting immediately on this limestone is a thin deposit of iron ore six inches thick. It varies little in thickness, and though of good quality is not available.

CHAPTER IX.

Relations of the Strata.

The studies of 1875 have led me to the conclusion that the numerous anticlinals within the district are extremely ancient; that some of them, at least, date back certainly to the middle of the Lower Barrens, and that the elevation of these axes was a gradual process, the result of lateral pressure and subsidence. This conclusion is supported by evidences of sub-aerial erosion, by the distribution of the limestones and by the variations in the intervals between the more important strata.

A great number of facts appear in the sections, which bear directly on the hypothesis proposed by me in 1872 to account for the origin of the Upper Coal strata, and thus far they serve to confirm the correctness of that hypothesis.

Though a discussion of these matters would, no doubt, be of interest, yet at present it could not be other than unsatisfactory. The season of 1876 will be spent by me in studying the same troughs on the eastern side of the Monongahela river, where they contain the eastern and north-eastern outcrops of the Upper Barrens and upper coals, as well as the lower groups of the Coal Measures. At some localities, also, the underlying rocks will be reached. This work will complement that of 1875, and in all probability, will afford the means of instituting direct comparisons of all parts of the great trough containing the upper coals. As all this material is essential to a proper discussion of the relations of the strata, I have thought it best to defer such a discussion until the results of 1876 can be obtained.

A brief examination of the relations of the oil obtained in this district will be found in the chapter on Economical Geology.

PART III.

THE GEOLOGY DESCRIBED BY TOWNSHIPS.

CHAPTER X.

Dunkard Township, Greene County.

In this township the exposed rocks extend from 370 feet below the *Pittsburg coal* to about 70 feet above the *Washington coal*, the higher strata being found on the northern and western borders and on a few high knobs in the center. The lower barren series is exposed only in the vicinity of the river and the surface rocks for the most part belong to the upper coal series. The dip throughout the township is toward the north west, but the rate is quite variable.

On Crooked run, in the south-eastern corner, where the *Pittsburg coal* attains almost its highest elevation above the river, the following section was obtained:

1. Sandstone	30′
2. *Pittsburg coal*	7′
3. Concealed	130′
4. Limestone	5′
5. Shale	15′
6. *Coal*	blossom.
7. Sandstone	95′
8. *Coal*	2′ 4″
9. Shale	4′
10. Clay shale	8′
11. Concealed to river	80

No. 7 is the first sandstone of the oil-wells on Dunkard creek and, as here exposed, is irregularly massive to shaly. It is persistent in the river hills, being traceable quite to Morgantown in West Virginia, though it shows marked variations in structure. At many localities it is extensively quarried for building purposes. During its deposition it cut out the underlying *coal*, which varies from 0 to 2 feet 4 inches in thickness. Fragments of coal are numerous in the lower portion of the sandstone, many of which are saucer-shaped, as though the material were soft when torn from its place. The *Pittsburg* is not opened in the vicinity of the section and its overlying sandstone is extremely coarse, almost conglomerate.

Opposite the mouth of Cheat river, the *Pittsburg Coal* is mined by Mr. A. Dillner, at 372 feet above the Monongahela river. It there shows the following structure and association:

1. Sandstone		30′ 0″
2. Shale		8′ 0″
3. *Pittsburg Coal.*		
1. *Coal*	4″	5′ 7″
2. Clay	1″	
3. *Coal*	5′ 2″	

This is an extraordinary degradation of the bed not often seen where the overlying rock is an argillaceous shale as in this case. I am inclined to suppose that the whole Roof-division was removed in some way, if indeed it ever existed, before the deposition of the shale, since the portion remaining does not show the ordinary thickness of the Lower division. No. 3 is only part of the upper bench. Mr. Field's opening, about a mile and a half further down the river and near Crow's ferry, shows the normal structure as follows:

1. Sandstone		35′ 0″
2. Shale with *coal* streaks		1′ 0″
3. *Pittsburg Coal.*		
1. *Coal*	6″	9′ 7″
2. Clay	3″	
3. *Coal*	8″	
4. Clay	1′ 2″	
5. *Coal*	7′ 0″	

The direction from Dillner's opening to this locality is almost north-west, and the fall of the *coal* is nearly 75 feet. At the mouth of Dunkard creek, three miles north north-west from Dillner's, the *Pittsburg* is opened and the following section is exposed:

1. Sandstone	40′
2. *Pittsburg Coal*	6′
3. Concealed	50′
4. Flaggy sandstone	45′
5. *Coal*	blossom.
6. Flaggy sandstone	25′
7. *Coal*	1′ 6″
8. Shale	30′
9. Limestone	4′
10. Concealed to river	60′

The dip to this point is somewhat more than 50 feet per mile, north north-west. The section shows a material thickening. The interval between the *Pittsburg Coal* and the limestone No. 9, being 150 feet, as against 130 feet opposite the mouth of Cheat River.

In the south-east corner of the township, openings upon the *Pittsburg* are quite numerous. The bed being worked by Messrs. L. Titus, Daniel Miller, George Brown, J. Van Varis and others. At Mr. Miller's opening, it shows its minimum thickness, the section being :—

1. Sandstone		40′ 0″
2. *Pittsburg Coal.*		
1. *Coal*	1′ 3″	6′ 11½″
2. Clay	1½″	
3. *Coal*	5′ 7″	

At the other banks the total thickness of the bed varies from eight feet four inches, near Zion church, to ten feet two inches at Mr. Titus' bank, half a mile east from Wiley post-office. The roof is usually thin and single, and yields a very pretty but extremely impure coal.

Along the Morgantown road the *Pittsburg* disappears under the hill about a mile north from the State line and near Mr. J. Dorr's, the *Redstone* is seen in the road. Its thickness is quite variable, being barely five inches at one exposure, while at another the blossom indicates a bed of not less than two feet. The *Sewickley* is first seen in the same vicinity and is mined by Mr. A. Lucas, about three-fourths of a mile farther north, where the following exposure occurs:

1. Shaly sandstone		30′ 0″
2. *Sewickley coal.*		
1. *Coal*	2′ 1″	5′ 5½″
2. Clay	½″	
3. *Coal*	3′ 4″	

This is a good coal. The rocks between the *Pittsburg* and the *Sewickley* are shown at a little distance farther north near Mr. B. Ross' residence, as follows:

1. *Sewickley coal*	5′
2. Shale	10′
3. Limestone	8′
4. Sandstone and shale	22′
5. *Redstone coal*	blossom.
6. Limestone	4′

7. Shale	20′
8. Sandstone	40′
9. *Pittsburg coal*	8′

Crossing to the west of the Morgantown road and still keeping to the south of Dunkard creek, we find the *Pittsburg coal* passing under Crooked run near Mr. Garlow's residence, where the exposure is quite imperfect, only six feet of the bed being seen. On this run, which follows somewhat irregularly the boundary between West Virginia and Pennsylvania, the *Sewickley* is mined by Mr. Scott John, about four miles from the river. It is roofed by twenty feet of sandstone and is in two divisions, two feet four inches and three feet respectively, separated by a thin clay parting. The interval between the *Sewickley* and the *Waynesburg*, as obtained in this vicinity on Mr. J. E. Taylor's property, just north from the run, is 260 feet. The *Sewickley* soon passes under the run, at about four and-a-half miles from the river, the *Waynesburg* is caught in the hills. The first openings in this coal observed along this run are those of Mr. J. Bowlsby on the State line, where the following measurement was made:

1. Sandstone		40′ 0″
2. Drab shale		5′ 0″
3. Dark shale		2′ 0′
4. *Waynesburg Coal.*		
1. *Coal*	1′ 0″	
2. Clay	0′ 3″	
3. *Coal*	2′ 4″	7′ 9″
4. Clay	0′ 8″	
5. *Coal*	3′ 6′	

No. 3 of this section is quite rich in impressions of plants, many of which are exceedingly well-preserved. The section of the coal is a typical one for the vicinity, though the relative thicknesses of the benches will not hold good, as in this respect the bed shows great and abrupt variations. Two openings on Smith's Run, the western boundary of the township, present the following differences:

1. *Coal*	1′ 2″	1′ 0″
2. Clay	0′ 2′	0′ 2″
3. *Coal*	2′ 1″	2′ 4″
4. Clay	2′ to 3′ 0″	1′ 4″
5. *Coal*	2′ 9″	2′ 6″

Sections taken in either of these banks at intervals of forty feet show greater variations than those in the measurements just given. The coal is far from being of first quality at any of the openings in this corner of the township.

Along Dunkard creek exposures are frequent, but until Bobtown is reached they are incomplete and few coal-banks are seen. At that village the Morgantown road crosses Dunkard creek near Maple's woolen mill, in the immediate vicinity of a vast number of oil-derricks. This is the central point of the Dunkard oil region. Here Syck's knob, rising almost 450 feet above the stream, shows a very imperfect section, but exhibits the *Uniontown* at 300 feet above the *Pittsburg*, which is worked directly opposite the bridge. The *Uniontown* is represented by a fissile black shale, underlaid by a ferruginous limestone, which weathers bright yellow. The *Sewickley* and *Redstone* are concealed, and the section from the *Pittsburg* to the creek is as follows:

1. Sandstone or sandy shale		30′ 0″
2. *Pittsburg Coal*		
1. Roof division	3′ 3″	
2. Clay	1″	
3. *Coal*	1′ 9″	
4. Clay	1″	
5. *Coal*	1′ 0″	9′ 1½″
6. Clay	⅛″	
7. *Coal*	7″	
8. Clay	½″	
9. *Coal*	2′ 5″	
3. Concealed		5′ 0″
4. Sandstone and shale		10′ 0″
5. Shale and iron ore		12′ 0″
6. *Coal*		1′ 6″
7. Limestone in bed of creek		

The roof-division is three feet three inches thick, and consists of two benches of coal, two inches and two feet respectively, separated by one foot of clay. The coal is very poor and is not removed. The main parting is reduced to the minimum, being only one inch thick. With one exception this is the only locality in the southern portion of the district where a parting is seen in the upper bench of the lower division. Here it is one inch thick, well-marked and persistent. The coal is of very fair quality, but contains an appreciable quantity of sulphur. The little coal below is said to be excellent and admirably adapted to blacksmiths' use. At this exposure the

Pittsburg sandstone is much degraded, being only a loose irregular sandy shale more or less contorted, but at a short distance farther up the stream it resumes its proper character. About half a mile above Bobtown we find the following opposite Mr. Hufty's house:

1.	Sandstone, seen	25′
2.	*Coal* and shale	2′
3.	*Pittsburg Coal*	9′

The sandstone is compact and in parts very clear and white, but owing to uneven structure it weathers into extensive cavities. Indistinct impressions of *Spirophyton* occur here. No. 2 of the section represents the Roof division of the bed. As its coal is fragmentary, I think that the overlying sandstone, during its deposition tore out the upper portion of the bed and mingled the coal and shale indiscriminately. This seems tne more probable, since at an exposure about one hundred yards below this the sandstone is much distorted at its base and contains fragments of coal; and in the same vicinity it rests directly on the lower division of the coal. The structure of the bed cannot be made out at the excavation near Mr. Hufty's house, as the face is covered with dirt and moss, but there is much pyrites present. A fine sulphur spring issues from the coal here. The flow of water is not large, but sulphuretted hydrogen is present in very considerable quantity.

Above this locality, on the road leading up the creek, there are no good exposures until within half a mile of Fairview, where an irregular opening on the *Pittsburg* is seen extending along the bank at the roadside for more than twenty feet. The overlying rock is the Pittsburg sandstone, whose under surface, where exposed by the removal of the coal, resembles a group of saucers arranged side by side. The upper division of the bed is from twelve to nineteen inches thick and consists of shale with numerous thin streaks of coal, which occasionally unite to form a layer several inches thick. The lower, of which only five feet can be seen, is badly broken up by hard clay "binders," which are persistent and often pyritous. Much of it is bony, and as a whole, the mass is of poor quality. The irregularity of the opening is due to the extreme dip, which is north 70° west, mag. at the rate of one foot in six. The Pittsburg sandstone as shown here is about twenty-two feet thick, coarse, incoherent and containing much feldspar in grains. Directly above the sandstone there is a richly bituminous shale, two or three feet thick, which contains some coal in thin streaks.

On the banks of the creek near Fairview the coal is mined somewhat extensively for local use. The section is

1. Sandstone seen		20′ 0	
2. *Pittsburg Coal.*			
1. Roof Division	1′ 6″	7′ 11½″	
2. Lower Divison	6′ 5½″		
3. Sandstone		20′ 0″	

The sandstone roof is quite irregular and encroaches upon the coal, having bowl-shaped excrescenses on the under side. The layers of the roof division for a foot or more coincide in direction with the base of the sandstone giving the whole a concentric structure. The main clay parting is absent, but in many places the surfaces in contact are slickensided. At no point, however, is there any difficulty in distinguishing the divisions, as the upper is always bony, burning well and making an intense fire but leaving a great proportion of ash. The lower division is said to improve in quality from the top downward so that, as the blacksmith at Fairview informed me, the best and purest coal is at the bottom of the bed. This is somewhat unusual, for at most localities the coal at the base is regarded as too poor to pay the cost of mining.

The dip is quite sharp, being one foot in ten, as was determined by careful levelings at three openings. It is, however, very irregular and the rate given is only the average obtained in a distance of 200 yards. Still there can be no doubt that a line of disturbance passes through this point and extends certainly to the Maple farm at Bobtown.

At a short distance south from Fairview, near the Center school-house, the *Sewickley coal* is seen at 110 feet above the *Pittsburg*. Its structure is as follows:

Sewickley coal.		
1. Bony *coal*	3″	5′ 2″
2. Parting	⅛″	
3. *Coal*	2′ 5″	
4. Parting	1″ to 2″	
5. *Coal*	2′ 4″	

No. 1 is a new feature in this township having been observed at none of the openings in the south-eastern portion. The coal throughout the bed is much better than that from the *Pittsburg*. The bottom bench is much prized by blacksmiths, but the middle bench is harder, less clean and contains more or less of pyrites in nodules. It is evident that the purity of the bottom is exaggerated, for after exposure it becomes covered with streaks of copperas. The *Pittsburg coal* is found in the bed of the creek about half a mile above Fairview, and in the same vicinity Mr. Steinrod Everly has an opening on the *Sewickley*. There the following section was obtained:

1.	Shaly sandstone	40′
2.	*Sewickley coal*	5′ 1½″
3.	Shale	12′
4.	Limestone	18′
5.	Shale	10′
6.	*Redstone coal*	1′ 6″
7.	Limestone	10′
8.	Sandstone	60′
9.	*Pittsburg coal.*	

Here the *Sewickley* shows no traces of the top or bony bench and is in two divisions, two feet and three feet respectively, separated by an inch of compact clay. At the mouth of Meadow run it is at the level of the creek and is mined by Mr. John Debolt. Its benches are two feet and two feet two inches, separated by one inch of hard clay. The *Waynesburg* is seen here on the hill 270 feet above the *Sewickley*. An excellent section of the interval is exposed on Meadow run below Davistown as follows:

1.	*Waynesburg Coal.*	
2.	Concealed	75′
3.	Sandstone and shale	20′
4.	*Uniontown Coal*	1′
5.	Limestone and calcareous shale	23′
6.	Sandstone and shale	60′
7.	Limestone and calcareous shale	53′
8.	Shaly sandstone	40′
9.	*Sewickley Coal*	4′ 3″

The *Uniontown coal* as here exposed is a richly bituminous shale with a cannel-like fracture, and contains numerous fish teeth and scales, all of them minute, together with occasional indistinct lamelli-branchiates. At a short distance farther up Dunkard, the concealed portion of this section immediately below the *Waynesburg* is exposed as follows:

1.	*Waynesburg Coal*	8′
2.	Shale and sandstone	40′
3.	Bituminous shale	2′
4.	Limestone	15′

Near the township line the *Waynesburg* is mined by Mr. W. M'Clure; it is in four benches, and shows a total thickness of eight feet two inches. Within a distance of forty feet the second and third benches vary from two feet three inches and two feet ten inches, to one foot six inches and one foot two inches, while in the same distance the intervening clay varies from one foot six inches to five feet. The coal, as is ordinarily the case with the *Waynesburg*, is poor and slaty.

Passing now to the north of Dunkard creek and beginning at the east end of the township, we find the *Sewickley* mined on the property of Mr. John South, where it shows the following section:

	Sewickley Coal.		
1.	*Coal*	1′ 11″	5′ 5½″
2.	Clay	2½″	
3.	*Coal*	3′ 4″	

The measurement is not complete, as the bottom of the bed could not be reached owing to the bad condition of the bank, and the thickness is probably not far from six feet. The clay parting is double and usually embraces an inch of bony coal. Its position is somewhat variable, as was ascertained by measurements made at several openings on South's run and on Deep run, the latter a tributary to Dunkard creek, entering near Bobtown. In this vicinity both benches of the bed yield excellent coal for fuel, but the lower bench is preferred for domestic use. The upper bench is hard and is mined with some difficulty; but it evidently contains much less pyrites than the lower one. The coal from all parts of the bed is quite free-burning and leaves a powdery white ash. The *Redstone* is represented here only by a bituminous shale.

On the farm of Mr. A. Garrison, in the north-east corner of the township, the *Waynesburg* is mined. The exposure at the bank is imperfect, and is as follows:—

1.	Sandstone		8′ 0″
2.	*Waynesburg Coal*, seen.		
1.	*Coal* and shale	11″	6′ 11″
2.	*Coal*	2′ 1″	
3.	Clay	2 0″	
4.	*Coal*, seen	2′ 0″	

At another opening, immediately adjoining, No. 3 is two feet eight inches thick, and No. 5 is exposed for two feet four inches; it is said to reach three feet. The coal is very poor, leaves a bulky ash and contains so much sulphur as to be very destructive to stoves. Openings on the same coal were seen in this vicinity on the farms of J. Garrison, J. M. Crumpshon and Mrs. Garrison. In the road near Mr. A. Garrison's house, the *Washington Coal* is exposed in the road at 160 feet above the *Waynesburg*, and is one foot six inches thick.

At Mr. N. Knott's opening in the *Waynesburg* on the Morgantown road, near the line of Greene township, the section is:

1. Sandstone		—
2. *Waynesburg Coal.*		
1. *Coaly* shale	10″	8′ 4″
2. *Coal*	1′ 9″	
3. Clay	1′ 9″	
4. *Coal*	4′ 0″	

The coal does not appear to be very pyritous and is quite open burning. For use in grates Mr. Knott prefers it to either the *Pittsburg* or *Sewickley.*

No other openings were found in this portion of the township. On the Davistown road near the Greene township line, the hills rise nearly 300 feet above the *Waynesburg*, but without affording any exposures. The *Waynesburg "A"* is not exposed in this interval; indeed it was not seen in this portion of the township. As one approaches Davistown, descending the east fork of Meadow run, he sees at the first fork in the road above that village, a slight blossom which probably belongs to this coal. That is the most south-easterly exposure of the bed. At Davistown the *Waynesburg* is mined somewhat extensively for local use, and the following sections were obtained at the opening back of the steam mill:

Waynesburg Coal	6′ 10″ to 7′ 0″
1. *Coal* and shale	0′ 7″ to 0′ 8″
2. Clay	0′ 1″ to 0′ 2″
3. *Coal*	1′ 10″ to 2′ 3″
4. Clay	1′ 11″ to 1′ 4″
5. *Coal*	2′ 5″ to 2′ 7″

These measurements are only fifteen feet apart, and exhibit imperfectly the variations. The top bench is well defined throughout the tunnels and is always of poor quality. The others yield good coal, which, for domestic purposes is thought to be fully equal to that from the *Sewickley* at the mouth of Meadow run. The Main Clay No. 4 is slickensided and interferes much with work, but seldom forms extensive horsebacks. No clay veins have been encountered in this vicinity.

This bed goes under the run about half a mile above Davistown, and the blossom of the *Washington* is seen at several places along the Garard's Fort road. On Lucas' fork, the latter coal is exposed near M'Clure's school house, and passes under the stream at a short distance above. On Bowen's fork it shows near Mr. Bell's residence. On Glade run it has been opened by Mr. M. Wildman, at whose bank it is composed of two benches, one foot six inches and two feet six inches respectively, separated by six inches of clay shale. The lower bench is quite good. The *Waynesburg Coal* is mined by Mr. L. M'Clure, about half a mile east from the line of Perry township, where the exposure is as follows:—

1. Sandstone		40′
2. Shale		6′
3. *Waynesburg Coal.*		
1. *Coal*	10″	
2. Clay	2″	
3. *Coal*	2′ 6″	8′ 0′
4. Clay	1′ 6″	
5. *Coal*	3′ 0″	

The coal does not differ in quality from that observed at other openings in the township.

Oil-Wells on Dunkard's Creek.

Perhaps the most interesting feature of the geology in this portion of the county is the occurrence of oil in considerable quantity along Dunkark creek. The general characteristics of this deposit and its relations to other localities within the county have been discussed in another portion of this report; so that it remains to give only such details as may be merely local in importance.

The productive oil territory begins about two miles above the mouth of Dunkard and extends along the line of the creek for barely two miles, including the Elliott, Maple, Garrison, Bailey and Ross farms. For the most part the wells were bored on the bottom and the curbs are only a few feet below the *Pittsburg coal.*

Elliott Farm.

Three wells were bored here, all of which were productive. One of them, the Bobtail, flowed for nearly three weeks, giving an average of 80 barrels per day; but the flow gradually diminished and at last ceased. Its total yield is not far from 5,000 barrels. Only one well is now in operation and that gives to the pump one or two barrels each day. The oil comes from the lower horizon about 400 feet below the Pittsburg coal. In each of the wells some oil was found at the upper horizon, but the quantity was very small.

Maple Farm.

Between this and the Elliott farm wells were bored at several localities, but they proved either insignificant or utterly barren. A great number of borings were made on the Maple farm and it was the main center of the oil excitment. The yield was extensive but in the absence of all records I am unable to estimate it even approximately. The Wiley well No. 1 has produced more than 5,000 barrels, and still yields a little each day. The Allegheny well at first flowed at the rate of 125 barrels per diem but the flow gradually decreased until the well ran dry. The Lone Star still yields a small quantity, two or three barrels a week. It flows on provocation, pumping being needed only to stir up the gas. The oil was obtained on this farm at two horizons, 175 and 440 to 460 feet below the *Pittsburg coal.* Salt water was found at various depths but no tests were made to determine its quality.

Garrison Farm.

This adjoins the Maple farm northward and includes Deep run, on which the Butler well was bored. This well, like the Lone Star flowed upon provocation, and yielded in all about 1,500 barrels. The flow was stopped by caving of the walls. No oil was obtained here at the upper horizon and very little water, either salt or fresh, was encountered in the boring.

Bailey Farm.

This adjoins the Maple farm along the creek. Only three wells were bored here. No. 1 found a good supply at the upper horizon and yielded at first from 30 to 40 barrels daily. It now gives only one or two. No. 2 has proved almost unproductive, and No. 3 is known as the "Drywell," as no oil and very little water have been found in it. Good brine, said to be ten degrees strong, was obtained from Nos. 1 and 2.

Ross Farm.

This is next above the Bailey farm on the creek. Of the fifteen wells put down here, twelve are on the "bottom" beginning at about ten feet below the *Pittsburg Coal.* Oil was obtained in large quantity at both horizons. The first well, known as Ross No. 1, flowed 200 barrels during the first day and then suddenly stopped as the walls had fallen in. A similar mishap befel several other wells which opened in an equally promising manner. Two are now in operation, which yield seven to eight barrels per week from the lower horizon.

Borings have been made at several localities along the creek above the Ross farm but they have proved uniformly unsuccessful. As no oil has been obtained at any point below the Elliott farm it is easy to define the limits, north and south, of the productive territory.

Oil from the upper horizon, 170 to 180 feet below the Pittsburgh coal is uniformly heavy, usually about 32 degrees. The only exception is the Bailey well No. 1, in which the gravity is 40 degrees. The lower horizon, in the Mahoning sandstone, yields a much lighter oil of 40 to 42 degrees gravity. This oil is much valued in the raw condition for cleaning wool, and when reduced by exposure or steaming to about 34 degrees it becomes a lubricating oil of excellent quality. By some of the inhabitants crude oil is used for illuminating purposes. Experience there leads me to assert that it is not well adapted to that use.

2. PERRY TOWNSHIP, GREENE COUNTY.

This adjoins Dunkard township on the west, and Dunkard creek flows irregularly near its southern border. The exposed section extends from nearly 400 feet above the *Washington Coal* to about fifty or sixty feet below the *Waynesburg.* The direction of dip is toward the north-west in the greater part of the township, but at the west it is slightly reversed owing to the influence of the Blacksville anticlinal.

The rate is never more than seventy-five feet to the mile, and for the most part it does not exceed fifty feet.

The *Waynesburg Coal* is exposed only in the south-eastern portion of the township, on Dunkard creek and Morris run, the latter being a tributary coming in at Mount Morris. On the former it is mined by Mr. G. T. Long, half a mile above the township line and shows the following section:

Waynesburg Coal.

1. *Coal*		6″	6′ 6″
2. Clay		3″	
3. *Coal*	2′	6″	
4. Clay		3″	
5. *Coal*, seen	3′		

A very similar section is seen at Mr. Donley's opening on the opposite side of the creek. But at Mr. W. M'Clure's, in the immediate vicinity, the clay No. 4. is one foot ten inches thick. At Mount Morris the Waynesburg sandstone is exposed, and the coal is probably 35 or 40 feet under the creek at that place. Ascending Morris run the strata are seen rising more rapidly than the bed of the stream, so that the coal soon comes to the surface and continues in sight to beyond the State line. The openings are very numerous and the exposures of the lower or more compact portion of the sandstone are very satisfactory. The bed shows less abruptness in variation here than usual, and the following section, obtained at Mr. Donley's bank may be regared as typical for this locality:

1. Sandstone			45′ 0″
2. Shale			8′ 0″
3. *Waynesburg Coal.*			
1. *Coal*		6″	9′ 10′
2. Clay		4″	
3. *Coal*	2′		
4. Clay	2′	6″	
5. *Coal*	4′	6″	

The characteristic plants of this horizon are found in Nos. 2 and 6. In these openings Nos. 3 and 5 are usually left in the bank, as they are too impure to be used profitably. No. 7 shows a thickness rarely equalled at any localities northward, and is much cleaner than one ordinarily finds it. A blacksmith at Mount Morris states that a layer one foot thick and near the bottom of the bed is preferable to the *Pittsburg* for smithing purposes.

On Colvin's run, which enters Dunkard below Mount Morris, as well as at the mouth of Morris run, the blossom of the *Washington* is seen, but is so insignificant that the bed must be very thin.

Crossing Dunkard creek just above Mount Morris we find, in the road, the *Waynesburg* "*a*," which is seen occasionally for more than half a mile. It is very thin, barely eight inches, and contains a good deal of sulphur. At the mouth of Bacon run it is below the surface, and the *Waynesburg* "*b*" is seen in the hillside at 35 feet above the road. This is six inches thick, and is enclosed between an overlying sandstone and an underlying limestone, Below the limestone and the *Waynesburg* "*a*" there is only shaly sandstone. The *Washington Coal* passes under this run about a mile above its mouth, but without yielding any satisfactory exposures. It is evidently thin, and no opening has been made upon it. The black shales below contain a little coal and some low grade iron ore. Limestones II and III of the upper barren series are seen farther up the run.

On Shannon's run, which comes in about a mile above Mount Morris, the *Waynesburg* "*b*" remains in view to the forks of the stream where it is seven feet above the road. It is eight inches thick and so compact that it stands out like a shelf, having withstood the effects of the weather better than did the sandstone above or the limestone below. This underlying limestone is about one foot thick and appears to be quite persistent in this township. On the south fork of the run the exposures are very fair, and the following section was obtained in going from the mouth of the stream to the summit between it and Headlee's run:

1. Concealed	20′
2. Limestone V	
3. Concealed	35′
4. *Jolleytown Coal*	blossom.
5. Shale	90′
6. Limestone (local)	4′ 6″
7. Sandstone	75′
8. *Washington "a" Coal*	3′
9. Shale	20′
10. Limestone III	3′-4′
11. Sandstone	40′

12. Shale	3′	
13. Limestone II	1′	6″
14. Shale	5′	
15. *Washington Coal*	blossom.	
16. Laminated Sandstone	10′	
17. Dark shale	8′	
18. Sandstone	23′	
19. *Waynesburg "b" Coal*		8″
20. Limestone	1′	
21. Sandstone	40′	

The section has been carried down to the mouth of the run where the *Waynesburg "a"* cannot be more than four or five feet below the stream. A similar section occurs on the North fork. The *Washington Coal* is not satisfactorily exposed on either of the forks, but its blossom is persistent and shows that the bed is thin. It passes under the South fork about half a mile above the junction. At a mile up this fork the *Washington "A"* is seen, apparently about three feet thick. Like the Limestone No. 6 it is local, having been identified only in Perry, Wayne and Whitely townships. The structure of the limestone is—

Limestone		8″
Calcareous shale	1′	6″
Limestone	2′	6″

The lowest layer is quite arenaceous, and weathered fragments are readily mistaken for sandstone. The *Jolleytown Coal* is seen only as a blossom, but is certainly very thin, not more than three or four inches. Between Headlee's and Rudolph's run there are no exposures, and the country rises nearly 400 feet above the *Washington Coal.* South from Shannon's run exposures are quite rare, but they are occasionally found on Dunkard creek just over the State line. At the mouth of Rudolph's run the *Washington "a"* is exposed near Mr. A. John's mill, where it is only a few feet above the surface of the stream. The section between this and the *Washington* is shown at a short distance farther down the creek, as follows:

1. *Washington "a" Coal*	3′	4″
2. Shale and sandstone	45′	0″
3. Limestone (III)	1′	6″
4. Shale	20′	0″
5. *Coal*	1′	6″
6. Limestone (II)	4′—6′	0″
7. Bituminous shale	2′	0″
8. *Washington Coal.*		

1. *Coal*	1′ 6″	}	5′ 8″
2. Clay	0′ 3″		
3. *Coal*	1′ 2″		
4. Clay	0′ 5″		
5. *Coal*	2′ 4″		
9. Laminated sandstone			10′ 0″
10. Black shale			3′ 0″

The *Washington "a"* coal is in two benches, one foot six inches thick respectively, which are separated by one foot eight inches of clay. Of this parting the top, four inches, are calcareous and contain occasional scales of fish. This bed is of no value, as easily appears from its structure. The Upper Bench is too impure, and the Lower is too thin to pay for working. The coal resting on limestone II is really a bituminous shale, but is very rich and shows a true cannel fracture; of course it is worthless. The section of the *Washington Coal* was obtained at Brown's mill, on Dunkard creek, only a few rods south from the State line. The bottom layer of the bed is said to be a very fair article for fuel and is mined somewhat extensively to supply the surrounding district in Pennsylvania and West Virginia. At the same time it contains a large percentage of ash, and a noteworthy proportion of sulphur. The other layers are quite inferior but are removed and mixed with the rest for sale. This operation has injured the reputation of the coal very materially. The laminated sandy shales underlying this bed contain innumerable fragments of vegetable matter and occasionally yield good impressions of leaves which belong chiefly to the genus *Neuropteris*.

3. Wayne Township, Greene County.

This lies directly west from Perry township. The exposures for the most part are very unsatisfactory and the only work at all clear was done on Dunkard creek which flows along the southern border. The numerous tributaries to this stream have a rapid fall and their beds, soon after leaving the creek, rise above the well marked strata and bring us into portions of the upper barren series where the details are very perplexing unless the exposures are frequent and connected. The section of the township is exclusively in the Upper Barrens and extends downwards barely to the *Washington coal*. This stratum is not exposed at any locality but its horizon is reached at Blacksville on Dunkard creek only a few yards beyond the State line. A slight anticlinal passes near Blacksville but through the rest of the township the dip is almost imperceptible.

The section along Dunkard creek as made out by Mr. White is as follows:

1.	Limestone V	1′ 6″
2.	Shale and sandstone	25′
3.	*Jolleytown coal*	1′ 10″
4.	Shale and sandstone	27′
5.	Gray shale	12′
6.	Sandstone and shale	71′
7.	Limestone	3′
8.	Shale	45′
9.	Sandstone	35′
10.	Shale	5′
11.	*Washington "a" coal*	4′ 3″
12.	Shale	4′
13.	Limestone	3′
14.	Shale	20′
15.	Sandstone	20′
16.	Shale	5′
17.	Limestone	1′ 6″
18.	Shale	20′
19.	Horizon of *Washington coal.*	

The *Jolleytown coal* is mined by stripping, near the western line of the township on land belonging to Mr. J. Cox, to supply Lantz' steam-mill with fuel. Probably 2,000 bushels have been taken out for this purpose. The coal is slaty and sulphurous but is of no little importance, as the nearest opening where better coal can be procured is that of Mr. Brown, eight miles away at the east. This little bed is found on Tom's run near Mr. J. Coen's but is not mined. Near the mouth of Hoover's run it is seen on the land of Mr. Kent, and farther up the same run it occurs on Mr. Eddy's property. At Mr. Lantz', about two-thirds of a mile below the mouth of Hoover's run, Nos. 4 and 5 have coalesced and formed a massive sandstone 70 feet thick, which is of irregular structure and weathers into fantastic forms. It shows this character at many localities on the creek and proves a very useful aid in working out the stratigraphy.

The limestone, No. 7, is quite persistent throughout this region and is evidently the same with that silicious rock found in Perry at about the same horizon. Mr. W. Spragg in the north-eastern portion of the township passed through it in digging a well. Half a mile below his house on Roberts' run it is seen in the road and is 18 inches thick with a carbonaceous shale resting on it. Mr. D. Spragg living in the same vicinity has burned it for lime and found that it yields a material of excellent quality. The coal, *Washington "a,"* presents the same characteristics here as in Perry township. It is seen at the mouth of Rudolph at 70 feet above Dunkard; on Roberts' run just above Blacksville at 40 feet above the stream, and on the road above Blacksville at 40 feet above Dunkard. The creek bed rises above it near Mr. Yeager's residence. This bed is of no economical value but owing to its peculiar structure and the persistence of its blossom it is a very serviceable guide to the stratigraphy. It is in two benches as in Perry township but is thicker, the upper being from 12 to 18 inches and the lower from 6 to 8 inches. The calcareous layer containing the remains of fishes is present and is a most valuable feature.

The upper series of limestones, that beginning with No. V, is not very satisfactorily exhibited in this township. No. V is occasionally seen along Dunkard, at from twenty-five to thirty-five feet above the *Jolleytown Coal*, as well as at several localities on Hoover's and Shepherd's runs. No. VII was recognized in the extreme northern portion, near the residence of Mr. C. Spragg, at the head of Hoover's run. On the high knobs in the centre of the township, there are many fragments of a limestone which, judging from physical characters alone, Mr. White thinks may be the No. X. No means exist whereby its true relations can be accurately determined, as there are no exposures in the vicinity of any locality where it was seen.

At Blacksville a boring was made for oil many years ago, but the reports respecting the section found in it are so conflicting as to be of little service. Dr. Strosnider of that village informed me that the succession is as follows:—

1. Debris	10′
2. Shale	8′
3. Sandstone	46′
4. *Coal*	7′
5. Sandstone	24′
6. *Coal*	9
7. Shale and loose sandstone	500′

Exposures in the vicinity amply show that there is some error in this section, for the *Washington "a"* is seen within one-fourth of a mile at only fifty feet above the creek. The curb of the well is not more than ten or twelve feet above the place of the *Washington Coal*, so that No. 4 of the section is most probably the *Waynesburg* "*b*," and No. 6 may be the *Waynesburg* "*a*." In each case the thickness must be greatly exaggerated, since neither of these beds ever exceeds one foot in this portion of the county. One of the workmen employed in putting down the hole, informed Mr. White that at 200 feet, after passing through 50 feet of hard white sandstone, a large bed of coal was reached; this is the *Waynesburg*. It is probable therefore, that the debris No. 1 of the section occupies the place of the *Washington Coal*, as along the southern line of the district the interval between the two coals is not far from 200 feet.

4. Gilmore Township.

For the most part, the exposures in this township are as unsatisfactory as those in Wayne. Dunkard creek flows irregularly through the southern portion, and is formed by the junction of two forks which may be designated the North and South forks. The section extends from the *Jolleytown Coal* in the south-east, to the highest rocks in the series in the north-west. The absence of the limestones which characterize the section farther north, and the readiness with which the rocks disintegrate, render accurate work exceedingly difficult. Near the western boundary is the high dividing ridge separating the waters of Dunkard and Fish creeks, which continuing northward, becomes in Jackson township the divide between Wheeling and Ten-Mile creeks. Here, as has been shown in a previous portion of this report, are found as high strata as any that occur in this district. From the divide great hog-back hills stretch out in all directions between the streams, and rise in some instances almost 600 feet above the adjoining valleys. Upon most of these one finds the great sandstone which has been mentioned previously as the highest stratum exposed over any considerable area in the southern part of the district.

At the extreme head of Dunkard creek, near Mr. P. Shough's residence, Shough's knob rises 125 feet above the great sandstone, but the strata are concealed by a thick coating of debris. On a similar knob, near Mr. J. Pestle's, the sandstone is exposed to a thickness of thirty feet, and forms bluffs containing many cavities which are favorite resorts of foxes when hard pressed by hunters. At thirty feet below this rock, Mr. A. Taylor, living on the ridge, found an impure nodular limestone while digging a well. Above it there is a thin bed of shale, containing impression of plants. Neither of these strata is exposed at any localities visited in the township.

In the same vicinity, but farther down the creek, the *Nineveh Coal* occurs one foot two inches thick on Mr. J. Taylor's property, at 275 feet below this limestone. Mr. Taylor mines it for his own use, and thinks it a very fair coal. It is a semi-cannel, and must be quite clean as blacksmiths have used it successfully. At twenty-eight feet below it is *Limestone X*, a dark blue rock, four feet thick, clean but somewhat earthy. No doubt it would yield a lime good enough for agricultural purposes, and as the soil here needs some such application the experiment is well worth trying. As usual there rests on this bed a dark calcareous shale, which contains ill-preserved remains of plants and fish.

On the main fork of Dunkard the *Dunkard Coal* is seen on Mr. Lee Garrison's property where the exposure is

1. Shale with iron ore		9′ 0″
2. Shale with plant impressions		3′ 0″
3. *Dunkard Coal.*		
1. *Coal*	5″	
2. Clay	2″	1′ 1″
3. *Coal*	6″	

It is mined by Mr. Garrison and is said to be very good. The association of plants in the overlying shale is quite interesting, and a number of specimens was collected and forwarded for study. About a mile below Mr. Garrison's place *Limestone IX* is seen in a ravine where it is dark dove colored and breaks with an irregular fracture, but is quite pure, and would probably yield a fair lime. In the same vicinity the *Upper Washington Limestone* occurs associated with the usual dark shale and a little coal. At the mouth of Negro run the relations of the several strata are shown as follows:

1. Concealed	200′ 0″
2. Shale with plants	2′ 0″
3. *Dunkard Coal*	1′ 2″
4. Limestone	0′ 6′

5. Sandstone	10′ 0″
6. Shale	15′ 0″
7. Limestone IX	1′ 6″
8. Shale and sandstone	28′ 0″
9. *Coal*	1′ 1″
10. Black calcareous shale	0′ 6″
11. Upper Washington Limestone VI	4′ 0″
12. Sandstone—not measured.	

No. 9 is the little coal just referred to. This aggregation of the carbonaceous particles into coal is rare in Greene county, having been observed at but few localities outside of this township. The Black Shale, No. 10, contains its accustomed fossils, the fish remains and minute bivalve crustaceans. The *Upper Washington Limestone* is very impure, dark-colored and somewhat brecciated. It passes under the creek just above this locality. No openings upon the Little coal were seen on this fork of the creek, but on the South fork it is mined by stripping by Mr. A. Taylor and Mr. W. T. White. It is eighteen inches thick, is quite sulphurous and contains much slate.

At Jolleytown, just below the junction of the forks of the stream, the *Jolleytown Coal* is seen crossing the road at the upper end of the village near the hotel. By the mill below the village it has been opened by Mr. J. Clovis, who finds it one foot eight inches thick, almost semi-cannel, and containing some slate and sulphur. About a mile below Jolleytown Mr. J. Lantz has opened it to secure fuel for his mill. The section at that locality is as follows:

1. Shale	20′ 0″
2. Limestone V	2′ 0″
3. Shale	24′ 0″
4. *Jolleytown coal*	1′ 9″
5. Shale and sandstone	25′ 0″
6. Shale	15′ 0″
7. Sandstone	4′ 0″
8. Shale to creek	12′ 0″

The coal is of the same quality with that obtained by Mr. Clovis. The rocks are rising eastward here the coal being 30 feet above the creek at Mr. Clovis' and 58 at Mr. Lantz'.

5. Springhill Township.

In this township the section embraces very nearly the same portion of the Upper Barren series as does that of Gilmore. For the most part the same difficulties exist to prevent satisfactory work, but along Fish creek, which crosses the township from east to west, the valley is so narrow and the enclosing hills so abrupt that a very satisfactory section was obtained. This has been given already in another portion of the report and need not be repeated here. The surface throughout the township is excessively rugged and many of the valleys eroded by tributaries to Fish creek are little other than deep ravines whose sides rise from 200 to 400 feet above the stream-bed. Hog-back hills stretch out from the divide as in Gilmore, and the Great Sandstone, near the top of the series, extends westward almost to the center of the township.

On the dividing ridge the Great Sandstone is a coarse grained rock, rather soft and inclined to wear to an irregular surface often presenting extensive cavities. Still it stands weathering fairly well and dresses easily so as to be a good building stone. The color is reddish brown.

On the south fork of Fish creek there are no satisfactory exposures for nearly 300 feet below this stratum. The *Nineveh Coal* is seen by its blossom only and, within a short distance of New Freeport, *Limestone* X is shown. It consists of several layers separated by shales, is somewhat brecciated and is evidently too impure to yield good lime. Above it for several feet are bituminous shales. Still farther down the stream is a sandstone irregularly bedded and somewhat flaggy, which underlies the limestone directly and is about 35 feet thick. Before reaching New Freeport it becomes a sandy shale and at that place rests on a thin coal.

On the north fork of the creek, about a mile and-a-half above the village, Mr. J. Whitlatch strips what appears to be the *Nineveh Coal.* It is one foot ten inches thick, and though containing much slate is a good fuel which finds a ready market in New Freeport.

At the lower end of that village a coal blossom is seen in the road associated with a peculiar nodular limestone which is more or less conglomerate. A coal was mined here at 115 feet above the creek. It is one foot thick but its relations were not fully determined. It is probably the *Nineveh Coal.* The fall of Fish creek is very rapid, so that although the dip of the rocks is north-westward one soon reaches much lower strata which continue in sight quite to the State line. At New Freeport a heavy sandstone is seen which becomes quite massive as it goes westward. Opposite the mouth of Herod's run the *Dunkard Coal* is exposed about ten feet under it and at 80 feet higher is a thin coal which was once worked by Mr. Anderson, although it is but 13 inches thick. This same bed is stripped by Mr. Burgess on Herod's run about a mile and-a-half above the mouth. At the mouth of Tice's run the *Dunkard coal* is exposed in the hill, and at 60 feet below it is another coal about eight inches thick.

On Laurel run a section was made by Mr. White which reaches from almost the top of the series down to the lower coal seen at Tice's. Although imperfect, I give it here:

1. Sandstone, massive	30′
2. Concealed	300′
3. Limestone X	4′
4. Concealed	40′
5. Limestone	5′
6. Concealed	75′
7. Massive sandstone	25′
8. Concealed	20′
9. Bituminous shale	2′
10. Limestone	6′
11. Concealed	50′

with traces of coal at the creek level. Above *limestone X* there is a very considerable mass of black shale.

The *Dunkard Coal* is easily followed down the stream to White's mill. Half a mile above that locality it has been mined by Dr. Owen and is somewhat more than one foot thick. At the mill Mr. S. White has been digging it for years to supply his mill. The quality is so good that the working is regarded as profitable although the bed is barely one foot thick. Occasional openings are seen below this to Wagon Road run where it passes under the creek. There the section is—

1. Sandstone	20′	0″
2. Bituminous shale	2′	6″
3. Limestone X	6′	0″
4. Shale and sandstone	25′	0″
5. Limestone	3′	0″
6. Shale and sandstone	60′	0″
7. Concealed	40′	0″
8. *Dunkard Coal*	1′	2″

No. 5 is evidently the limestone of which fragments are found on the South fork of Fish creek, at half a mile above New Freeport. From the mouth of Wagon Road run for nearly a mile, everything is concealed, but at two-thirds of a mile the stream turns sharply southward and brings the rocks up again, so that at Mr. Ferguson's place the *Dunkard Coal* is twenty feet above the creek. On a little run coming in opposite that house, the following exposure is shown:—

1. Bituminous shale		2′ 0″
2. Limestone X		6′ 0″
3. Shale and sandstone		132′ 0″
4. *Dunkard Coal*		1′ 1″
1. *Coal*	6″	
2. Clay	2″	
3. *Coal*	5″	
5. Concealed to creek		21′ 0″

The coal is no longer mined and the measurements are given on the authority of Mr. Ferguson, who has burned it for several years, and says that it does very well. The coal is double at no other exposure on Fish creek, though that is its chief characteristic on Dunkard and Ten-Mile. Below Mr. Ferguson's, the coal is again carried under the stream and the massive sandstone overlying it remains in sight, descending westward, almost to Hutchinson's mill, where the dip is reversed and the rocks rise to the west. Before reaching that hill the following section was obtained on Raccoon run, which differs somewhat from that seen near Mr. Ferguson's house:—

1. Sandstone	40′ 0″
2. Shale	20′ 0″
3. Bituminous shale	2′ 6″
4. Limestone X	5′ 0″
5. Shale and sandstone	30′ 0″
6. Calcareous shale, hard	3′ 0″
7. Shale and sandstone	20′ 0″
8. Massive sandstone, to creek	40′ 0″

Just below the mill we find:—

1. Shale	30′ 0″
3. Bituminous shale	2′ 0′
3. Limestone X	6′ 0″
4. Shale	10′ 0″
5. Sandstone	70′ 0″
6. Concealed	70′ 0″
7. *Dunkard Coal*	1′ 2″

Limestone X has become exceedingly impure and contains much carbonaceous matter. It is in several layers separated by shale, more or less bituminous. It is fast losing its character as a limestone and very probably disappears as such not very far beyond the State line in West Virginia. The limestone which appears at twenty-five to thirty feet below this in several sections obtained farther up the creek has entirely disappeared, having become more and more earthy in each section westward. The heavy sandstones of this section are the same with those which are so striking a feature all the way down the creek from the mouth of Herod's run. They are admirably adapted to building purposes, as they dress easily and resist the weather well. They show bold bluff faces. The *Dunkard Coal* has been burned at Hutchinson's mill, and as obtained there is said to contain little shale and to be almost entirely free from sulphur.

A comparison with the general section shows that Springhill township is hopelessly without coal. At the lowest point, geologically speaking, which is a little way above White's mill, the *Waynesburg* is at least 650 feet, and the *Pittsburg* 1,050 feet below the surface, while at many localities on the ridge, the latter coal is nearly 1,700 feet underneath. The only dependence for the greater portion of the township is the *Dunkard Coal*, while in a small area at the east on the North fork of Fish creek, the *Nineveh Coal* is available. It is fortunate that these beds are accessible in the bottoms of many streams so that they can be mined by stripping, since they are so thin that mining in the ordinary way would be impracticable.

6. MONONGAHELA TOWNSHIP.

In this township the section extends from 220 feet below the *Pittsburg Coal* at Greensboro', to the *Washington Coal* in the north-western part. The *Pittsburg*, *Sewickley* and *Waynesburg*, are well exposed and are of workable thickness. The first of these goes under the river near Gray's landing, but the others remain above its level along the whole river face.

About three-fourths of a mile above Greensboro', a steep bluff on the river affords a detailed section of the upper portion of the Lower Barren series, as follows:—

1. Concealed	50′	0″
2. Dark shale	8′	0″
3. *Pittsburg Coal*	9′	0″
4. Fire-clay	4′	0″

5. Calcareous shale		5′ 0″
6. Sandstone		20′ 0″
7. Shale		5′ 0″
8. Sandstone		5′ 0″
9. *Coal*		1′ 6″
10. Limestone		12′ 0″
11. Shale		25′ 0″
12. Massive sandstone		35′ 0″
13. Sandy shale		15′ 0″
14. *Coal and Fire-clay*		0′ 11″
15. Limestone and shale		7′ 10″
16. Shale		2′ 6″
17. Flaggy sandstone		10′ 0″
18. Fossiliferous shale		0′ 6″
19. Sandy shale		3′ 6″

The sandstone No. 6 is a persistent stratum usually very compact, but here yielding easily to the weather and wearing into cavities. The iron ore above it is the same with that so well known in south-western Fayette, as the Oliphant blue lump. I feel some hesitation in attempting to identify the Little Coal No. 9, with any farther up the river in West Virginia, but I think it most likely to prove the same with that which in another connection, I have termed the *Little Pittsburg*. Ordinarily it is a very fair coal, but here it appears to be decidedly impure. The dark calcareous shale, No. 16, is very rich in well preserved bivalve shells.

From this point the river flows eastward and the strata rise, so that at Greensboro', the *Pittsburg Coal* is somewhat more than 200 feet above the line of low water. In the vicinity of this village that coal is quite extensively mined. In a ravine near the old glass works, the following section was obtained:—

1. *Sewickley Coal*		5′ 6″
2. Sandstone and shale		12′ 0″
3. Bituminous shale		0′ 6″
4. Limestone		3′ 0″
5. Shale		10′ 0″
6. Limestone		5′ 0″
7. Shale		15′ 0″
8. Bituminous shale	*Redstone Coal,*	8′ 0″
9. *Coal*	*Redstone Coal,*	1′ 6″
10. Bituminous shale	*Redstone Coal,*	5′ 0″
11. Limestone		16′ 0″
12. Shale		15′ 0″
13. *Pittsburg Coal*		14′ 0″
14. To the river		146′ 0

On the river road leading from Greensboro' to Mapletown,

the *Sewickley* is mined at somewhat more than a mile from the former village. Other openings were seen on the direct road between the two villages. The thickness varies from five feet to five feet six inches, and the character of the coal is the same throughout, somewhat more open burning than that from the Pittsburg and less sulphurous. This coal finds a ready market in Greensboro', and by many is preferred to the *Pittsburg*. In this vicinity the *Redstone* is everywhere a mass of richly bituminous shale, including midway a rather variable seam of bony coal. It is well shown on the direct Mapletown road just back of Greensboro', as well as at several localities on the river road. About twenty years ago a company was organized to mine this material for the manufacture of coal oil, but the discovery and enormous production of petroleum promptly brought the work to an end.

The interval between the *Sewickley* and *Redstone* in this section is only forty-five feet, and between the latter and the *Pittsburg* only thirty-one feet. This is in marked contrast with sections obtained in the adjoining township of Dunkard, not more than three miles away, where the intervals are sixty and fifty feet respectively. The Great Pittsburg sandstone has changed to a mass of shale only fifteen feet thick, thus approaching the conditions seen just over the line in West Virginia, where the interval between the *Pittsburg* and *Redstone* varies from fifteen to twenty feet.

Exposures of the Roof division of the *Pittsburg Coal* are not frequent in this neighborhood, but the whole bed is fully exposed at two localities along the river road to Mapletown, at one of which it was formerly worked. Here the roof is enormously expanded as is shown in the following section:—

1. Dark shale	6″	6′ 8″
2. Coaly shale	8″	
3. *Coal*	1′ 9″	
4. Clay with streaks of coal	2′ 10″	
5. *Coal*	2″	
6. Clay	8″	
7. *Coal*	1″	
8. Clay, main parting		0′ 10″
9. *Coal*, lower division seen		6′ 0″

The coal in the Upper division is bright and handsome but is heavy, and leaves a great bulk of white ash. At less than half a mile above Greensboro', this division contains only two feet of coal. On the river hill the ore below the coal occurs in moderate quantity, and the *Little Pittsburg* (?) is seen thirty feet below the coal and resting on its limestone. About half a mile above Gray's landing a section shows:—

1. Sandstone	40′	0″
2. *Sewickley Coal*	6′	0″
3. Shale	8′	0″
4. Limestone	15′	0″
5. Shale	10′	0″
6. Limestone	8′	0″
7. Shale	15′	0″
8. *Redstone Coal*	5′	0″
9. Calcareous sandstone	25′	0″
10. Shale	15′	0″
11. *Pittsburg Coal*	10′	0″
12. To river	30′	0″

The peculiar feature in this section is No. 10, which is the same as No. 11 of the Greensboro' section. Here it is a sandstone with a large proportion of calcareous matter, whereas there it is a limestone of fair quality. At a short distance below this locality the *Pittsburg* disappears, and at Gray's landing is under the river; while back of the distillery and 90 feet above the water level the *Sewickley* is mined, showing the following structure:

1. *Coal*	3′	
2. Clay	2′	6″
3. *Coal*	2′	
Total	5′	2½″

The coal is good throughout, but the lower bench is somewhat superior. Below this locality, for nearly half a mile, the exporsures are not good, but at about one-fourth of a mile above the mouth of Whiteley creek the following very interesting section was obtained by Mr. White:

1. *Sewickley Coal.*				
2. *Coal*	2′	6		
3. Sandstone	15′	0		
4. *Coal*	0′	5		
5. Shale	2′	0″	33′	6″
6. *Coal*	0′	1″		
7. Sandstone	12′	0′		
8. *Coal*	1′	6″		
9. Shale and sandstone			5′	0″
10. Limestone			15′	0″

11.	*Redstone Coal*	1′ 6″
12.	Bituminous shale	7′ 0″
13.	Calcareous sandstone	12′ 0″
14.	*Pittsburg Coal* in bed of river.	

This breaking up of the *Sewickley* is shown also on Whiteley creek as will appear from the references made in the description of the country along that stream. About half a mile below the mouth of Whiteley the *Waynesburg* is caught in the river hills and is seen on the land of Mr. Geo. Evans, where it shows five benches as follows:

1.	*Coal*	0′ 2″
2.	Clay	0′ 8″
3.	*Coal*	0′ 6″
4.	Clay	0′ 1″
5.	*Coal*	0′ 8″
6.	Clay	0′ 3″
7.	*Coal*	2′ 0″
8.	Clay	0′ 6″
9.	*Coal*	2′ 6″
	Total	7′ 4″

Below it the section is exposed to the Upper Division of the *Sewickley*, as follows:

1.	Waynesburg coal.	
2.	Shale and sandstone	40′
3.	Limestone	6′
4.	Shale	45′
5.	*Uniontown Coal*	2′
6.	Limestone	6′
7.	Shale and sandstone	65′
8.	Limestone	40′
9.	Sandstone	25′
10.	*Sewickley Coal*	2′ 6″

At Mr. Minor Gray's opening, half a mile north-west from the last, the *Waynesburg* shows only three benches, 1 foot 3 inches, 2 feet, and 2 feet respectively, and is 285 feet above the river. Throughout this region it is known as the "Horseback coal," and yields a poor fuel, owing to the large proportion of slate and sulphur. The separating clays inevitably become mixed with the coal, notwithstanding the utmost care, and render the quality still worse. At Mr. Gray's opening the Waynesburg sandstone is exposed to the thickness of 45 feet. About half a mile below the mouth of Whitely creek the river turns toward the east, so that at Hatfield's ferry the *Sewickley* is found 40 feet above the stream and 2 feet 6 inches thick, while at 240 feet higher the *Waynesburg* is seen 6 feet thick, and overlaid by 40 feet of sandstone. The interval is wholly concealed. At the mouth of Little Whitely, the northern boundary of the township, the *Sewickley* is in the bed of the river, and is exposed at low water.

The *Waynesburg* is mined on the valley fork of Little Whiteley by several persons, beginning about a mile from the main stream. Measurements made at three of these openings give as follows:

1. Shale	10′ 0″		
2. *Coal*	1′ 0″	1′ 0″	0′ 8″
3. Clay	0′ 1″	0′ ½″	0′ ½″
4. *Coal*	2′ 2″	2′ 1″	1′ 11″
5. Clay	0′ 10″	0′ 8″	1′ 0″
6. *Coal*	2′ 9″	2′ 11″	3′ 2″
7. Clay	concealed	concealed	6′ 0″
8. *Coal*	concealed	concealed	0′ 5″

Owing to the prevalent mode of mining most of these banks are in very bad condition. As the coal is conveniently exposed along the streams, an opening is pushed in only until the water becomes troublesome through defective drainage, or until some timbering is necessary to support the roof, when it is deserted and a new one is begun at a distance of only a few yards. The old one soon falls in and shuts off all access to the coal behind. The coal from this locality is said to be quite good, and comparatively free from sulphur. The shale overlying the bed contains fine impressions of plants. At 35 feet below the coal the limestone is seen 8 feet thick, the interval being filled with sandstone. Near the Valley schoolhouse the *Uniontown* appears in the road as a bituminous shale 3 feet thick, and holding about midway a thin coal of poor quality. On the next fork above, this coal is seen as a black shale resting on a bright ferruginous limestone, and the *Waynesburg* occurs 95 feet higher up, but is unopened. On the ridge between Greene and Monongahela townships the *Washington Coal* is exposed in the road near the corner of Cumberland township, but as there is no opening upon it, its thickness cannot be ascertained.

On Whiteley creek the *Sewickley Coal* is well exposed up to Mapletown, as the general course of the stream is almost north and south, and the fall is very nearly the same with the dip. At the fork in the road near Minor's distillery one of the lower divisions of the coal is seen five inches thick, and overlaid by sandstone. It is probably the second of the river section, obtained near the mouth of the creek. On the road, coming in here from the south-east, the *Uniontown* and *Waynesburg* are both exposed by their blossoms. At an opening in the upper division of the *Sewickley*, the coal is three feet thick, and of by no means superior quality. On the road between this and Hartley's mill the *Uniontown* is exposed near Mrs. Conant's residence.

At Hartley's mill the excavations in the *Sewickley* are numerous and extensive, showing well the structure of the bed. Two measurements are as follows:

Sewickley Coal.		
Coal	0′ 9″	0′ 7″
Clay	0′ $\frac{1}{8}$″	0′ $\frac{1}{4}$″
Coal	0′ 2″	0′ $7\frac{1}{2}$″
Clay	0′ $\frac{1}{2}$″	0′ $\frac{1}{3}$″
Coal	1′ 11″	1′ 10″
Clay	0′ 1″	0′ $\frac{1}{2}$″
Coal	0′ $7\frac{1}{2}$″	0′ 1″
Bituminous Clay	0′ $2\frac{1}{2}$″	0′ 2″
Coal	1′ 1″	1′ 2″

A comparison of these measurements, with the river section of Mr. White, shows that at that locality the bed has been broken up simply by the thickening of the partings, and this matter is still more clearly set forth in the section of the bed at Mapletown. At Hartley's mill the top and bottom benches are evidently very fair clean coal, but the middle benches contain much sulphur, and the exposed surfaces are coated with copperas. The whole is easily mined, and for fuel is preferred to the *Waynesburg*, owing to the much smaller proportion of ash. The roof is a laminated sandstone, full of obscure vegetable fragments, and is very insecure. This is a serious drawback and prevents many from availing themselves of this coal.

At Mapletown the openings in this bed are very numerous, but most of them are worked only during the winter. The insecure roof falls readily and chokes up the entry, so that in the summer or autumn one has much difficulty in procuring exact measurements. At the opening belonging to Mr. Debolt, the following section was made:

Sewickley Coal.

1.	*Coal*	2′ 3″
2.	Clay	0′ ½″—1″
3.	*Coal*	0′ 4″
4.	Clay	0′ 2″
5.	*Coal*	2′ 6″

No. 5 shows a thin parting near the top. No. 4 is a richly bituminous shale here as in nearly all the openings examined. The bottom bench is said to be much inferior to the top in respect both of sulphur and ash.

In that portion of the township lying south from Mapletown, the *Waynesburg* and *Uniontown* are occasionally seen. The latter is a bituminous shale, and the former is not mined. The country is too high to permit either the *Pittsburg* or the *Sewickley* to be available.

7. Greene Township.

Whiteley creek crosses this township from east to west, and affords some very satisfactory exposures. The *Sewickley Coal* passes under the creek near the, eastern boundary, and at Vance's mill is 60 feet below the stream. A curious feature here is, that in a boring made for oil the *Pittsburg* was not reached until at a depth of 220 feet below the *Sewickley*, whereas at no locality where the two coals are exposed so as to admit of direct measurement was the interval found to be more than 110 or 115 feet. This extraordinary increase would be incredible were it not for the circumstantial manner in which the details respecting the boring were given, and the occurrence of the oil-bearing strata at the proper distance below the lower coal.

This well was bored by Mr. G. Vance, who obtained oil at 120, 368 and 395 feet below the *Pittsburg Coal.* When the lowest horizon was reached, oil of light specific gravity, about 40 degrees, flowed at the rate of two barrels per diem; but soon, as in so many other instances in this region, the walls caved and the flow ceased. An attempt to clean it out was made, but was unsuccessful, resulting in the loss of two sets of tools which, with 100 feet of rope, still remain in the well. Notwithstanding all these obstructions, there is an energetic flow of gas, and oil is obtained whenever pumping is resorted to. Oil from the top horizon was heavy, and that from the other was light.

Still following up the creek, the *Uniontown Coal* is found at the roadside where the Morgantown road crosses. On this road the *Waynesburg* is mined by Messrs. South and Keener, but at the time of examination, the banks were closed by debris so that no measurements could be made. They are said, however, to differ little from the bank belonging to Mr. Knott, on the same road in Dunkard township. Turning off to the west from this road at the Willow Tree tavern, the *Uniontown Coal* is occasionally seen in the road, and at about a mile and a quarter from the tavern, there is an opening upon the Waynesburg which gave the following measurement:—

Waynesburg Coal.

1.	Clay shale	1′ 3″
2.	*Coal*	0′ 6″
3.	Clay	0′ 2″
4.	*Coal*	2′ 2″
5.	Clay	0′ 2″
6.	*Coal*, seen	1′ 11″

At a little distance farther up the hill, the *Waynesburg "a"* and the *Washington* are seen in the road at 80 and 180 feet above the *Waynesburg*. The former appears to be not far from one foot six inches thick; no estimate of the latter could be made. It is accompanied by its black shale about twelve feet below. Somewhat further on, Mr. Samuel Minor's opening in the Waynesburg is seen, which exhibits a still more striking thinning out of the partings than that just given.

1.	Sandstone		40′ 0″
2.	Shale		1′ 6″
3.	*Waynesburg Coal.*		
4.	*Coal*	0′ 4″	
5.	Clay	0′ 1½″	
6.	*Coal*	2′ 3″	5′ 8¾″
7.	Clay	0′ 0¼″	
8.	*Coal*	3′ 0″	

Here the partings have almost disappeared. The section seems to be fully characteristic of the bed throughout this opening.

Following up Whiteley creek, the *Waynesburg Coal* is seen falling westward, until at Garard's Fort it is scarcely 40 feet above the stream. The openings are numerous, but they show no material differences from the sections already given. The bed goes under at a short distance above the village, and the

Washington comes down to the road near Mr. H. Lantz's residence, beyond which it is almost constantly in sight to the township line.

North from the creek, on the road to Waynesburg, one is compelled to carry his horizon almost wholly by means of the barometer, as the covering of debris is sufficiently thick to conceal everything except an occasional petty exposure of sandstone. The line of the synclinal has not been fully determined and it cannot be, so that the relative horizons along this road cannot be given with exactness. At Murdock's store, the *Washington Coal* must be near the surface.

In the northern part of the township, one frequently finds fragments of *Limestones* V and VI, on the Hillsides. The *Washington Coal* is seen near the head of the run which enters Whiteley at Garard's Fort, and on the same stream, near Mr. Long's residence, the *Waynesburg* "*a*" is seen with *Limestone* I above it.

8. Whiteley Township

In but one other township of this county is the work so unsatisfactory as in this. The whole surface is covered by debris so that one can neither obtain sections nor trace strata to any considerable distance. To render the matter more complicated the synclinal, southeast from the Waynesburg axis, passes through the township, but owing to the want of exposures its precise course cannot be determined, and the geologist is unable to rely upon calculations for dip to aid him in ascertaining his position. A few observations were secured along Whiteley creek in the eastern portion of the township, but no very definite information could be obtained west from Newtown, east from which the synclinal passes.

The *Washington Coal* and its black shales are seen along the creek, showing very little dip until near the Methodist church below Kirby's school house. There the dip suddenly increases so that within a few hundred feet the coal is deeply under the surface and *Limestone* III is seen at the roadside. From this point the dip slackens and at the mouth of Dyer's run the coal is not more than 70 feet below the road, while at a short distance up the run the *Washington* "*a*" *Coal* appears at the roadside. It continues in sight along the run to where the stream is crossed by the road to Newtown. It is again exposed half a mile farther up on Mr. J. Shriver's place.

On the road passing north from Newtown a limestone is seen at 80 feet above the creek road in that village and at 80 feet higher is a second limestone, which seems to be Limestone V of the upper barren series. If this identification be correct the *Washington "a" Coal* should not be far from the creek level at Newtown, and the *Waynesburg* is not more than 230 feet underneath. The lower limestone observed in the road is probably the same with that found in nodules about a mile above Waynesburg, and at several localities in Perry and Wayne townships.

About a mile above Newtown a limestone is seen. No complete exposures occur, but the quantity of fragments indicate that the bed is not far off. From its lithological character I am inclined to regard this as *Limestone* III, which is about 50 feet above the *Washington Coal.* This is the more probable, as the marked flattening of the dip east from Newtown shows that the synclinal must pass very near that village. This limestone, or rather *Limestone* III, has been found on Dyer's run near its head, on Mr. Zimmerman's property, and on Mr. J. Shriver's place about a mile farther down the stream. Betwen these points it has a very sharp south-east dip. It is more than probable, therefore, that the two limestones are the same. Two miles above Newtown this limestone is seen in place.

From this last locality the creek and the road ascending it, rise so rapidly as to bring the higher rocks in sight, so that within one-eighth of a mile from the western boundary of the township, *Limestones* V and VI are reached. The limestone observed near Newtown, at 80 feet below No. V is not exposed here and very probably is not present, or if present is very thin or earthy. Along the western line, the coating of debris is so thick and so admirably distributed as to conceal everything and to render even fragmentary exposures of sandstone rare.

9. Cumberland Township.

This lies north from Monongahela township, and Muddy creek flows irregularly across it from west to east. The section extends from the *Sewickley Coal* to *Limestone* V. At the mouth of Little Whiteley creek, the coal is in the river bed.

At M'Kann's ferry, a little below the mouth of that creek, the following section was obtained:—

1.	Waynesburg sandstone	40′ 0″
2.	*Waynesburg Coal*	concealed.
3.	Shale	35′ 0″
4.	Limestone	10′ 0″
5.	Shale and sandstone	50′ 0″
6.	*Uniontown Coal*	blossom.
7.	Limestone	8′ 0″
8.	Shale and sandstone	45′ 0″
9.	Limestone	50′ 0″
10.	Concealed to river	22′ 0″

At one time the *Waynesburg* was mined here by Mr. W. Flenniken, but the opening has been deserted and no definite information can be secured respecting the coal. Following up the creek, the *Uniontown Coal* is exposed near the first fork in the road, where it rests as usual upon a bright ferruginous limestone and is about three feet thick. As far as one may judge from the blossom alone, it seems to be a coal of fair quality, but no openings have been made upon it to test its value. In the immediate vicinity, the *Waynesburg* is seen in the hill at 95 feet above the *Uniontown*, and three-fourths of a mile north at the forks in the road. The *Waynesburg* "*a*" is shown apparently two feet thick. On the next road crossing the stream, and about midway between it and the Presbyterian church, the *Waynesburg* is mined by Mr. G. W. Conner, at whose bank the following measurements were made:—

1.	*Coal*	1′ 1′	1′ 1″
2.	Clay	0′ 1″	0′ 1″
3.	*Coal*	1′ 6″	1′ 10″
4.	Clay	1′ 8″	0′ 10″
5.	*Coal*	2′ 8″	2′ 9′

The variations in thickness were observed at an interval of ten feet. The coal is said to be of very fair quality, and is mined extensively to supply the surrounding section of country. Between this opening and the creek, the blossom of the *Uniontown* is seen at the roadside. At Ceylon post-office, nearly a mile and a half farther up the creek, the *Waynesburg* is worked by a number of persons. Mr. D. C. Stevenson's bank shows the following section:—

Waynesburg Coal.		
1. *Coal*	0′ 10″	
2. Clay	0′ 1″	
3. *Coal*	2′ 0″	2′ 2″
4. Clay	0′ 1″	0′ 10″
5. *Coal*	3′ 0″	

The coal is said to be very good, and blacksmiths use it without much difficulty, though they prefer the *Pittsburg*. There is less ash than usual, and the partings are so thin as to be no drawback in working.

Some years ago a boring was made here by Mr. D. C. Stevenson to the depth of 700 feet, the curb of the well being 36 feet below the *Waynesburg* coal. At 288 feet the *Pittsburg* was reached, giving the interval between the coals as 324 feet, which is much less than at any locality south from this where direct measurements have been made. Oil was obtained at 460, and between 600 and 700 feet. These answer to the first and second horizons of the Dunkard wells. Brine was found in close proximity to the oil. The lower stream was strong and yielded excellent salt, but no thorough test was ever made to determine the strength of the brine, nor has any one attempted to utilize it by the manufacture of salt.

On the river, about half a mile below the mouth of Little Whiteley, the *Waynesburg* is opened in the hills by Mr. D. Parker, and near him by Mr. John Hewston. The section at the former locality is as follows:

1. Sandstone		30′
2. Shale		8′
3. *Waynesburg Coal.*		
1. *Coal*	1′	7′ 4″
2. Clay	2″	
3. *Coal*	2′ 1″	
4. Clay	1′ 1″	
5. *Coal*	3′	
4. Imperfectly exposed		35′
5. Limestone		10′ 0″
6. Sandstone and shale		45′ 0″
7. *Uniontown coal*		1′ 6″
8. Shale and sandstone		40′ 0″
9. Limestone		50′ 0″
10. Concealed to river		10′ 0″

No. 8 is imperfectly exposed and it is quite probable that the ferruginous limestone is concealed. Mr. Parker has used coal from the Uniontown, and he pronounces it a very superior article, well adapted to smiths' use. At a little distance up the hollow from Mr. Parker's opening the *Waynesburg "b"* was once opened on Mr. Walter's property and proved to be of excellent quality.

On the other side of the hill a fine section can be obtained on the road beginning at the schoolhouse above Mr. Hewston's residence and descending to the river bluff. It is this—

1.	*Washington Coal*	blossom.
2.	Concealed	50′ 0″
3.	*Waynesburg "b" Coal*	1′ 6″
4.	Imperfectly exposed	40′ 0″
5.	*Waynesburg "a" Coal*	1′ 6″
6.	Sandstone, Waynesburg	70′ 0″
7.	Shale	10′ 0″
8.	*Waynesburg Coal*	5′ 10″
9.	Fireclay and shale	15′ 0″
10.	*Coal*	1′ 0″
11.	Shale	20′ 0″
12.	Limestone	8′ 0″
13.	Shale and sandstone	45′ 0″
14.	*Uniontown Coal*	2′ 0′
15.	To river	80′ 0″

At one time the *Washington Coal* was worked near the school house, but the opening has long been deserted, and I have been unable to get any definite information about it. The bed is said to have been double, about two feet thick, and to have yielded very bad coal. The *Waynesburg* is opened in an astonishing way on Mr. Hewston's place. The road passes down a little run which has worn for itself a narrow ravine, exposing the coal on both sides for nearly 100 yards. The whole face of the bed on both sides of the run is pitted with shallow openings which are so close to each other and so carelessly guarded, that the fall of the overlying rock and the consequent destruction of the property seems to be inevitable. At this locality the main partings, those marked Nos. 2 and 4, in the section of Mr. Parker's opening, vary from two to six inches and rarely exceed three inches. They are excessively hard and render mining quite difficult. The coal contains much ash and the upper bench shows pyrites in such quantity that leaching the slack for copperas might not prove unprofitable. This bench has six inches of bony coal on top, which is occasionally replaced by drab shale.

The shale over the coal is literally crowded with impressions of plants, most of which are admirably preserved. The Waynesburg sandstone is variable; at Mr. Hewston's works it is flaggy throughout, but at a little distance north on the other side of his farm, the lower portion is massive and loose grained, yielding readily to the weather and breaking down into a clean white sand. Irregular cavities have been eroded in it, some of which are of large size. East from the school house, the *Washington*, *Waynesburg* "*b*" and "*a*" are exposed in the road.

About a mile below Mr. Hewston's, along the river, the *Waynesburg* is mined by Mr. J. C. M'Clary, at 175 feet above the river. The section there, is:—

1. Sandstone		40′ 0″
2. Shale		8′ 0″
3. *Waynesburg Coal.*		
1. *Coal*	1′ 2″	
2. Clay	0′ 2″	
3. *Coal*	2′ 0″	6′ 10″
4. Clay	1′ 0″	
5. *Coal*	2′ 6″	

Just a little way below this locality borings for oil were made, but were unsuccessful and no records were preserved. In each case, however, the *Pittsburg Coal* was reached at about 180 feet, making the interval between it and the *Waynesburg* not far from 340 to 350 feet in this region. This is very near the bottom of the synclinal southeast from the Waynesburg axis. At the mouth of Muddy creek the *Waynesburg* is mined by Mr. E. Flenniken at 240 feet above the river. A rise of nearly 45 feet per mile from the last opening. The same coal is mined by Mr. J. S. Fuller at one mile below the mouth of Muddy creek. The sections at these localities are as follows:

1. Sandstone.		
2. Shale	0′ 8″	8′ 0″
3. *Waynesburg Coal*	6′ 8″	5′ 10″
4. Shale	40′ 0″	45′ 0″
5. Limestone	8′ 0″	8′ 0″
6. Shale and sandstone	45′ 0″	40′ 0″
7. *Uniontown Coal*	2′ 0″	1′ 6″
8. Limestone	8′ 0″	6′ 0″
9. Shale	40′ 0″	35′ 0″
10. Limestone	65′ 0″	70′ 0″
11. *Sewickley Coal*	3′ 0″	2′ 6″
12. Sandstone to river	30′ 0″	40′ 0″

At Mr. Flenniken's opening the three benches of the coal

are 8, 18 and 28 inches thick, while at Mr. Fuller's they are 9, 24 and 30 inches. The lower benches yield a very fair fuel, which, however, is burdened with ash and sulphur. The shale overlying the coal is rich in impressions of plants, of which a number were collected for determination. No test has been made to ascertain the value of the *Uniontown*, which throughout this township is a coal, and not a merely bituminous shale. In these sections the *Sewickley* reappears, but as a bituminous shale. The underlying sandstone has thickened up while the one which is above it at more southern exposures has disappeared, so that the coal now lies directly under the great limestone. This is its place in the greater part of the district. A noteworthy feature of these sections, as compared with others obtained higher up the river, is the rapidly increasing thickness of the lower division of the great limestone.

Following up Muddy creek we find at a little east from the stream and about a mile from its mouth the *Waynesburg* worked by Mr. A. Burwell. The mode of opening here is quite as bad as that seen at any place, for within a distance of sixty yards ten holes were observed, of which eight had already caved. This condition is especially reprehensible here as the coal is accessible only in a little ravine where the face has for the most part been already ruined. The coal is in three benches 11, 25 and 30 inches thick, and the main parting between the second and third benches varies from 7 to 18 inches. The coal is easily mined, but contains so much slate as to make the ash bulky and annoying. Above this no openings were found before reaching Glade run, which enters the creek two-thirds of a mile below Carmichael's. About three-quarters of a mile east from Carmichael's the coal is extensively mined by Mr. J. Guseman and Mr. A. Grover. At these banks the following measurements were made, 1 and 2 at the former and 3 and 4 at the latter:

Sandstone.				
1. Shale	6′ 0″	4′ 0″		
2. *Coal*	0′ 10″	0′ 11″	0′ 10″	0′ 10″
3. Clay	0′ 2″	0′ 3″	0′ 2″	0′ 3″
4. *Coal*	1′ 6″	2′ 4″	1′ 9″	2′ 2″
5. Clay	2′ 0″	0′ 6″	3′ 0″	0′ 4″
6. *Coal*	2′ 8″	2′ 8″	3 0″	3′ 0″

At Mr. Guseman's bank the main clay parting varies from 1 inch to 2 feet, but usually is not more than 6 inches. The coal is very good, much better than this bed usually yields. The poorest rests directly upon the main parting, and the best is immediately under it. Below this, for about one foot, the coal is quite inferior. As a whole it is open-burning, easily mined without blasting and has not a large amount of visible pyrites. The bank has a good reputation and is well worked. The overlying shale contains many fine impressions of plants and one *Calamodendron* was seen exposed for a length of six feet. At Mr. Grover's bank the general character of the bed is the same as at the last, but the main parting varies at the expense of the overlying bench and at one spot near to the mouth it has removed all the coal except two inches. The under surface of that bench is much slickensided. On a little tributary coming in nearly half a mile above these banks the coal has been worked extensively and at the same time so carelessly that the numerous openings have neither fallen in or become choked with debris. At the head of this run there are banks operated by Messrs. Stevenson, Barnhart & Wyckoff, in all of which the main parting varies from one-eighth of an inch to ten inches, while at Mr. Bailey's openings, farther down the stream, the main parting averages about one foot four inches, and frequently contains a two inch layer of coal near the top. The bottom bench is exposed here to the thickness of two feet nine inches, but the whole is clearly not shown. The sandstone comes down directly upon the bed and is so shaly as to form a very insecure roof.

Just below Carmichael's at Horner's dam, there is an interesting exposure which exhibits a very complex division of the bed, and also the thin coal below, observed on Mr. Hewston's property. The section is—

1. Sandstone		15′ 0″
2. Shale		5′ 6″
3. *Waynesburg Coal.*		
1. *Coal*	1′ 0″	9′ 1″
2. Clay	0′ 2″	
3. *Coal*	1′ 9″	
4. Clay	2″ to 0′ 10″	
5. *Coal*	2′ 2″	
6. Clay	3′ 0″	
7. *Coal*	0′ 3″	
8. Clay	0′ 4″	
9. *Coal*	0′ 5″	

4. Shale, with iron ore	15′ 0″
5. *Cannel coal*	0′ 6″

On the opposite side of the stream the coal is mined. The second bench is bony throughout and requires to be broken into small pieces before it can be burned. The third bench is better, in that it is softer and contains less ash, but it has much sulphur in lumps, binders and films. The effect of so much sulphur is disastrous to grate bars. This is the most southeastern locality at which the shales underlying the coal are ferriferous, but this character prevails to a considerable distance northwestward, though the quantity is never sufficient to repay the cost of mining.

Above Horner's dam the coal is exposed along the creek for more than half a mile or almost to Kerl's mill, and an excellent exhibition of its variations is afforded. The upper benches are excessively impure and, where sheltered by overhanging rock, are coated by copperas which is usually tinged red from partial decomposition of the salt. The lowest bench seems to contain rather less of pyrites, but is very far from being a clean coal. It is stripped out of the creek near the saw-mill where it is said to be quite free from sulphur. It is probable that the changed pyrites has been leached out by the water. The upper parting is constant in its thickness and averages two inches. But the main parting varies abruptly. For the most it is only from four to six inches, but within ten feet it may increase from the minimum to three feet. When thin it is extremely hard.

The shale overlying the coal is in lenticular masses and varies at the expense of the sandstone above just as the main parting in the coal varies at the expense of the bench above it. Within 100 yards it will have increased from zero to 10 or 13 feet, and have returned to zero again. It seems to have been cut out during the deposition of the sandstone so that a more accurate statement would have been that the sandstone varies at the expense of the shale. At one foot above the coal is a layer one foot to eighteen inches thick which contains immense numbers of beautifully preserved specimens of leaves. A few of these were collected and forwarded for examination. The number of species is small, but the beauty of the specimens makes the locality an important one to those interested in the study of fossil botany.

Though the sandstone frequently rests directly on the coal, it rarely forms a horse-back. Its lower portion is soft, occasionally cross-bedded and is uneven in structure, so that it weathers into cavities, whose surface is usually honeycomed. The upper portion is irregular and cross-bedded so that its exposed surface is generally shattered. For the most part the cross-bedding dips north-west, but it is not regular and occasionally dips to the north-east. The *Waynesburg Coal* passes under the creek at Carmichael's.

Following a little branch of the creek entering from the west half a mile above the village, we ride on the Waynesburg sandstone almost to the head of the stream. The frequent undulations in the road bring into sight the *Waynesburg* "*a*," but the hills on each side are covered by debris and show no exposures of any higher rocks. At the head of the run, the road ascends the ridge dividing Cumberland and Jefferson townships and exposes both the *Waynesburg* "*b*" and the *Washington*. On top of the ridge *Limestone* III is seen. Taking from this point the Ridge road to Muddy creek, no exposures are found in the road, but at about a mile and a half from the starting point, the *Waynesburg* "*a*" is mined by Mr. M'Cree, in a hollow south from the road; it is said to be three feet six inches thick, and of excellent quality. At the time of examination, the working was filled with water and no measurement could be made. This coal is exposed in the road, half a mile from Muddy creek.

From the schoolhouse on that creek, nearly two miles above Carmichael's, the *Waynesburg* "*a*" continues in sight, rising westward with the creek quite to the township line. Wherever this coal occurs the *Waynesburg* will be found about 80 feet below.

On the Ridge road leading from Muddy creek, at the township line, to a tributary of that stream coming in at Mr. Doulin's, the *Washington Coal* is seen near the schoolhouse on the Summit, and the *Waynesburg* "*b*" near the old steam saw-mill. At Mr. J. Cregg's the *Waynesburg* "*a*" was once opened, but the difficulty of mining rendered it expensive, and the work was abandoned. This coal is mined a little farther on by Mr.

R. Doulin, who finds it three feet thick, and a caking coal of rather superior quality. No specimens could be obtained other than from the outcrop, as the opening was in bad order being worked only in the fall to supply fuel for the winter.

Returning to the river we find the *Waynesburg* worked by Mr. A. Noble, two miles below the mouth of Muddy creek. In a ravine leading from his bank to the river the following section is exposed:

1. Sandstone	30′	
2. Shale	8′	
3. *Waynesburg Coal*	5′	8″
4. Shale and sandstone	36′	
5. Bituminous shale	2′	
6. Limestone	9′	
7. Shale and sandstone	45′	
8. *Uniontown Coal.*	2′	
9. Limestone	8′	
10. Sandstone	30′	
11. Limestone	76′	
12. *Sewickley coal*	2′	
13. Shaly sandstone	40′	
14. Limestone to river	3′	

The *Waynesburg* is in three benches, six, twenty-two and thirty-three inches thick, while the partings are only one-half and six inches. The *Sewickley* is represented by a richly bituminous shale with a cannel-like fracture. At Mr. J. Davidson's opening in the *Waynesburg*, opposite Davidson's ferry, the coal shows a somewhat different arrangement as follows:

1. *Coal*	1′	2″
2. Clay	1′	0″
3. *Coal*	3′	6″
Total	5′	8″

The section below the coal here does not differ from that just given, except in the development of the Great Limestone, which makes the section between the *Uniontown* and *Sewickley* thus:

1. Limestone	6′
2. Sandstone	28′
3. Limestone	90′

Portions of the lower division are in high repute for fluxing purposes, and have been extensively quarried for use in the manufacture of glass and iron. At Baird and M'Clure's opening in the *Waynesburg* near Rice's landing there are only two benches, fourteen and thirty inches thick, separated by two feet of clay. The coal is of the usual quality. Between the bank and the river the *Uniontown* and *Sewickley* are seen, both of them coals and two feet thick. The two divisions of the Great Limestone are 70 and 80 feet thick, separated by 35 feet of sandstone. One mile south from Davidson's ferry the *Waynesburg* is mined by Mr. Geo. Hewitt, at whose bank the coal shows its normal structure, thus:

1. Sandstone	40′	0″
2. *Coal*	0′	5″
3. Clay	0′	1″
4. *Coal*	2′	0″
5. Clay	0′	2′
6. *Coal*	3′	0″

One hundred feet inside the tunnel No. 5 thickens up to three feet, to the great detriment of No. 4. In quality No. 6 is always preferable to No. 4, containing less ash and sulphur; but at the same time it is a very poor article. On Pumpkin run the *Uniontown Coal* was once worked near the works of the Vesuvius Manufacturing Company, where it showed the following structure:

1. *Coal*	1′	6″
2. Flaggy sandstone	10′	0″
3. *Coal*	0′	6″
4. Ferruginous limestone	8′	0″

No. 1 is said to be good though sometimes a little slaty. No. 2 is evidently a thickening up of one of the partings as No. 3 rests directly upon the upper division of the great limestone.

By means of oil-borings it is ascertained that the interval between the *Pittsburg* and the *Waynesburg* coals at the mouth of Pumpkin run is not far from 370 feet.

10. Jefferson Township.

This is the northern township fronting on the river. Its northern and northwestern boundary is Ten-mile creek, and its section extends from the *Pittsburg* to Limestone X of the upper barren series. The *Pittsburg* is reached at the north of Ten-mile, and the *Waynesburg* is available only on the river face and on the creek, being deeply buried in the greater part of the township. The general structure is easily worked out as the upper limestones are frequently exposed, but details are to be obtained only with difficulty as sections are rarely seen.

On Bush run, entering the river about one mile below Rice's Landing, the *Waynesburg Coal* was found at 285 feet above the river and six feet thick. The *Uniontown* and *Sewickley* are imperfectly exposed, but seen to be not far from two feet thick. The Great Limestone is well shown in two divisions, 8 and 85 feet, and separated by 35 feet of shale and sandstone. At a mile farther down the river a fine section was obtained in the river hill, beginning at Mr. Jacob Bowser's opening. It gives the details of the whole upper coal series, thus :

1.	Waynesburg sandstone, seen	20′
2.	Shale	3′
3.	*Waynesburg coal*	5′ 3½″
4.	Shale	40′
5.	Limestone	6′
6.	Sandstone and shale	45′
7.	*Uniontown coal*	1′ 6″
8.	Limestone	6′
9.	Shale and sandstone	38′
10.	Limestone	82′
11.	*Sewickley coal*	1′ 9″
12.	Sandstone	40′
13.	Limestone	25′
14.	Sandy shale	30′
15.	*Redstone coal*	1′ 6″
16.	Flaggy sandstone	15′
17.	Massive sandstone	30′
18.	Top of *Pittsburg coal*, in the river.	

Which makes the interval between the *Pittsburg* and the *Waynesburg* 360 feet. The latter coal is in three benches, four, fifteen and thirty-six inches, with partings two to six inches thick; the quality is as usual. The *Uniontown* and *Sewickley* show coal as given, unaccompanied by shale, but no effort has been made to ascertain their character. The *Redstone*, which is seen here for the first time below Gray's landing, is represented only by a bituminous shale. No limestone occurs in the interval between it and the *Pittsburg*, nor does the sandstone appear to be calcareous. At a short distance below this, the *Pittsburg* rises above the river, and is mined by several persons. On the property of Mr. L. Vernon, the exposure is:—

1.	Sandstone	30′ 0″
2.	*Pittsburg Coal.*	
	1. Shale with coal	1′ 0″
	2. *Coal*	1′ 2′
	3. Clay	0′ 10′
	4. *Coal*	7′ 0′

The sandstone is massive, coarse and very hard. Its lower layers contain lumps of coal, some coarse pebbles and rude fragments of vegetable stems, the whole looking as though its deposition was accompanied by sufficient disturbance to tear off part of the old swamp. It is clear, however, that not all of the coaly fragments found in the base of this stratum belong to the same bed. Some of them are crushed or fractured in such a manner as to show that they had been consolidated before removal, while others are saucer shaped. The latter were probably transported but a short distance and belonged to the unconsolidated *Pittsburg* marsh; whereas the others may have been torn from the outcrop of some bed belonging to the lower coal series exposed at the east along the line of the Allegheny mountains.

At this opening the upper or roof division is bony, and is not removed in mining. The lower division shows two thin partings separated by three inches of coal at thirty inches from the bottom. The coal is good, but contains an appreciable quantity of sulphur. At an opening in the same coal on Ten-Mile, three-fourths of a mile below Clarksville, the two divisions are one foot six inches, and six feet seven inches thick, with the clay parting one foot one inch. The thin partings in the lower division are only two inches apart. The overlying sandstone is forty-five feet thick and no limestone appears directly above it.

Near Clarksville, an old opening in the *Waynesburg* is seen on the hillside, but it affords no opportunity for measurement. Near Burson's school-house, somewhat more than a mile from that village, Mr. Shape's bank shows:—

1. Sandstone.
2. *Waynesburg Coal.*

1. *Coal*	1′ 0″	6′ 4
2. Clay	5″	
3. *Coal*	1′ 3″	
4. Clay	1′ 2″	
5. *Coal*	2′ 6″	

In the same vicinity, Mr. David Bowser mines this coal. He has the partings only two and six inches. At both openings the coal is quite inferior. On Pumpkin run one mile north of the last, at Mr. J. Price's bank, the bed is somewhat thicker but no better.

Along Ten-Mile creek no banks were seen in operation above Clarksville, until Jefferson was reached, but in the vicinity of that village the workings are quite extensive and the annual consumption is said to be not far from 110,000 bushels. On a little run about a mile below Jefferson, Mr. John Rex has a number of openings, from which a large amount of coal is taken during the fall. At the time of examination only two of these afforded a full exposure of the Lower bench, but I was assured that it varies little from three feet. The comparative measurements at five openings are annexed:—

Sandstone....	35′				
Shale........	1′ 0″	3′ 0″	1′ 0″	1′ 4″	0′ 10′
Coal.......	0′ 9″	0′ 10″	0′ 8″	0′ 11″	1′ 3″
Clay.........	0′ 3″	0′ 2″	0′ 3″	0 2″	0′ 6″
Coal........	1′ 11″	2′ 0″	1′ 4″	1′ 11″	1′ 7″
Clay..........	0′ 8½″	0′ 6″	2′ 2″	1′ 5″	1′ 5″
Coal........	2′ 11″	2′ 6″	2′ 10″	2′ 11″	2′ 3″

The shale usually contains streaks of coal. No. 3 is not worked and consists of coal and shale, each in thin streaks from one-half to one inch thick. No. 5 is hard and slaty and like No. 3, contains much pyrites. No. 7 is a fair coal, but is by no means so good as the *Pittsburg*, and is inferior to that obtained from some other openings in the neighborhood. Only the lower division of the sandstone is exposed here, the extreme thickness being forty feet. For fifteen feet at the base this is cross-bedded, very soft and weathers to a rounded surface. The underside of overhanging cliffs is frequently honeycombed. The upper portion is irregularly bedded and the exposed face has a shattered appearance. This part has been quarried for flagging, and the other for building purposes. The latter dresses easily and is a handsome stone, but does not withstand the weather. In this respect the rock differs at different localities, for a house built of it near Rice's landing is now nearly one hundred years old, and the stone is in perfect condition.

Just back of the college grounds at Jefferson this coal has been opened a great deal, but most of the openings have been suffered to fall in. At the one last made the bed shows a total thickness of ten feet nine inches, with four benches, five, eleven, twenty-one and thirty-two inches respectively. It is highly probable that the whole thickness of the bed is not exposed, as the bottom bench rests on a carbonaceous shale. The lower clay parting contains numerous poorly preserved impressions of leaves, and the overlying shale has a thin layer of low grade iron ore. The coal does not appear to be very good. In Mr. L. L. Minor's opening at the upper end of the village, the bed varies from seven feet six inches to eight feet four inches, and shows three benches, eleven, twenty-two and thirty-five to forty-five inches thick. The bottom bench has a band of hard black clay two inches thick at ten inches from the base, and its coal is irised to a distance of ninety feet from the mouth of the pit. It is soft, richly bituminous, cakes readily when thrown on the fire, and is regarded as little inferior to the *Pittsburg* obtained at Millsburg, on the river. The amount of pyrites is considerable, but it occurs in lumps which is an unusual condition in this bed. Some binders of this mineral were seen, but they are not numerous. The upper benches are not so good.

Near Houlsworth's mill the coal shows three benches, ten, twenty-four and twenty-eight inches respectively, and separated by two and eighteen inches of clay. The lower parting contains streaks of coal. On the hill above the mill, the blossom of the *Washington Coal* is seen with ten feet of laminated sandstone between it and the black shale below.

A very fine section was obtained by Mr. White on the run heading near Dowling's school-house, about a mile and a half south from Jefferson, and emptying into Muddy creek. It is—

1.	Limestone X	5′
2.	Shale and sandstone	70′
3.	Limestone VIII	2′ 6″
4.	*Coal*	blossom.
5.	Shale	30′
6.	Limestone VII	3′
7.	*Coal*	blossom.
8.	Shale	40′
9.	*Coal* (?)	blossom.
10.	Limestone VI	5′
11.	Shale	30′
12.	Limestone V	4′
13.	Concealed	220′
14.	*Washington Coal Bed*	4′ 2″

No. 1 is extremely hard and of a blue color. The Upper Washington limestone, VI of the series, and limestone V show their usual characteristics. The former being flesh-colored and weathering white, while the latter is coarse and brecciated. The *Washington* coal was once opened at M'Crie's mill, near the mouth of this stream, where its section was—

1.	*Coal*	1′ 3″	4′ 2″
2.	Clay	8″	
3.	*Coal*	2′ 3″	

The lower division of the bed is good, but the upper is impure. The *Waynesburg* has been opened on Coal Lick, but the only banks now in operation are in Franklin township.

11. Morgan Township, Greene County.

This adjoins Jefferson at the west, and is separated from it by the South Fork of Ten-mile creek. The north-eastern boundary is the north fork of the same stream, which separates it from Washington county. The exposed section extends from the *Sewickley* coal to limestone X of the upper barren series. The *Waynesburg* is accessible on both forks of Ten-mile creek.

Along the north fork no openings in the coal were found. Many years ago there was one on the property of the Messrs. Hawkins, in the extreme northern corner of the township, but it has long been deserted and no one knows anything about the coal farther than that its quality was somewhat inferior. The section from this bank to the creek is as follows:

1.	*Waynesburg Coal Bed.*	
2.	Concealed	35′
3.	Limestone	11′
4.	Sandstone	35′
5.	Limestone	13′
6.	Shale	10′
7.	Limestone	20′
8.	Sandstone	18′
9.	Limestone	6′
10.	Concealed	34′
11.	Limestone	33′
12.	Shaly sandstone	10′
13.	Limestone	15′
14.	Clay	2′
15.	*Sewickley Coal Bed*	1′ 6″
16.	Dark fissile shale seen	14′

No. 4 is imperfectly exposed at the base, so that the *Uniontown* is concealed. The interval, No. 10, is probably wholly occupied by limestone as there are frequent partial exposures of that rock. The peculiar feature of this section is the separation of the great limestone into four divisions, having in all a thickness of not far from 120 feet, whereas there are ordinarily only two divisions with an extreme thickness in this county of 96 feet. About a mile and a half W. S. W. from this locality there is an opening in the *Waynesburg* near the Bollenfield school house, which shows:

Coal	1′ 6″	5′ 11″
Clay	1′ 5″	
Coal	3′	

On Ten-Mile this coal is exposed in the forks of the road near Mrs. Cox's residence on the State road, half a mile from Clarksville. The blossom of the *Waynesburg* "*a*" is seen at 65 feet higher, that of the *Waynesburg* "*b*" at 115 feet, and that of the *Washington* at 160 feet. The *Waynesburg* "*a*" is associated with a ferruginous limestone.

At the mouth of Casteel run the *Waynesburg* is 190 feet above the creek, and the interval contains somewhat more than 80 feet of limestone. The coal is not caught in the hills here, but a deserted opening occurs about one-third of a mile south. The first bank on this run is that belonging to Mr. W. Steward, at the mouth of Bacon Street run, nearly a mile and a half up from the creek, where the section is—

1.	*Coal* and shale	1′ 4″	9′
2.	Clay	1′ 1″	
3.	*Coal*	0′ 11″	
4.	Clay	0′ 2½″	
5.	*Coal*	1′ 8″	
6.	Clay	1′ 3″	
7.	*Coal*	2′ 9″ to 3′ 7″	

No. 6 contains a thin streak of coal, and shows but slight variations in thickness, the extremes being 10 and 17 inches. On Bacon Street run the *Waynesburg* "*a*" is seen one foot four inches thick, and resting on a heavy limestone, which yields an excellent lime and is much used for that purpose. The *Washington Coal* is reached by the run just below Mr. Keys' house, and remains in sight to the head of the stream. It is shown on the State road beyond, accompanied by Limestones II and III.

At the mouth of this run the ferriferous shale underlying

the *Waynesburg* and the limestone below are exposed in the road. At one time the ore was mined here by stripping, but the enterprise proved unprofitable and was abandoned. At a little distance farther up Casteel the coal is worked on Mr. Cox's property, where the following measurement was obtained:—

1. *Coal* and shale	5′ 0″	12′ 4″
2. *Coal*	0′ 10″	
3. Clay	0′ 3″	
4. *Coal*	1′ 11″	
5. Clay	1′ 10″	
6. *Coal*, seen	2′ 6″	

The distance between this and Steward's opening is barely half a mile, yet the coal is 30 feet lower than at that bank, showing a comparatively rapid dip for this region to the northwest. The coal goes under the run just below Mr. J. Greenlee's house. Near the Casteel schoolhouse the *Waynesburg "a"* is seen in the road, but beyond that everything is concealed until the Baptist church, at the head of the run, is reached, where limestone X is exposed in the road. Nodules of the Upper Washington limestone are scattered on the hill-sides at half a mile above the schoolhouse.

At the mouth of Hews' run, the first entering Ten-Mile above Casteel, the coal is 120 feet above the creek. At a short distance up the run Mr. Hews has an opening in which the coal shows the same general structure as at Cox's on Casteel, and has a total thickness of eight feet. The coal is open-burning, but quite variable in quality; some of it is good, but the greater part is slaty and sulphurous. It contains much mineral charcoal in layers, some of which are one-fourth of an inch thick. At a short distance below the Center schoolhouse on the same run is Mr. S. C. Orr's bank, in which the bed shows three benches, seven, twenty-three and forty inches, and has a total thickness of seven feet eleven inches. The main parting varies wholly at the expense of the bench above it, and the combined thickness of the two is quite constant. At one spot in the entry the coal is only five inches, while the clay is thirty-eight inches. Stringers of coal pass down into the clay and frequently show complex sub division. The bottom bench is hard, open-burning and contains much ash. Some portions of it must be broken into small fragments before it is thrown upon the fire; otherwise it will not be reduced to ashes. The bed goes under the run about 150 yards below the schoolhouse. On this run the Waynesburg sandstone is 65 feet thick, and is an excellent building stone. Blocks of any desired size can be procured, as there are massive layers 8 to 10 feet thick. The *Uniontown Coal* is shown at several localities 95 feet below the *Waynesburg*, and is represented by a black bituminous shale. Nine feet below the latter coal is the ferriferous shale, containing ore in lumps weighing from one to three pounds, which many years ago was mined and taken to the old furnace at Clarksville. But the quantity was not sufficient to render the enterprise a profitable one.

On the ridge between Hews' and Casteel runs the *Washington* is exposed at Mr. Orr's house, and on the other side at Mr. Steward's. Below it are the blossoms of the *Waynesburg* "*a*" and "*b*," the former associated with its limestone. Near the head of Hews' run the *Washington* is shown in the road accompanied by Limestone II, but the exposure is imperfect and everything else is concealed.

Directly opposite Jefferson, on the property of Mr. D. W. Rogers, the *Waynesburg Coal* has been wonderfully slaughtered, for within a distance of less than 200 yards there have been twenty openings, all of which have been permitted to fall in. Farther up the stream, on the hill followed by the Waynesburg road, the coal is mined, but owing to the bad condition of the bank no information could be obtained. On Bell's run which enters the creek about a mile above Jefferson, the coal has been opened in many places, but a number of the entries have been closed by debris. At an opening near the distillery the following section was obtained:

1. Sandstone		20′ 0″
2. Shale		4′ 0″
3. *Waynesburg Coal Bed*		
1. *Coal*	1′ 3″	
2. Clay	3″	
3. *Coal*	2′ 0″	7′ 11″
4. Clay	1′ 3″	
5. *Coal*	3′ 2″	
4. Fire clay		3′ 0″
5. Sandstone		6′ 0″
6. Ferriferous shale		12′ 0″
7. Flaggy sandstone		10′ 0″
8. Clay shale		8′ 0″
9. Limestone		7′ 0″

Nodules of iron ore are numerous in No. 10, but not in quantity to be of any economical value. At a short distance farther up the run the benches are six, fifteen and forty-two inches respectively, and the total thickness of the bed is six feet eleven inches. The top bench is bituminous shale with cannel-like fracture, and is the material commonly known throughout this vicinity as cannel coal. The ferriferous shale shows about the same amount of ore as at exposures previously noticed. The *Waynesburg* passes under the run beyond Mr. Bell's house, and at less than half a mile farther up the *Waynesburg* "*a*" is exposed resting on a limestone. The blossom of the *Washington* appears at the roadside just below Mr. Crayne's house near the head of the run.

At a little way east from Harry's school house the *Waynesburg* is mined on the Jesse Bell estate, just below the State road. It is in two benches, nineteen and thirty-two inches thick, separated by a clay parting of nineteen inches. This is the ordinary form of the bed from this point westward. The top bench is absent or represented only by a bituminous shale. At this opening the lower bench has a clay parting at fourteen inches from the bottom.

On Ruff's creek several banks belonging to Mr. C. C. Harry are seen about one-fourth of a mile from the State road. Measurements were made at two of these, but the others were in such condition as to render examination impossible. The sections are—

1. Shale	6′ 0″	5′ 6″
2. Bituminous shale	1′ 0″	0′ 10″
3. *Coal*	1′ 2″	1′ 11″
4. Clay	1′ 10″	1′ 7″
5. *Coal*	2′ 10″	2′ 8″

Here as in other instances previously given, the parting varies at the expense of the overlying bench of coal. It contains many thin streaks of coal, which in nearly all cases can be traced into the bench above. No. 5 shows a layer of clay at nine inches from the bottom. Between these openings and Martinsburg the road rises above the place of the *Washington Coal*, but everything is concealed except the *Waynesburg* "*a*." Just below Martinsburg, on Ruff's creek, the *Waynesburg* is mined by several persons. The structure is the same as at Harry's openings, being in two benches with a bituminous shale resting on the upper one. The parting occasionally becomes three feet thick, and varies wholly at the expense of the overlying bench. The lower bench is quite constant, shows no clay streak and yields a fairly good coal. Very little pyrites is visible but the quantity present must be considerable, for the prevailing cause of complaint against the coal is the sulphurous odor given off during combustion. The ash is said to be large. The bed goes under the creek before reaching Martinsville.

On a little run coming in at Martinsville from the north, the *Waynesburg* "*a*" is seen near the first cross-road associated with its limestone. The latter must be good, for at the time this locality was visited, preparations were making to burn a large kiln for lime. Three-fourths of a mile farther on, the blossom of the *Washington Coal* is indistinctly exposed. At the head of this run is the dividing ridge between Ruff's creek and Casteel run. Just west from Mr. Fulton's house on this ridge the blossom of the *Jolleytown Coal* shows in the road, and at 40 feet above it is limestone V. The Upper Washington limestone VI, of the Upper Barren series, is next seen, and at 40 feet above it is the blossom of a coal underlying a dark limestone. Beyond this there are no exposures until the highest point of the ridge is reached, within half a mile of the township line, where there is a dark blue limestone associated with a much inferior sandy iron ore. This is evidently limestone X, and at fifty feet below it is a coal six inches thick.

On Ruff's creek the blossom of the *Waynesburg* "*a*" is shown near Heaton's school-house, and at three-fourths of a mile farther up, that of the *Washington* is in the bluff alongside the road. This coal goes under the creek just before reaching the township line.

12. Washington Township, Greene County.

This adjoins Morgan on the west. Except on Ruff's creek and the high ridge marking the Washington county line, exposures are very rare. Even along these lines no sections can be obtained, and the only source of information is an occasional exposure of one of the characteristic limestones. The section extends from limestone III to limestone X, of the Upper Barren series, an interval altogether devoid of interest. Upon

Craigs', Craynes', Johns' and Overflowing runs, as well as on Bates' fork of Ten-Mile creek, everything is utterly concealed.

On Ruff's creek, the *Washington Coal* goes under just east from the township line and limestone III is seen in the road at the line. Two-thirds of a mile farther up the stream the Middle Washington limestone V of the series appears, and is well exposed on a run coming in there from the south, where it has a mass of black shale resting upon it. On the same run the Upper Washington limestone occurs, the interval between it and the one below being 98 feet. From this locality to the mouth of Boyd's run everything is concealed, but at a short distance up that run Mr. George Huffman, Jr. mines a coal thirty inches thick, embedded in a mass of dark shale which is fifteen feet above the Upper Washington limestone. The coal is a semi-cannel and is said to be in favor as a fuel. It has much ash and sulphur. This is merely a local manifestation and does not occur at any other locality, although the shale is persistent. Limestones VI and V are exposed in the road below the coal; VI is a handsome rock and yields excellent lime.

Near Mr. Ross's house on Ruff's creek this limestone is seen in the road, with a thick black shale resting upon it. The dip, owing to the influence of the Pin-hook anticlinal, is slightly eastward, so that for nearly two miles the limestone rises as rapidly as the creek bed. It is exposed at the old mill and at the mouth of John's run, and continues in sight to the forks of the stream near Mr. Huffman's house, beyond which it passes under the creek.

At the head of Boyd's run Limestone X is exposed in the road at 180 feet above Limestone VI. It was seen at several localities along the ridge west from that point.

13. Franklin Township, Greene County.

This is of irregular shape, and lies west from Whiteley, Jefferson and Morgan, and south from Washington township. It is crossed from west to east by Ten-Mile creek. Its section extends from 150 feet below the *Waynesburg Coal* to 250 feet above Limestone VIII; but above that limestone everything is concealed. The exposures of the remainder of the section are numerous and satisfactory, exhibiting completely the structure of this portion of the series. Near the county poor house on Coal Lick the *Waynesburg Coal* is mined, and shows the following at two openings:

1. Sandstone seen	30′ 0″	20′ 0″
2. Shale	8′ 0″	15′ 0″
3. Bituminous shale	0′ 6″	0′ 4″
4. *Coal*	1′ 11″	1′ 8″
5. Clay	2′ 9″	3′ 8′
6. *Coal*	3 8″	4 0′

This is the typical structure for the township. The bed is rarely found triple, unless indeed the bituminous shale No. 3 may be regarded as representing the top bench with the parting below obliterated. An interesting feature here is the variation of the parting without material effect upon the overlying bench of coal.

In the eastern portion of the township the Waynesburg is opened at numerous localities. A deserted bank was seen opposite the old round school house. At a short distance farther up Ten-Mile creek Mr. U. L. Green mines the coal at the Dexter coal works, where the bed has a structure similar to that shown at the next locality. The Waynesburg sandstone is well exposed here, standing out as a bluff nearly 40 feet high. It shows much cross-bedding, varies from flaggy to massive, from fine to coarse-grained. The color is almost white and ferruginous matter is present in but small quantity. Not far from these works are the Excelsior works, operated by Mr. Thomas Sayers, where the bed is six feet one inch thick, and shows two benches of coal fourteen and forty-two inches. The lower bench is hard and splits out in blocks two or three feet long and three or four inches wide and thick; the upper bench is softer. The coal throughout is open-burning and contains much ash. Attempts have been made to coke the "slack," but without success.

At Smith's coal works, immediately adjoining the last, the bed has its normal structure and shows three benches fourteen, six and forty-two inches, the total thickness being six feet. In this bank the upper parting is persistent, though at times it diminishes to a mere knife-edge. The main parting frequently contains thin streaks of coal which, in nearly all cases, come from the overlying bench. The coal from this opening resembles that from the Excelsior works.

Where the road crosses the creek below Dodysburg, the coal is worked on both sides of the stream. Measurements made at two openings here are as follows:

1.	Shale		3′ 0″
2.	Bituminous shale	0″	8″
3.	*Coal*	1′ 6″	1′ 8″
4.	Clay	1′ 3″	1′ 3″
5.	*Coal*	3′ 8″	3′ 5″

These measurements, however, give very little idea of the true condition of things. The parting No. 4 shows abrupt and extreme variations, which here are always at the expense of the lower bench. In the entry belonging to the Ten-Mileworks, No. 3 varies from one foot to one foot six inches; No. 4 from one foot to four feet, and No. 5 from one foot to eleven feet. Precisely similar changes occur in the opening on the other side of the creek. As the better coal comes from the lower bench these variations are grievously annoying and add much to the cost of mining. This bench is sometimes cut down for a number of yards, so that the available thickness is less than fifteen inches. The miners receive no pay for the removal of this "horseback," and the temptation to throw some of it into the coal on the dump can hardly be resisted; especially since a little shrewdness prevents the detection of the trick until after the coal has been delivered to the consumer. These mines are worked quite extensively to supply the village of Waynesburg, where the annual consumption is somewhat more than 500,000 bushels per annum. The coal is hard but is taken down without blasting. The upper bench is badly cut up by binders of clay, which in some parts are barely one inch apart, and the lower bench is not much better. The coal therefore has a large proportion of ash, but in this respect is no worse than that from the banks already noticed, and the proprietors claim that it is superior in that it contains only a small percentage of sulphur.

The *Waynesburg Coal* passes under the creek at a short distance above these works or just below the little village of Dodysburg. Near the distillery an inclined shaft was sunk upon it, but it is not worked to any extent. The Waynesburg sandstone is well exposed along the creek to Hook's mill immediately below Waynesburg, where it passes under the stream. It is about 75 feet thick and is divided into two portions by a sandy shale, which frequently changes into a flaggy sandstone, like the upper portion of the stratum. The lower division, which is not far from 40 feet thick, though sometimes slightly conglomerate, is usually a coarse-grained and rather soft rock, which dresses easily and makes a neat but not durable building stone. The face of a wall becomes pitted, and the corners soon lose their sharpness.

At Hook's dam the *Waynesburg "a"* is seen in the following section:—

1.	Sandstone and shale	25′ 0″
2.	Limestone and shale I	13′ 0′
3.	*Waynesburg "a" Coal Bed*	1′ 6″
4.	Shale	10′ 0″
5.	Waynesburg sandstone seen	8′ 0″

It is quite probable that the *Waynesburg "b"* is only a few feet above the top of this exposure. The coal in this section is good, but owing to its thinness it has never been mined.

On Purman's run, which enters the creek at the lower end of Waynesburg, the *Washington Coal* is seen near the first bridge, where it seems to have been opened at one time. An opening was found at one-eighth of a mile above the bridge, but when visited it was closed and no measurement could be made. Directly opposite the first locality, the following characteristic section was obtained:—

1.	Shale, with iron ore		7′ 6′
2.	Limestone III		3′ 0″
3.	Sandstone		18′ 0″
4.	Black shale		3′ 0″
5.	Limestone II		3′ 6″
6.	Shale		6′ 0″
7.	*Washington Coal Bed.*		
	1. *Coal*	0′ 8″	
	2. Clay	0′ 5″	2′ 4″
	3 *Coal*	1′ 3″	
8.	Fire-clay and shale		2′ 6″
9.	Arenaceous shale		12′ 0″
10.	Black shale		8′ 0″

The black shale No. 4 contains fish teeth and scales in great numbers, associated with innumerable specimens of bivalve crustaceans, and occasionally there occurs a fragment of a leaf, which shows evidence of long maceration in water. The shale, is very fissile, splits readily into very thin layers and has a distinctly cannel-like fracture. It is exceedingly rich in bituminous matter. No. 9 is the thinly laminated sandstone which invariably accompanies the coal, and is literally crowded with fragments of vegetable matter, among which an imperfect leaf is sometimes found. No. 10 is only a dark fissile shale, with barely sufficient carbonaceous matter to give it color, for on exposure to the air it becomes gray. The coal is quite inferior.

At the first fork in the road ascending this run, limestones V and VI are exposed in a little bluff at twenty-two feet apart. The former is coarse and brecciated on top, but the lower layer is probably good enough to make a strong lime. The Upper Washington limestone, VI, shows the characteristic snow-white surface and the flesh-colored interior. It is seen in the bed of the east fork at the road-crossing, half a mile farther up. At the head of this stream the road is 620 feet above the *Waynesburg Coal.*

Where the road leading down this run crosses the street in Waynesburg, limestone III is exposed in the bluff. Fragments of the Upper Washington limestone are found in the superficial deposit at the cemetery near the village.

At the mouth of Smith's creek opposite Waynesburg, the *Washington Coal* is seen with the laminated sandstone and the black shale below.. Ascending this creek which flows from the south, the rocks are seen rising somewhat more rapidly than the creek bed for nearly two miles and a half. At the first cross-road there is an old opening in the *Washington Coal*, but its condition is too bad to admit of measurement. The coal is double and seems to be in all about three feet thick. Between it and limestone II, there is a small deposit of iron ore apparently quite good. A little way below this, the sandstone underlying the black shale is quarried; it is seventeen feet thick and the lower portion is a good building stone. Immediately under it is the *Waynesburg* "*b*," one foot thick, which is said to be an excellent coal and well suited for blacksmiths' use.

On property belonging to Mr. E. Kent, about a mile and a half from Waynesburg, the *Washington* is mined by stripping. The exposure is very imperfect owing to the debris and water in the excavation. The bed consists of two benches, fifteen and twelve inches thick, separated by four inches of clay. On the opposite side of the stream it is mined in the same way by Mr. J. M. Bell, but there was no exposure at the time the locality was visited. The upper division of the coal is slaty, but the lower bench is said to be quite good. The iron ore above the coal was seen here on both sides of the creek. Near the brick school-house the dip of the rocks is reversed, and at the old saw mill the coal comes down nearly to the road and continues in sight to the forks of the road, about two-thirds of a mile farther up. There it passes under the hill and as one ascends to the ridge, he sees the upper limestones coming out in their regular order of succession. On the ridge between Smith creek and Sand run, the *Washington Coal* is exposed at half a mile from Ten-Mile, and limestone VI, at two miles from that stream.

At three miles above Waynesburg, near the second bridge crossing Ten-Mile creek, the following section was obtained on property belonging to Mr. J. Yeador:—

1.	Concealed	350′
2.	Limestone VII, seen	2′
3.	Shale and sandstone	40′
4.	Black shale	2′
5.	Limestone VI, seen	3′
6.	Shale and sandstone	32′
7.	Limestone V, seen	2′ 6″
8.	Sandstone and shale, to creek	45′

The total interval between Limestone V and the *Washington Coal*, as ascertained on the hill opposite Waynesburg, is not far from 240 feet. At Waynesburg, Limestone III is thirty feet above the coal, and at the upper part of the village the interval between that limestone and V is found, by direct measurement, to be 200 feet. At the first fork in the road west from the village, the Upper Limestone V is seen in the hill, with the *Jolleytown Coal* below it. The same coal is exposed for nearly 200 yards in the road east from Hill's school-house, and is there seen to be 30 feet below Limestone V. At the school-house the limestone is in the road, and at two-thirds of a mile farther up limestones are well exposed.

The three limestones given in the section are nicely shown at the first bridge above the mouth of Brown's fork, where fragments of No. VIII are seen at 25 feet above VII. Limestone VI, or the Upper Washington, is easily distinguished by its peculiar white exterior and its flinty fracture, as well as by the fact that overlying it is a dark coal-like shale containing leaves of plants. It is very pure, and is the chief source of lime. On the Yeador property there seems to be a coal near the horizon of Limestone VIII.

14. Centre Township, Greene County.

This adjoins Franklin on the west. At Rogersville, about two miles from the eastern boundary, Ten-Mile creek is formed by the junction of Gray's and M'Courtney's forks. The former flows through the township in a south of east direction from Richhill, while the latter comes in from Jackson, and follows a north of east course. Just below the junction the creek receives an important tributary from the south, known as Pursley creek. The exposed section extends from the *Jolleytown Coal* to the top of the upper barren series as found in this State. The lower portion of the series is frequently shown, but the upper part is not seen in detail at any exposure.

At the mouth of Pursley creek Limestone V is about 60 feet above the stream, but it rises rapidly southward, so that at the fork in the road below Oak Forest it is 120 feet above the creek. On the west fork of Pursley, which comes in not far above that village, it soon is at the road side, and within a mile from the village Limestone VI is in the road. Upon the latter there is a black shale eight feet thick, rich in bituminous matter, which, in its lower portion, shows many impressions of long slender leaves belonging probably to *sigillaria menard*, and in its upper layers it has vast numbers of bivalve crustaceans. Above it is a dismal reddish shale fifty or sixty feet thick, which contains at its base layers of iron ore. This is well exposed on the property of Mr. J. Hoge, two miles from Oak Forest, near whose house the following section was obtained:

1. Sandy shale	40′	
2. Limestone		4″
3. Clay shale	2′	
4. *Iron ore*		2″
5. Clay shale	1′	
6. *Ore*		8″
7. Clay shale	10′	
8. *Ore*		6″–10″

9. Clay shale	5′	
10. *Ore*		3½″
11. Clay shale	1′	6 ″
12. *Ore*	1′	
13. Dark shale	7′–8′	
14. Limestone VI.		

The ore is a carbonate, which is said to yield 51 to 54 per cent of metallic iron. The layers are not absolutely continuous, but very nearly so; the ore occurring in masses weighing from 2 to 75 or even 100 pounds, which are closely packed together. The character of the associated rock is such, as to render it doubtful whether drifting for the ore would be profitable. Fortunately this will not be necessary for some time after the work begins. At Mr. Hoge's house two runs unite, each having a broad "bottom." That from the west exposes No. 8 of the section in such a way that a very large quantity can be obtained by stripping off a thin cover, while that from the south in like manner renders Nos. 10 and 12 easily accessible throughout an area of several acres. Another run from the west, entering at a short distance below Mr. Hoge's, exposes the ores, but the available area is much smaller than on the runs just referred to. If the analyses by the State Chemist confirm those made by private parties, there can be no doubt that this deposit will prove to be of some value. It does not occur on the south fork of Pursley creek.

On the road leading up the Little Run just below this locality limestone X is seen at 265 feet above limestone VI. There are no exposures in the interval.

On a little stream entering Ten-Mile at Rogersville from the south, some ore is found below limestone V on the property of Mr. Joshua Knight, but it is in small quantity, and is of no economical importance. Some of the residents on this run have mistaken limestone V for iron ore, and have cherished great expectations which they are unwilling to give up. They should understand that the deposit is worthless, and that any time spent in developing it is utterly wasted.

Along Ten-Mile from the mouth of Pursley to Rogersville the exposures are quite satisfactory. On Mr. G. Fry's land half a mile from the eastern boundary of the township, the following section was procured, which shows the lower portion of the series as it occurs here:

1. Concealed	278′	
2. Sandstone	50′	
3. Limestone X	7′	
4. Sandstone	70′	
5. Limestone VIII	5′	2″
6. Sandstone	19′	
7. Limestone VII	2′	6″
8. Sandstone	31′	
9. Dark shale	4′	
10. Limestone VI	7′	6″
11. Shaly sandstone	20′	
12. Limestone V	3′	
13. Shale and sandstone	35′	
14. *Jolleytown Coal Bed*	1′	
15. Shale to creek	10′	

These intervals are much smaller than at any other localities in the township. On Mr. D. W. Fry's land, less than half a mile away, the interval between limestones VII and VIII is 35 feet, and between them is a coal which seems to be directly under VIII. This coal is seen on Mr. Buchanan's land near by, where it is 20 inches thick; at both places it has been opened to a slight extent, but it is too thin to pay for working though the quality is much superior to that of the *Waynesburg*. It does not appear in the section given above. Limestone VIII attains its greatest thickness here and is a compound stratum in four layers, as follows:

Limestone	1′	
Shale		6″
Limestone	1′	8″
Limestone	2′	

Limestone VII is a dark blue rock with black calcareous shale resting on it, which shows indeterminate fragments of vegetable matter. The dark shale over limestone VI is similar to that seen on Pursley creek, but the arenaceous shale above it does not contain the iron ore. The *Jolleytown Coal*, as exposed in the road, varies abruptly in thickness, the extremes being 2 and 12 inches. It is a semi-cannel of fair quality.

The lower limestones are seen all the way to Rogersville along the creek. At that village limestone V is 50 feet above the road at the bridge, and the upper Washington lies 22 feet higher with its black shale resting upon it. At 100 feet above the latter there is a line of dark limestone fragments associated with a coal blossom, so that here we have VIII at a considerably increased distance above VI. VII is not exposed. Nodular ore occurs below V, but none is associated with VI. The *Jolleytown Coal* is concealed.

About half a mile above Rogersville, where the two forks, Gray's and M'Courtney's, unite, limestone V is fifty feet above the road. On Hargus' creek, which enters M'Courtney's fork less than half a mile above this point, the fall of the stream is greater than the rise of the rocks, so that at the fork in the road near Hopkins' mill limestone V is barely twenty feet above the road. Thence to the school-house the shale underlying that limestone is constantly in sight. It has a good deal of iron ore, which, if collected so as to be within a vertical range of three feet would no doubt be of some value, but distributed as it is throughout twelve or fifteen feet of shale it is practically worthless. At the school-house limestone VI is in the road.

About a mile up a little tributary coming in here, the Upper Washington limestone is exposed with the black and ferriferous shale. Fragments of limestone VII were observed here, but the rock could not be found in place. Nearly half a mile further up the stream the ore occurs in the "bottom" on the Adams' farm, but I was unable to determine how much is available for stripping, and could find no exposure which would give a definite idea respecting the amount of ore present. Near the head of this run limestone X was seen in the road, and at 30 feet higher the blossom of the *Nineveh Coal*. At 100 feet higher another limestone was seen which may be limestone XI, though the interval is somewhat greater than at any locality where the two strata have been positively identified.

On Hargus' creek above the school-house, the rise of the stream for more than two miles is about the same with that of the rocks, and the Upper Washington limestone continues in sight up to Mr. Jesse Orndorff's house, where it passes under the stream. The black shale continues present all the way but with diminished thickness. Careful search was made for the iron ore above this stratum, but at no place was it found aggregated so as to be of economical value. The shale is always ferriferous, but the ore is disseminated and rarely occurs even as nodules. Still it may be present at some localities, and it would not be amiss to make test openings on the broad bottom near Mr. Orndorff's residence, as, if present there, it would be very valuable owing to the large area for stripping. From Mr. Orndorff's to the head of the stream, there are no exposures and the ridge at the crest is on the horizon of limestone X.

On M'Courtney's fork, about half a mile above the mouth of Hargus creek, a boring was made on the property of Mr. John Pettit, in which a thick coal bed is said to have been struck at the depth of 140 feet. The curb of the well is at 60 feet below limestone V, which is exposed near by. The interval between the coal and limestone VI is 220 feet, which shows that the coal is the *Washington.* This interval on Dunkard creek is 300 feet, and Waynesburg 280, and in Northern Richhill township only 135 feet. The coal is said to be seven feet thick, but this is not all coal as can be seen by reference to any of the descriptions of the *Washington Coal* in this report. The quality of the bed is always poor, so that it would be unadvisable to sink a shaft for it, as some of those residing in the vicinity had proposed to do.

On Mr. Pettit's property a coal occurs twenty inches thick, and directly underlying limestone VII. The limestone is only a calcareous shale which contains much carbonaceous matter, and is nearly eight feet thick. The coal is bright and burns well, though it is slaty and contains a considerable proportion of sulphur. It has been mined by stripping in the run. Quite a tunnel was driven in on it, the owner having the impression that the bed would thicken up under the hill; of course the operation was rewarded with failure. This impression seems to prevail here, so that it may be well to state that it is wholly erroneous. When a sound roof is reached, the character of the coal is fully shown. No marked increase of thickness is to be looked for unless a horseback should occur at the mouth of the entry, and of this the workman can easily satisfy himself.

The Upper Washington limestone has been burned for lime here, and is the only source of supply for the whole region. It comes down to the road a little way above the first fork, where, though very thin, it shows its characteristic mode of weathering, and has the following structure:—

1.	Limestone	1′ 3″
2.	Shale	3′ 0′
3.	Limestone	0′ 10

The upper layer is a beautiful rock, blue on fresh fracture and weathering almost snow white. The lower layer weathers to a dingy white, and is dull flesh colored on the fresh fracture. The shale separating the two layers is drab and slightly calcareous. This stratum continues in sight for nearly two-thirds of a mile. The black shale is from two to five feet above it, the interval being filled with drab shale. The iron ore of Pursley creek is altogether absent.

Along this creek limestone VII is first seen near the second fork in the road, and remains above the road to just above Wood's mill. It is in three layers separated by thin shales, in all thirty inches thick. The rock is dark blue, somewhat laminated and contains much earthy matter. Four feet below it is a layer of blueish limestone eight inches thick, weathering very white and bearing much resemblance to the upper layer of VI. Between VI and VII only sandy shale is seen. Near Wood's mill there is a calcareous shale at thirty feet above VII, which may represent limestone VIII. From this locality to the township line only shales and sandstones are exposed in the roads, and the hill-sides are so deeply buried under debris that everything is concealed.

On Gray's fork, the limestones V, VI and VII are seen disappearing in succession, but limestone VIII is not exposed at any point above the junction of this stream with M'Courtney's fork. On Clover run, limestone IX crops out near the dwelling house on Mr. M. Johnson's property, where it is six or seven feet thick and has a dull, slightly ferruginous surface. At sixty feet above it is a compact sandstone twenty to twenty-five feet thick, which stands out on the hill-sides in bluffs. In these it appears to be massive, but where exposed in the road it is found to be made up of layers each about an inch thick. Between it and the limestone the blossom of the *Dunkard Coal* is seen.

Along the creek the sandstone is conspicuous. At the mouth of a little run above Strawn's mill Limestone VII is in the creek, and at 100 feet higher Limestone IX is imperfectly exposed on the hill-side. Thirty-five feet above it is the *Dunkard Coal*, which has been mined on the run by stripping. The sandstone seems to rest directly on the coal here. At the mouth of

Scott's run, this limestone is fully shown in the forks of the road where it is a mass of limestone and calcareous shale ten feet thick. At a short distance higher the *Dunkard Coal* is seen resting on a ferruginous limestone, IX*a*, and is mined, by stripping, on this run about half a mile from the creek. At a little way farther up the creek this coal has been stripped on property belonging to Mr. Joseph Scott. There is an old opening here on the south side of the stream in which the coal is about 20 inches thick, and a very fair semi-cannel. As in the southern townships it is double, and rests on a limestone. It is so exposed that it can be mined at many localities by stripping, and considering the quality of the coal one is surprised to find that so little advantage is taken of its accessibility; the more so since the nearest point whence coal can be procured is Waynesburg.

The massive sandstone above the coal is in sight quite to Rutan post-office. About one-eighth of a mile farther up the creek, and at 120 feet above the *Dunkard Coal*, Limestone X is exposed in a field near a spring, and at 35 feet above it the *Nineveh Coal* is in the road. The hill rises above this for two hundred feet, but everything is concealed except a fragmentary exposure of sandstone near the coal. In a gully this sandstone is found to be similar to that above the *Dunkard Coal* and has about the same thickness. Both of these rocks yield excellent building stone. Beyond this locality to the Baptist church, half a mile below the township line, there are no exposures. In the meantime the dip has changed so that, in the road at the church, Limestone X is exposed with its dark shale resting upon it. From the church to the line everything is shown, but Limestones VIII and IX do not appear. They have evidently thinned out.

Turning north at the Baptist church, the following succession is found in the road:

1. Red shale	70′
2. Limestone XIII	—
3. Concealed	50′
4. Limestone XII	—
5. Concealed	75
6. Limestone XI	—
7. Concealed	80′
8. Limestone X	—

The exposures are imperfect, and afford no opportunity for measuring the limestones, or for determining the character of the rocks in the intervals. On the ridge, XIII is easily traced to Hopewell church, beyond which it is in the road at Mr. Babbitt's blacksmith shop, where it is five feet thick. Under it is a compact sandstone for 40 feet, and the total interval to XII is 60 feet. Midway between this point and Hunter's cave, nearly a mile farther east on the ridge, the top of the series is reached. At 80 feet above XIII, the interval being filled by red shales and sandstones, there is a constant line of limestone fragments, which I have designated as Limestone XIV. This stratum has not been found *in situ*, but the horizon of the fragments cannot be far from the true place of the bed. Above this to the top of the hill, 60 feet, is concealed.

On the road leading from Hunter's cave to Sargent's mill on Brown's fork of Ten-Mile, the *Nineveh Coal* with its sandstone and limestone X are exposed, but everything else is concealed. The same exposure occurs on the road from Hopewell church to Brown's fork.

15. Morris Township, Greene County.

This lies directly north from Centre. Except along Brown's fork of Ten-Mile creek, the exposures are very unsatisfactory and for the most part are utterly worthless. The section extends from the top of the series to limestone V.

At Sargent's mill, in the extreme south-east corner of the township, the Upper Washington limestone comes down almost to the creek, and just below it is V, of which fragments are scattered in great numbers on the hill-side east from the mill. No. VI is about eight feet thick, in several layers and accompanied by its black shale. It reaches the creek at the mouth of a little run one-third of a mile farther up the stream, where it shows the same characteristics as at the mill. Above the black shale there is a good deal of iron ore, but not enough to be of any economical importance. In the hill, at nearly 170 feet above the limestone, is a heavy sandstone which proves to be that overlying the *Nineveh Coal.* Just beyond this locality limestone VII is seen in the road and continues in sight almost to the mouth of Brushy Fork. VIII and IX do not occur on this stream.

In a little run below Mr. George Lightner's residence, and half a mile above the mouth of Brushy Fork, the following section was obtained:—

1.	Sandstone	60′
2.	Limestone XI	2′ 6″
3.	Shale and flaggy sandstone	42′
4.	*Nineveh Coal Bed*	1′ 10″
5.	Shale and sandstone	36′
6.	Carbonaceous shale	1′
7.	Limestone X	2′
8.	Shaly sandstone, to road	45′

The coal consists of shale four inches, and coal twenty inches. An attempt has been made here to mine it by drifting along its face, but this seems to have been abandoned after the removal of about one hundred bushels. The bed is too thin to be worked in this way. It has been mined by stripping on a run entering a creek at Mr. Lightner's residence, where it is said to be twenty inches thick. At the mouth of this run, limestone X is in the road. The black shale above it contains minute teeth of fish, and occasional impressions of leaves. The sandstone above the *Nineveh Coal* is shaly here.

At the mouth of Lightner's run below Nineveh, the coal is exposed in the bank at the road-side. Overlying it is shale on which rests a flaggy sandstone. At Nineveh, limestone XI is seen in a bluff about twelve feet above the creek, and is in four layers, thus—

1.	Nodular limestone	0′ 4″
2.	Calcareous shale	0′ 6″
3.	Compact limestone	1′ 6″
4.	Shaly limestone	0′ 6″

It weathers to a light blue color and is apparently good. Below it to the creek is argillaceous shale containing many nodules of carbonate, but so scattered through the mass as to be unavailable. About one-eighth of a mile above the village, XI passes under the creek, and XII is shown in the road just beyond. The northerly dip continues to half a mile north from Nineveh, carrying this limestone below the road, but at that distance it is reversed and XII soon comes up again, rising about as rapidly as the stream. At two miles north from Nineveh, this limestone is exposed in a bluff near Mr. Auld's house, where it is twelve feet thick, but is made up of shale and limestone about equally divided. The limestone layers vary from shaly to compact, and in color from flesh to light blue and almost black. Beyond this point to the summit everything is concealed. The interval between limestones XI and XII, is filled with sandstone. A thin coal is said to occur at thirty feet above XII, but I was unable to find it at the locality to which I was referred.

On Brushy Fork, the *Nineveh Coal* and limestone X were seen. The coal is not mined. Exposures are entirely wanting on the road from Nineveh to Prosperity, though one occasionally sees limestone fragments resemble XI. On Bates' Fork, the Upper Washington limestone is exposed for some distance in the vicinity of Hopkins' mill. Where the Waynesburg and Washington railroad turns off from the stream, a cut has exposed limestone X, while the *Nineveh Coal* is shown just beyond and the sandstone above it stands out in bluffs on both sides of the creek. Another limestone was seen about a mile farther up, but could not be identified.

16. JACKSON TOWNSHIP, GREENE COUNTY.

This lies directly west from Center and north from Gilmore. In its central portion is the high ridge separating the waters of Dunkard, Ten-Mile and Wheeling creeks, which, as already intimated, extends into Gilmore and Aleppo townships. The section reaches from the top of the series down to *Limestone* IX. This is the most perplexing part of the whole column and its character in this township is not such as to make work in it at all simple. The limestones which in northern Center are so well marked are either absent or concealed, and the other rocks yield so readily to the weather that connected sections do not exist.

On M'Courtney's fork of Ten-Mile the *Limestones* V,VI,VII and VIII are seen disappearing in succession before reaching the western boundary of Center township. In Jackson that stream flows along the strike, or nearly so, until within two miles of White Cottage. Near the Methodist church, below that place, the *Dunkard Coal* has been opened by Mr. Johnson, who, by stripping, has taken out between two and three hundred bushels to supply the church. Just above White Cottage, near the watering trough, the *Nineveh Coal* was once worked. It is only six inches thick, but is of good quality, and all that could be obtained by stripping was removed. *Limestone* X is not exposed here.

About 250 feet above the *Nineveh Coal* there is a dark limestone which is imperfectly exposed in the road. Fragments of it obtained on the cultivated hills at White Cottage show it to be a tough impure rock, containing crystals of blende and occasional minute scales of fish. This is *Limestone* XIII, the same with that found on the ridge in western Gilmore. Attempts to burn it into lime have not been successful. On the ridge above White Cottage, the great sandstone, already referred to as occurring in Gilmore and Springhill townships, is seen at thirty feet above the limestone. It is hardly so compact here as there and shows less ability to withstand the weather. The exposed surface is ordinarily somewhat shattered, though I have been informed, since visiting this locality, that the rock is in massive cliffs not far from the head of South Wheeling creek in this township.

On South Wheeling creek no exposures were found until near the line of Richhill township, where the highest two limestones of the Richhill section occur. I am unable to identify these limestones satisfactorily. At twenty-five feet above the upper one there is a thin coal bed.

On Wood's run, which enters M'Courtney's fork near the township line, Mr. White obtained the following section:

1.	*Nineveh Coal Bed*	0′	8″
2.	Dark shale	2′	4″
3.	Concealed	95′	0″
4.	Massive sandstone	20′	0″
5.	Shale	10′	0″
6.	*Dunkard Coal Bed*	1′	8″
7.	Limestone IX "a"	3′	0″
8.	Flaggy sandstone	20′	0″
9.	Shale	20′	0″
10.	Limestone IX, seen	6′	0″

The *Dunkavd Coal* is worked by Mr. Wood, who finds it:

Coal	0′	11″
Clay	0′	1″
Coal	0′	8″
Total	1′	8

According to Mr. Wood, the coal is preferable to that from the Waynesburg. On the same run, this bed is stripped by Mr. Bayard, on whese property the Upper Bench is one foot thick. On the Wheeling creek side it is mined by Mr. John Scott, who has it thinner as follows:—

1.	*Coal*	0′ 8″
2.	Clay	0′ 1″
3.	*Coal*	0′ 5″
	Total	1′ 2″

The quality here is the same as at the other localities. The underlying limestone in every case is quite light colored. In the section given above, *Limestone* IX is dark colored. It is in numerous thin layers separated by thinner layers of shale, some of which are bituminous. A similar condition occurs at the exposure of this stratum on Gray's fork, but there the bituminous matter is proportionately less in quantity.

17. Aleppo Township, Greene County.

This lies directly west from Jackson and north from Springhill. It is drained by the South or Aleppo fork of South Wheeling creek, which flows northward through the township. The section extends from the top of the series to about one hundred feet below the Upper Washington limestone.

Along the creek Mr. White made out a detailed section which condensed, is as follows:—

1.	Concealed	60′
2.	Limestone	4′
3.	Shale	25′
4.	Bituminous shale	2′
5.	Shale	30′
6.	Sandstone	30′
7.	Concealed	300′
8.	*Nineveh Coal Bed*	1′
9.	Shale	25′
10.	Limestone and shale X	8′
11.	Shale and sandstone	40′
12.	*Coal*	1′ 2″
13.	Limestone	2′
14.	Sandstone	30′
15.	Limestone	8′
16.	Shales and sandstone	70′
17.	Limestone	2′
18.	Shale	15′
19.	*Dunkard Coal Bed*	2′
20.	Limestone IX "*b*"	1′ 6″

21.	Shaly sandstone	25′
22.	Limestone IX "*a*"	2′
23.	Shales and sandstone	115′
24.	Dark calc. shale	2′
25.	Limestone VI	6′
26.	Shale and sandstone	40′
27.	Limestone V	2′
28.	Shale and sandstone	60′
29.	Shale	10′
30.	Sandstone	25′

As will be seen by reference to Richhill township, this shows some difference from the section obtained along the Dunkard fork, but is much more like the typical section than that one is.

The highest members of it are reached only in the southern portion of the township, on the ridge dividing the waters of Fish creek from those of South Wheeling. The limestone No. 2, is dark blue, somewhat flinty, and is reported by Mr. White as observed at but two localities, one on land belonging to Mr. Bosworth, at the head of Walnut fork, and the other at the graveyard by Windy Gap church. At the same localities the bituminous shale No. 4, occurs. The Great Sandstone is frequently seen on the ridge and shows the peculiarities so characteristic of it in Gilmore and Springhill. It is soft, somewhat coarse grained, dresses easily and is a handsome building stone.

The *Nineveh Coal* was seen only on the property of Mr. Lee Moffatt, near the head of Hart's run. *Limestone* X is dark blue, and some of its layers contain much carbonaceous matter. The dark shale above it holds teeth and scales of fish as well as imperfect impressions of leaves, which belong to *neuropteris* and *pecopteris*.

The coal, No. 12, is mined on Hart's run by Mr. Perry White, and lower down by Mr. Jackson Hinerman. At both exposures it is about fourteen inches thick, and rests on a dark brecciated limestone. A coal occupying a similar position was found near the township line at Mr. Smith's property, on the South fork of Wheeling creek. This bed seems to be persistent in the northern portion of the township. The thick limestone No. 15, is also persistent, and it was seen on nearly all the little streams. It is earthy, weathers to a dirty white and breaks with an irregular fracture.

The *Dunkard Coal* is worked on Hart's run by Mr. Jesse

Hinerman, at half a mile from the West Virginia line, where it shows the following structure:—

1. Red shale			20′ 0″
2. *Coal*	10″	}	1′ 10″
3. Clay	4″	}	
4. *Coal*	8″	}	
5. Limestone IX *b*			4′ 0″

The coal is semi-cannel but is quite clean, and Mr. Hinerman, who has used it, finds that it answers very well for smithing purposes. On Mud Lick, a tributary to Wheeling creek, Mr. Levi Murray has opened this bed by drifting. In his entry, which was carried 50 yards before it was abandoned, the coal is somewhat thicker than usual, showing—

1. *Coal*	11″
2. Clay	3″
3. *Coal*	10″
Total	2′ 0″

It is stripped on Mr. Lough's farm on Dunkard fork, where it is a little thinner than at the last exposure. It is a semi-cannel with much slate and sulphur.

The Upper Washington Limestone is reached only on the creek near the township line. It is quite pure, and has been burned for lime, yielding an excellent article.

18. Richhill Township, Greene County.

This is the most north-western township of Greene county, and is separated from Washington county by Hunter's fork of South Wheeling creek. The Dunkard fork flows north-westwardly through the south and south-eastern portion, and is joined by its south or Aleppo fork at Ryerson's station. In the eastern part Gray's fork of Ten-mile rises, and is separated by a low divide from streams flowing into Dunkard and Hunter's forks. The Washington anticlinal passes through this township, bringing up the Waynesburg coal on the waters of both forks of Wheeling creek. The section extends from the top of the series to the *Waynesburg Coal*, a greater vertical range than is exhibited by any other township in the county.

Along Hunter's fork no openings in the *Waynesburg* were seen, although it is quite extensively mined on the opposite side of that stream, at rather more than two miles from the West Virginia line. The limestone underlying the coal is exposed at a short distance below these openings. On Mill run, which enters the creek near Clouse's mill, the following section was obtained:—

1.	Limestone III.		
2.	Sandy shale	15′	
3.	Limestone		8″
4.	Sandy shale	12′	
5.	Dark shale	3′	
6.	Limestone II		6″
7.	Dark shale	5′	
8.	*Washington Coal Bed*		6″
9.	Laminated sandstone	6′	
10.	Concealed	20′	
11.	Buff limestone I*b*	8′	
12.	Concealed to creek	10′	

The *Washington Coal* shows some strange variations in this county, but this is the only vicinity where it approaches total extinction. The overlying *Limestone* II, is so thin at no other exposure. *Limestone* I*b* is a new member of the series in Greene county, and is found only on Hunter's fork, though it is quite persistent throughout the western portion of Washington. At a short distance above Clouse's mill, the coal and *Limestone* II are seen in the bluff and exhibit the same features as in the section on Mill run.

On Owens' run, which enters the creek about a mile above Clouse's mill, the following section was obtained in a ravine on Mr. S. A. Houston's property:—

1.	Upper Washington limestone VI	8′
2.	Interval not fully exposed	69′
3.	Limestone	0′ 8″
4.	Clay	7′
5.	Middle Washington limestone, IV	3′
6.	Shaly sandstone	35′
7.	Limestone III, (in run.)	

The blossom of a thin coal is occasionally seen at twenty-five to thirty feet above *Limestone* VI. This rock weathers white, but inside is dingy blue to dull slate color. The interval No. 2, contains a fine sandstone ten to fifteen feet thick, which is an admirable building stone. When first taken out it is quite soft but soon hardens on exposure. This is the only locality, aside from Ruff's creek, at which the Middle Washington limestone has been recognized in Greene county. *Limestone* III is traceable from the mouth of the run to Mr. Houston's residence, where it passes under. It is tough and slate colored, but is said to burn into good lime. Above that point *Limestone* IV is shown in the road for nearly a mile, and before the sum mit of the dividing ridge is reached *Limestone* VI appears for a short distance. On the other side, on Gray's fork, the road again shows it near the church, from which it is constantly in sight to Graysville, dipping somewhat more rapidly than the stream and passing under at the village.

Above the mouth of Owens' run, the limestones of the last section are in sight quite to the township line, and for the most part the *Limestone* IV is above the road.

At the West Virginia line on Dunkard fork of South Wheeling creek, the *Waynesburg Coal* is twenty feet above the stream and is said to be available to the junction of that stream with Hunter's fork. A deserted opening was seen at Crow's mill, near the State line; thence the coal rises about as rapidly as the creek to within a short distance of the mouth of Crabapple, and the openings are numerous. One, which is half a mile above Crow's mill, shows—

Coal	1′ 4″
Clay	2′ 2″
Coal, seen	2′ 7″
Total	6′ 1″

At another worked by Mr. John Logston, somewhat farther up the stream, the following exposure occurs:—

1. Sandstone		30′
2. Shale		2′
3. *Waynesburg Coal Bed.*		
1. *Coal*	1′ 4″	
2. Clay	1′ 8″	6′ 2″
3. *Coal*	3′ 2″	

The top bench is quite slaty and is not mined, but the lower bench is very good and compact. It contains little pyrites.

Half a mile below the mouth of Crabapple, the road leaves the creek and after passing over a high hill, returns to it again at Crabapple. On the eastern slope of this hill, the *Washington Coal* is exposed in the road at 130 feet above the *Waynesburg* with the black shale below it, and *Limestones* II and III above it. At the mouth of Crabapple the *Waynesburg* is mined extensively, to supply a large region, coal being drawn from this locality to all parts of Richhill and even into Centre. Mr. Lyon's opening shows—

Coal and shale	0′ 3″
Clay	1′ 4″
Coal, seen	2′ 10″
Total	4′ 5″

In Mr. Keys' bank one hundred yards farther down the creek, the upper bench is wanting and the sandstone rests directly on the clay, which is slickensided. The lower bench is three feet three inches thick. In this vicinity the clay is quite troublesome and frequently cuts out the coal very badly. At a bank opposite that of Mr. Keys and now deserted, the coal was found six feet thick at the entrance, but it was soon cut out altogether by a clay horseback several yards thick, behind which there were but three or four feet of coal, which moreover was of poor quality. No horsebacks of sandstone were reported from this locality. On Crabapple there have been many openings, but most of these are deserted probably because they were directed down the dip. At Mr. Daniel Young's bank, the section is—

1. Dark shale with *coal*	4″
2. Drab clay	1′ 6″
3. Black clay	3″
4. *Coal*	2′ 9″

Above this there are rude openings until at one-third of a mile from the main creek the coal goes under. On the latter it disappears barely half a mile above Crabapple.

In this vicinity, and indeed all the way to the West Virginia line, the coal is very good, and far superior to that obtained from the *Waynesburg* at any place else in Greene county. It is hard, compact, not very heavy, and does not yield a bulky ash. It is said to be open-burning, and the appearance is such as to confirm the statement. It would bear shipment well. Unfortunately it contains too much sulphur to be employed in the furnace, as the pyrites is visible in considerable quantity. At the same time the quantity is much less than is usually found in this bed, and the coal is an excellent article for ordinary fuel. As may be supposed, it is in high repute.

The Waynesburg sandstone is 65 feet thick, and for 25 feet at the base is massive. This portion is soft, almost loose, and breaks down easily under influence of the weather, so that large cavities are formed whose walls are often honeycombed. Some of these seen on Crabapple are 30 to 40 feet deep, and contract behind in such a manner as to render it probable that they are connected with others farther in. On Crabapple, at the Jacksonville road, the following section was obtained:—

1.	Concealed	115′
2.	*Coal*	1′ 6″
3.	Concealed	50′
4.	Black shale	2′
5.	Sandy shale	2′
6.	*Waynesburg "b" Coal Bed*	1′
7.	Shale	24′
8.	*Waynesburg "a" Coal Bed*	blossom.
9.	Shale	10′
10.	Sandstone	65′
11.	*Waynesburg Coal Bed*	4′

The relations of the coal No. 2 are somewhat uncertain. It is 162 feet above the Waynesburg, which is very nearly the interval between the *Washington* and *Waynesburg* exposed in Franklin township; but less than a mile south of west from this place the interval between these coals was found, by direct measurement, to be but 130 feet. The distance between the coal of the section and No. 6, which is without doubt the *Waynesburg* "*b*," is too great to permit us to regard it as the *Washington*, for no such interval has been observed between those beds north from Dunkard creek. And besides, the intervals show a decided decrease north-west. For these reasons I think the *Washington* is in the concealed interval, and that the black shale No. 4 is the shale which underlies that coal.

Nearly two miles farther up the stream, on a little run coming in just below an old saw-mill the *Washington Coal* is mined, but when visited the opening was in such shape that no definite information could be obtained. At the base of eight feet of black shale two feet of coal were seen, which is a cannel with numerous thin streaks of bituminous coal. It shows a good deal of sulphur, and the ash is evidently large. But it is doubtless a very fair fuel.

On the road leading from this opening to Jacksonville there are no exposures for 180 feet, when a light colored limestone, evidently VI, is seen; and at 130 feet higher, at the north end of Jacksonville, limestone X shows, accompanied by its black shale. Ascending the road leading northward from the coal-bank to the ridge separating the waters of Crabapple from those of Owens and Mill runs, one finds *Limestone* III at 20 feet above the coal, and the Upper Washington at 120 feet higher. This limestone is in the road at the head of Mill run.

Returning to the main creek, we find the *Waynesburg Coal* passing under the stream about half a mile above the mouth of Crabapple, and the lower portion of the overlying sandstone remains above the stream until within a mile of Ryerson's station. On the road between the station and Crabapple, the *Washington Coal* is mined by Mr. George Woodruff, who has it only two feet thick with a clay parting two inches. It is easily followed to Ryerson's station. At that place Mr. S. H. Barnhart has an opening in the *Waynesburg* "*a*" which shows—

Shale	10′ 0″
Coal and black shale	0′ 5″
Coal	1 2″

The coal is very poor, for though there is little sulphur there is a great proportion of ash. In burning, it soon forms a compact mass which does not go to pieces or reduce in bulk. There is said to be "no heat in it," which is highly probable. Ascending the South fork, this coal is found at the level of the stream, at one-third of a mile above the village. The road there ascends a hill and shows a thin *coal* associated with a ferruginous limestone at forty feet above the *Waynesburg* "*a*," and at fifty feet higher another coal is seen also accompanied by a limestone but very imperfectly exposed. The upper of these beds is the *Washington*, and comes down to the creek about a mile and a half from Ryerson's station, near the mouth of a run entering from the west. At a little distance up this run *Limestone* III is seen.

Returning to the station, Dunkard fork of South Wheeling creek is again reached. Just above the village, the *Washington Coal* is exposed in the road, and near Baldwin's mill a deserted opening in it is seen. On a run coming in at the mill, there are many openings which, with one exception, have been deserted. The only one now worked shows the following section:—

Coal	0′ 6″
Clay	1′ 9″
Coal	1′ 6″
Total	3′ 9″

An exposure in the road seems to prove another bench below No. 4. The coal is inferior owing to the presence of much slate, but otherwise it is quite good being sufficiently free from sulphur, to be used by blacksmiths. Forty feet higher is a nodular limestone two feet thick, with a *coal* four feet above it, having this structure:—

Coal	0′ 2″
Clay	0′ 6″
Coal	0′ 2″

At forty feet higher, there seems to be another *coal* resting on one foot of limestone, and at eighty feet farther up the hill is a terrace-like bench on which are great numbers of large fragments of limestone, light colored and of coarse texture. Near the school-house a partial exposure occurs, which gives as follows—

1. Concealed	250′	
2. Shale	10′	
3. *Coal*	0′	1″
4. Clay shale	50′	
5. Sandstone	20′	
6. Shale	10′	
7. *Coal*	1′	
8. Shale	15′	
9. Sandstone, to creek	20′	

Immense blocks of the limestone seen at Baldwin's mill are scattered about in the lower part of No. 1, and at thirty feet above the little coal there is abundance of black shale. The coal No. 2, occupies the place of the *Jolleytown Coal*, and the shale below it contains much iron ore. No. 9 is a very massive rock and forms steep bluffs along the creek here. At somewhat less than half a mile farther up the stream, the limestone comes down to the road and is exposed in a run, as follows:—

1. Limestone, seen	1′	
2. Shale	25′	
3. Black shale	2′	
4. Iron ore	0′	3″
5. Limestone VI	6′	

Above this everything is concealed to the top of the hill, or nearly 300 feet. No. 5 is undoubtedly the Upper Washington limestone, though differing in some important respects from that rock as it ordinarily appears. Here it is coarse and more or less conglomerate; in this resembling *Limestone* V; but the presence of the black shale, the interval, 203 feet, between it and the *Washington Coal* and the occurrence of *Limestone* V under it on the South fork in Aleppo township, show that this can be no other than VI. On the Jacksonville road which leaves the creek at say two and one-half miles from Ryerson's station, a light blue limestone is seen at 150 feet above No. 1 of the last section, having a coal blossom about twenty feet below it. At the mouth of a run entering the creek near this road, the first limestone above VI is seen in the bluff fourteen inches thick, and two hundred yards beyond it comes down to the road and shows a thickness of two feet. It is moderately pure, breaks with a clean fracture and is almost flesh colored. Barely one-third of a mile farther, still ascending the stream, another limestone is exposed in the bluff at fifteen feet above the creek. This is an exceedingly compact rock, two feet thick, showing a ragged fracture, weathering rusty yellow, but of a beautiful flesh color on the fresh surface. I find it impossible to obtain a direct measurement between this and the limestone first below it, as it is not exposed on the little run near the Jacksonville road, and where the other passes under, the debris is so deep that although great fragments of the upper rock are seen, no determination of its place can be made. This difficulty is encountered all along the creek, so that owing to the utter impossibility of making direct measurements, most of the intervals have been calculated. Somewhat less than half a mile above where the third limestone of this series appears in the creek bluff, the fourth is exposed on a little run coming in just below Stall's mill, where it shows—

1. Limestone	1′ 2″
2. Calcareous shale	4′
3. Limestone	1′

The whole mass is light-colored, and the limestone is too earthy to make lime, as may be seen in an old kiln at the mouth of the run. Twelve feet of argillaceous shale rest upon it. Between this and the third limestone no direct measurement could be obtained, but by following the sandstone underlying the former the interval is determined to be not far from 100 feet. There seems to be another limestone in this interval, for fragments were seen which do not belong to either the third or fourth of the series, and not far from these there is much dark shale. If the estimated interval between the two limestones be correct, then that between the second and third is 50 feet, taking the measurement on the Jacksonville road as the basis. This is not far from correct.

This light-colored fourth limestone comes down to the creek at fifty yards below Stall's mill, where, on the opposite side of the stream, there is an impure limestone 35 feet above the fourth, showing this arrangement—

1. Limestone		10″
2. Clay		10″
3. Arenaceous limestone	1′	6″

and resting on sandstone. Some carbonaceous shale is associated with it, and at 30 feet higher there is a line of limestone fragments.

The latter, the sixth of the series on this creek, comes down to the road at one-sixth of a mile above the mill, where it is seen to be a dark-blue somewhat slaty rock, with dark shale resting on it. It shows—

1. Limestone		8″
2. Clay	1′	0″
3. Limestone	1′	3″

This is clearly *Limestone* X of the general section. It is 40 feet above the fifth limestone, and the interval is filled with sandy shale. At 90 feet higher, the interval being occupied chiefly by reddish sandy shale, the seventh limestone appears in a field almost opposite the mouth of Long run, and just above the school-house. This stratum comes down to the creek at a short distance beyond Adams' mill, where it has a thickness of eight feet, and is made up of several layers separated by calcareous shale. Thirty feet higher in the bluff at this place the eighth limestone is exposed with a thickness of three feet.

The direction of dip changes here from south-east to north-west, and the last two limestones continue in sight along the creek up to Woodruff's mill in Jackson township. And at several places a thin coal was seen at a few feet above the upper one. Above the eighth limestone there are no exposures, though the total thickness of rocks thus concealed is very nearly 350 feet.

In tabular form the succession of these limestones is as follows:—

1.	Concealed		350
2.	Eighth limestone		2′
3.	Interval		30
4.	Seventh limestone		8′
5.	Interval		90′
6.	Sixth Limestone X		2′ 11″
7.	Interval		40′
8.	Fifth limestone		3′ 2″
9.	Interval		35′
10.	Fourth limestone		6′ 2′
11.	Interval	100′?	
12.	Third limestone	2′	150′
13.	Interval	50′	
14.	Second limestone		1′ 4″
15.	Interval		27
16.	First Limestone VI		6′
17.	Interval		203′
18.	*Washington Coal Bed.*		

CHAPTER XI.

Washington County.

19. East Bethlehem Township.

This, the south-eastern township of the county, borders on the river and has for its southern boundary Ten-Mile creek and the North fork of that stream. The section extends from the Waynesburg sandstone to eighty feet below the *Pittsburg Coal.*

At the mouth of Ten-Mile, the *Pittsburg Coal* is exposed at a few feet above the river, and at Dog Hollow, three-fourths of a mile up the creek it is seen ten feet from the water, underlying forty-five feet of sandstone and showing no roof division. It is seven feet one inch thick, the partings being two feet ten inches from the bottom. On the hill above this, there is an opening in the *Waynesburg* belonging to Mr. S. G. Gayman, from which to the creek, the following section was obtained:

1.	Waynesburg sandstone	35′
2.	*Waynesburg Coal Bed*	5′ 8″
3.	Concealed	125′
4.	Sandstone	10′
5.	Limestone	80′
6.	Shale	5′
7.	*Sewickley Coal Bed*	2′
8.	Sandstone	40′
9.	Limestone	25′
10.	Sandstone and shale	25′
11.	*Redstone Coal Bed*	blossom.
12.	Sandstone	50′
13.	*Pittsburg Coal Bed*	7′
14.	Concealed	12′

The *Waynesburg* is here in four benches, six, fourteen, nineteen and nine inches respectively. The coal is very poor, containing a good deal of sulphur and leaving much ash and cinder. It is opened by many of the farmers in this portion of the township, especially by those living on the ridge. Though very poor it is easily reached on the hilltops and is mined because of the difficulty of hauling coal from the river at Frederick or Millsborough. Between school-houses seven and eight, openings were seen belonging to Messrs. Henry Brister, A.

Bowser, E. Buckingham and N. Barnard, which gave the following measurements —

Waynesburg Coal.

1.	Shale	0′ 0″	1′ 2″	0′ 2″	0′ 0″
2.	*Coal*	0′ 5″	0′ 8″	0′ 9″	0′ 5″
3.	Clay	0′ 3″	0′ 5½″	0′ 8″	0′ 6″
4.	*Coal*	1′ 4″	1′ 3″	1′ 4″	1′ 6″
5.	Clay	1′ 0″	0′ 11″	1′ 2″	1′ 4″
6.	*Coal*	1′ 6″	1′ 8″	2′ 5″	2′ 0″
7.	Clay	0′ 0½″	0′ 0½″	0′ 5″	0′ 5″
8.	*Coal*	0′ 8″	0′ 10″	——	0′ 3″

At No. 4, Mr. Barnard's opening, there are two layers of bituminous clay four and three inches thick, and separated by four inches of drab clay, at one foot below No. 9 of the section.

At Clarksville, the *Sewickley Coal* is fifteen feet above the creek. Along the North fork it remains in sight as far as Hawkin's mill, where the Waynesburg axis crosses the stream. The shaly sandstone underlying it becomes more and more laminated until, at barely a mile from Clarksville, it is a fissile shale with numerous vertical fissures. At three-fourths of a mile above that village, the following section was obtained :—

1.	Sandstone	35′
2.	Limestone	10′
3.	Concealed	40′
4.	Limestone	10′
5.	Sandstone	8′
6.	Limestone	15′
7.	Sandy shale	7′
8.	Limestone	50′
9.	Shale	2′
10.	*Sewickley Coal Bed*	1′ 10′
11.	Dark sandy shale	40′
12.	Limestone	10′

This section is of interest as showing sub-divisions of the Great Limestone similar to those observed in a section taken in Morgan township of Greene county. The shale under the *Sewickley* has great numbers of indistinct vegetable remains.

From Hawkins' mill the strata dip north-westward quite rapidly; the *Sewickley* soon disappears, and at the mouth of Plum run, one-fourth of a mile east from the township line, the *Waynesburg* is only 180 feet above the creek. On that run the coal goes under at the old mill, say two miles from the creek. At one time it must have been worked to a very considerable extent, for the deserted openings are very numerous. None are now in operation, as the farmers choose to bring their coal from the river in preference to burning the Waynesburg. A bank on Mr. J. Berkheimer's property, at the head of the East fork of this run, is still worked somewhat for domestic use, and shows this structure—

1.	*Coal*	1′
2.	Clay	1′
3.	*Coal*	2′ 10″
4.	Clay	4″
5.	*Coal*	5″
	Total	5′ 7″

The main bench has a thin parting near the middle. The coal is said to be rather better than is usual in this vicinity, and leaves a by no means bulky ash.

Along the river exposures are quite frequent. At Millsborough the following section was obtained:

1.	Limestone	25′
2.	Sandy shale	30′
3.	*Redstone Coal Bed*	1′
4.	Sandy shale and sandstone	15′
5.	Massive sandstone	30′
6.	Shale	8″
7.	*Pittsburg Coal Bed*	8′ 6″

Here the *Redstone* is represented by bituminous shale, but at the lower end of Fredericktown it consists of coal 6 inches, almost midway in a mass of bituminous shale 5 feet thick, and is separated by 45 feet of sandstone from the *Pittsburg* below. The variations in the lower coal are very slight, nearly all the openings showing the following structure:

1.	*Coal*	1′ 2″
2.	Clay	9″
3.	*Coal*	6′ 6″
	Total	8′ 5″

In the vicinity of Fishpot run, which enters the river a mile below Fredericktown, the whole thickness of the bed is about 8 feet, the only variations being in the roof and the clay. In all the openings two thin partings occur in the lower division, and are separated by from two to four inches of coal. Above these the coal is bony, and contains so much sulphur as to be useless to smiths, and it is apt to clinker. Below the partings the coal for two feet is very pure, and is in high repute among blacksmiths. The bottom, for from six to ten inches, is of uncertain quality, and is not mined. When not slaty or sulphurous, it is so brittle that it cannot bear handling. Half a mile above the mouth of Fishpot run the *Little Pittsburg ?* shows one foot of fairly good coal, and at fifteen feet below it is another coal of the same thickness. Both beds are represented by bituminous shale at the mouth of Fishpot, where the section is—

1. Sandstone	25′	
2. *Pittsburg Coal Bed*	8′	
3. Shaly sandstone	22′	
4. *Little Pittsburg Coal Bed*	1′	
5. Limestone	3′	
6. Sandy shale	6′	
7. Limestone	1′	6″
8. Shale	4′	
9. Bituminous shale	1′	
10. Limestone and shale	15′	
11. Arenaceous shale	30′	

On the run, half a mile from its mouth, the *Pittsburg* is exposed with the roof only four inches thick, and the clay six inches, while the lower division is 6 feet 9 inches. It is overlaid here by 50 feet of sandstone.

Near the mouth of this run, salt is manufactured by Register & Bair, who obtain brine from a depth of 585 feet below the *Pittsburg Coal*, in sufficient quantity to produce four barrels of salt per day. The boring was made nearly fifty years ago, and no record exists. In 1858, another boring on the land of Mr. J. Wright in the same vicinity, was made to the depth of 650 or 660 feet below the *Pittsburg Coal*, but was abandoned as the supply of brine was very small. The record of this well is as follows:—

Record of J. Wright's salt well.

1. Debris, &c	40′	
2. Shale	22′	6″
3. Limestone	2′	6″
4. Shale	9′	
5. Limestone	2′	
6. Shale	21′	
7. Sandstone	8′	
8. Shale	17′	
9. Sandstone, [1st Dunkard oil sand]	42′	
Depth of well, recorded as	170′	
10. Black shale	11′	
11. Clay and shale	14′	

12. *Coal* 1′ and fire-clay	7′
13. Shale	8′
14. Sandstone	4′
15. Shale	18′
16. Sandstone	88′
17. *Coal* 1′ and black shale	10′
18. Fire-clay	10′
19. Sandstone, [U. Mahoning S. S.?]	21′
Depth of well	373′
20. *Coal* 6′′ and black shale	15′
21. Sandstone, [L. Mahoning S. S.?]	44′
22. Dark shale	33′
23. Light shale	32′
Depth of well	487′.
24. Sandstone	33′
25. *Coal* and shale	7′
26. Dark shale	9′
27. Sandstone, [Piedmont S. S.?]	73′
Depth of well	604′

No. 9 of this section is the first oil-sandstone of the Dunkard oil region. I am inclined to regard Nos. 19, 20 and 21 as the Mahoning Sandstone, and No. 27 as the one which, in this region has been commonly termed the Tionesta [Piedmont] sandstone. This is saliferous in the Wright well, although the brine is present in but small quantity. The brine used by Regester & Bair is obtained from this rock.

At a short distance below the mouth of Fishpot the river turns and flows to the south, so that within three miles the *Pittsburg Coal* is only five feet above low water, and at one mile farther, is probably fifty feet under. There the course of the river is again changed to north, and at dam No. 5 the coal is not more than twenty-five feet below low water, having been reached by piles driven for the dam, while at Brownsville it is at low water level. Two openings in the *Pittsburg*, below the mouth of Fishpot, show the lower division varying from six feet six inches to seven feet, while the roof maintains a constant thickness of one foot. The latter is bony and is not removed from the mine. Two miles below Fishpot the blossom of the *Sewickley* is seen, and at ninety-five feet below it is the *Redstone*, represented by five feet of bituminous shale. The *Pittsburg* is forty feet below the latter, and has a total thickness of eight feet. Between the *Redstone* and *Sewickley*, there is a mass of limestone twenty-five feet thick.

On Fishpot, the *Waynesburg* is worked by Mr. N. Baker, at two miles and a half from the river. His opening shows—

1. Sandstone (seen)	30′
2. *Coal*	0′ 10″
3. Sandstone	2′ 6″
4. *Coal*	0′ 4″
5. Clay	0′ 10″
6. *Coal*	2′ 6″
7. Parting	0′ 1″
8. *Coal*	0′ 6″

This is the section as shown at the mouth of the bank, but at a short distance inside No. 3 thins away and No. 4 becomes one foot four inches. Here is one of the few instances where the parting varies at the expense of the bench below. No other openings are now worked on the run. Returning to the river, we find at three miles below this run, an exposure on the land of Mr. Elias Crouch, which gives—

1. Limestone, seen	65′
2. *Sewickley Coal Bed*	blossom.
3. Shaly sandstone	35′
4. Limestone	30′
5. Shale	25′
6. *Redstone Coal Bed*	2′
7. Shaly sandstone	10′
8. Massive sandstone	30′
9. *Pittsburg Coal Bed*	8′
10. Concealed, to river	5′

One mile above this locality the limestone is twenty-five feet, while the sandstone and shale associated with it are each five feet thicker. At lock No. 5 the condition is the same and the *Sewickley Coal* is three feet. On Mr. Richardson's property here this coal is shown and is said to be very good. Near Mr. Henry Howard's residence, one mile above Brownsville, it is well exposed and has this structure—

Coal	3′ 6″
Black sandy shale	2′ 0″
Coal	0′ 8″
Total	6′ 2″

Above the coal and separated from it by four feet of shale, is the Great Limestone exposed for forty feet, while below it are sandstone twenty-five feet and limestone thirty feet.

Along the National pike the blossom of the *Waynesburg* "*a*" is seen at several places east from Centreville, and in the cut just west from that village. The *Waynesburg* is reached near the pike, on the properties of Messrs. J. Horton and J. B. Welch, where it was formerly worked.

20. West Bethlehem Township, Washington County.

This lies directly west from East Bethlehem. The North fork of Ten-Mile crosses it in the southern portion, and Daniel's run with its tributaries flow southward through it to the creek. In its northern part a fork of Pigeon creek cuts very deeply, while at Hillsborough, on the National pike, is probably the highest point in the whole district. The Pinhook axis passes across the north-eastern corner and the synclinal east from it crosses the south-east corner of the township. The section extends upward to nearly 550 feet above the *Waynesburg Coal*, reaching a stratum 425 feet above the *Washington Coal.*

Just above Zollarsville, on Ten-Mile, the *Waynesburg* is mined by Mr. W. H. Ulery, at whose opening it is fifty feet above the creek, and shows the following structure:—

Coal	1′ 1″
Clay	1′ 0″
Coal	2′ 10″

The overlying sandstone is exposed to the extent of thirty feet and is rather shaly, having a shattered surface, and the shale between it and the coal varies from zero to ten feet. The Upper Bench of the coal is said to be much better than the lower, for it contains little ash and sulphur, while the latter leaves a bulky ash and has so much sulphur as to clinker badly. Half a mile above Zollarsville the coal passes under the creek, and at a mile above the following section was obtained:—

1. Limestone IV		6′
2. Shaly sandstone		40′
3. Black shale		1′
4. Shale		30′
5. Limestone III?		4′
6. Concealed		15′
7. Shale		3′
8. *Washington Coal Bed.*		
1. *Coal*	0′ 8″	
2. Clay	0′ 6″	
3. *Coal*	0′ 6″	3′ 3″
4. Clay	0′ 5″	
5. *Coal*	1 2″	
9. Laminated sandstone		9′
10. Black shale		1′ 4′

11.	*Little Washington Coal Bed*	2′	
12.	Shaly sandstone	25′	
13.	*Waynesburg Coal Bed* "*b*"	0′	8″
14.	Sandstone	12′	
15.	Limestone	2	
16.	Sandstone	25′	
17.	Shale	8′	
18.	Limestone I	7′	
19.	Shale	3′	
20.	*Waynesburg Coal Bed* "*a*"	3′	6″
21.	Waynesburg sandstone	55′	

The *Waynesburg* is but a few feet below the base of the section. The *Washington* is not mined anywhere in the vicinity, but the exposures show that it is very bad coal and of no economical value whatever, if anything else can be had for fuel. The little coal below it is much thicker here than at any other locality in the district, but is poor, being made up of thin layers of coal and shale interstratified. The *Waynesburg* "*a*" is very thick but seems to be of little value, and no openings have been made to test it.

The *Washington* continues in sight to the township line at the mouth of Little creek, or as it may be termed the Pinhook fork of Ten-Mile, where it is 35 feet above the stream. Following up this fork the rocks are seen rising and the strata of the section re-appear in succession until, at about one-third of a mile below the village of Pleasant Valley or Pinhook, the *Waynesburg Coal* comes up and is worked. At two miles below Pinhook, in a ravine leading from near Mr. T. M'Ginnis' residence to the creek, a section is exposed similar to that obtained at Zollarsville, but showing *Limestone* II, 21 feet 6 inches thick, and resting directly on the *Washington Coal*, with *Limestone* III five feet thick and sixteen feet above the last. The interval between the *Washington* and *Waynesburg* "*a*" *Coals* already shows a very marked decrease, being only 80 feet here and 95 feet at Zollarsville. Near Mrs. Weaver's there is an imperfect section exposed, part of which is of interest as showing the intervals as follows:—

1.	Limestone VI	12′
2.	Concealed	20′
3.	*Jolleytown Coal Bed*	blossom.
4.	Concealed	120′
5.	*Washington Coal Bed*	blossom.
6.	Concealed	80
7.	*Waynesburg* "*a*" *Coal Bed*	blossom.

The first opening in the *Waynesburg* near Pinhook is that belonging to Mr. David Bane, about one-fourth of a mile below the village; the next is at Pinhook on Brush run, and belongs to Mr. J. C. Frazier. At a little way above this the coal goes under the run, but at one-eighth of a mile farther up Mr. J. C. Frazier shafted for it, and reached it at 12 feet from the surface. The measurements obtained at these several banks in the order given are as follows:—

Waynesburg Coal Bed.

1. *Coal*	1′ 2″	1′ 3″	1′ 2″
2. Clay	1′ 2″	1 0″	0′ 10″
3. *Coal*	2′ 11″	3′ 1″	3′ 8″
4. Clay	0′ 3″	0′ 3″	0′ 2″
5. *Coal*	0′ 6″	0′ 10″	0′ 6″

Other openings occur in the same vicinity, but they will be referred to in the description of Amwell township.

The quality of the coal varies little in the different banks. The top bench is poor and is not mined as it aids materially to strengthen the roof, which is usually a blue shale and far from being safe. The main bench is a fair coal, though by no means equal to the *Pittsburg* at Frederick or Millsborough, as it contains much sulphur and ash, making a good deal of clinker in a forge fire. It is quite free-burning and comes out nicely in blocks with but little waste in slack. The bottom bench is said to be superior to the others. This bed is extensively mined to supply a large area in the southern portion of Washinton county which is entirely destitute of coal. The quantity available here is large, but the line of outcrop is short, extending down the creek below the village for less than one-fourth of a mile, and reaching not so far above on Brush run and Little creek. The hills rise abruptly from the streams, so that shafting is impossible except at great expense. Though inferior in quality this coal is much prized since no other is within reach.

On the road leading up Brush run to the National pike, a little coal 20 inches thick is seen near the latter road, along which it is frequently exposed for two or three miles. It is the same with that in the Weaver section and is from 20 to 50 feet below the Upper Washington limestone, maintaining the former interval quite constantly in the greater part of the township. In the immediate vicinity of Hillsborough the remains of a massive sandstone stratum are found in huge blocks lying in the fields on top of the hills. This is not far from 525 feet above the *Waynesburg Coal*, as exposed on Pigeon creek, but I am unable to determine its place in the series. It is nearly 250 feet above the Upper Washington limestone, but in this portion of the Upper Barren series the intervals show such extreme diminution in thickness northward that I do not feel justified in attempting to identify it with any similar stratum in the Greene county section, especially since tracing is utterly out of the question. On the second road west from Hillsborough, leading to Pigeon creek from the pike, the following section was found which, though imperfect, is very suggestive when compared with sections in Greene county exhibiting the same portion of the series.

1.	Limestone VI	—
2.	Interval	150′
3.	*Washington Coal Bed*	—
4.	Interval	40′
5.	*Waynesburg "b" Coal Bed*	—
6.	Interval	40′
7.	*Waynesburg "a" Coal Bed*	—
8.	Interval	45′
9.	*Waynesburg Main Coal Bed*	—

On Pigeon creek the Pinhook axis brings up the *Waynesburg Coal*, which is mined by Messrs. S. Tombough, Whitfield and D. M. Letherman, at whose banks the following measurements were made:—

Waynesburg Coal Bed.

1.	*Coal*	1′ 0″	1′ 11″	10″
2.	Clay	1′ 0″	1′ 3″	1′ 0″
3.	*Coal*	2′ 0″	2′ 3″	2′ 6″
4.	Clay	0′ ½″	0′ 2″	0′ 1″
5.	*Coal*	0′ 3″	0′ 2″	0′ 2″

The top bench is slaty but soft, and though leaving a bulky ash does not clinker. The middle and lower benches are very hard, show many binders of pyrites and clinker badly. At the same time coal from these contains much less ash than that from the Upper bench. Below Mr. Letherman's the coal is, for the most part, in the creek bed until at Mr. Nichols the creek falls more rapidly than the coal, which is thus again available. Here the bed shows an unexpected increase, being 6 feet 1 inch thick, and the Middle bench has 3 feet 6 inches of coal. At a little south from this locality the coal is mined on land belonging to Judge Hart, where it is in three benches, 9, 41 and 9 inches thick. In all these openings the character of the bed is the same. The Waynesburg sandstone rests directly upon it, and frequently cuts out the top bench—a condition which is of by no means common occurrence in this bed, though it frequently happens with the *Pittsburg*. A measurement on Judge Hart's property shows the interval between the *Waynesburg* and *Washington* coals to be 165 feet, which, if correct, is a very rapid increase, for at only three miles west it is but 125 feet. *Limestone* II is thick here; has a dark gray color, and is largely used for lime. The *Waynesburg* "*b*" is 30 feet below the *Washington*, and is 15 inches thick.

At Mr. John Dage's, about a mile and a quarter north-west from Hillsborough, the *Waynesburg* "*a*" was once opened and found to show 15 inches of very fair coal. Above it is a lime stone several feet thick.

On Daniel's run and its tributaries exposures are not numerous, and there are no available coals. At the head of the run the *Waynesburg* "*a*" is seen in the road near Mrs. Densor's residence, where it appears to be about two feet thick. Near Mr. D. Hildebrand's the *Washington Coal* shows in the road, and on Little Daniel's run this coal has been mined by Mr. S. Mowl, who has it 2 feet 6 inches thick.

21. Amwell Township, in Washington County.

This lies directly west from West Bethlehem, and Ten-Mile creek flows across its southern border. It is crossed by the Pinhook axis, and the section is almost the same with that of the last township.

The exposures along Little creek, so far as it forms the eastern boundary of this township, have been mentioned in the last section. Above Pinhook the stream lies wholly in Amwell, and heads near the National pike. In the vicinity of Pinhook or Pleasant Valley the *Waynesburg* is mined by Mrs. Cooper, H. and A. M'Crury, C. Chamberlain, S. L. Hughes, Amos Wise, J. A. Moninger and Frederick Ferrel. The bed shows much variation, as appears from the measurement taken at the several openings.

Waynesburg Coal.	1.	2.	3.	4.	5.	6.	7.
1. *Coal*.........	1′ 1″	2″-9″	1′ 4″– 11″	1′ 0″	1′ 0″	4″– 5″	1′ 0″
2. Clay...........	1′ 0″	1′ 2″	1′ 6″– 10″	1′ 0″	1′ 0″	1′ 2″–10″	1′ 0″
3. *Coal*.........	3′ 0″	2′ 10″	2′ 7″–3′ 4″	3′ 2″	3 3″	2′ 10″	2′ 8″
4. Clay...........	6″	6″	2½″	3″	3″	?	?
5. *Coal*.........	......	6″	5″	7″	6″	6″	......

The changes in the upper bench seem to depend upon the presence or absence of shale above it, for when the sandstone comes down upon the coal the latter is always more or less cut out. No. 3 is a strong coal, excellent for raising steam, but the unanimous opinion of those who use it for that purpose is that on the whole it is bad, containing so much sulphur as to leave a great deal of clinker, and to be very destructive to grate bars. No. 5 is said to be clean enough for use in smithing, and is taken out whenever the clay above is thin.

The Waynesburg sandstone is forty-five feet thick, and at a mile above the village the *Waynesburg* "*a*" appears above it. The *Washington Coal* is seen at somewhat more than a mile farther up the creek. Both of these are thin, and no openings have been made in them. Limestone II is exposed near Mr. S. L. Slusher's residence, and appears to be not less than twenty-five feet thick. The strata descend northwardly to Hughes' mill, where the dip is reversed. From this point the Upper Washington Limestone shows in the road for nearly three miles, but beyond that the rise of the surface is rapid to the dividing ridge between the waters of Ten-Mile and Chartiers creek, so that the limestone is not seen again until within a mile of Washington on the National road. Above it there are no exposures. Below it to the *Washington Coal* is about one hundred and fifty feet, and to the *Waynesburg* in this vicinity is not far from three hundred feet.

On the first road east from Martinsburg, passing south from the pike, the section extends downward from forty-two feet above Limestone VI to Limestone II, at Read's mill. The exposures are very poor, but Limestone VI is seen to be fifteen feet thick, with a coal blossom at twenty-five feet above it. The interval to the *Washington Coal* seems to be not far from one hundred and eighty feet, according to the measurement made by Mr. White. The lower limestone is fifteen feet thick, and has bituminous shale resting on it. East from this the exposures along the pike are far from being satisfactory, but at the eastern boundary of the township the following section was obtained:

1. Limestone	2
2. *Coal*	Blossom.
3. Shale	20
4. Limestone VI, seen	8′
5. Concealed	45′
6. *Jolleytown Coal Bed.*	
7. Shale	40′
8. Limestone IV, seen	5′
9. Concealed	40′
10. *Coal?*	
11. Limestone and Shale II	25
12. *Washington Coal Bed*	3′
13. Concealed	45
14. Limestone I	12
15. *Waynesburg "a" Coal Bed*	2′

Here the total interval between the *Washington Coal* and the middle of Limestone VI, is but one hundred and fifty-five feet, so that the true distance between the two strata is not more than one hundred and forty-five feet, and is probably a little less. In this section the *Washington Coal* shows:—

1. *Coal*	0′	6″
2. Clay	2′	0′
3. *Coal*	1′	6″

It is not mined. The *Waynesburg "a"* is imperfectly exposed, but is not less than two feet.

From the mouth of Little Creek the *Washington Coal* is constantly above Ten-Mile creek to the western boundary of the township, as the strata rise steadily in that direction. The coal has been opened at Ten-Mile village on land belonging to Mr. S. Tharp, where it shows:—

1. *Coal*	1′	0″
2. Clay	0′	3″—7″
3. *Coal*	0′	11″
4. Clay	0′	4″—6″
5. *Coal*	0′	8″
Total	3′	6″

This coal is miserable stuff and most of it is little better than a rich bituminous shale, although when freshly mined it is quite bright. It has been used for fuel by one or two families, but they now think it more profitable to haul their coal from Pinhook, five miles away, than to burn this, which is at hand. Underlying the coal is the thinly laminated sandstone, full of macerated fragments of vegetable matter and containing thin layers of argillaceous shale which show very handsome impressions of plants. An incomplete exposure on the hill here gives the interval between the coal and Limestone VI as one hundred and eighty-four feet, and Limestone II is nearly seventeen feet thick. It includes a layer of bituminous shale which blocks out almost like coal, and contains bivalve crustaceans, together with macerated fragments of vegetable matter. On the road leading north from that village the coal disappears at one mile from the creek and does not reappear until within sight of Pinhook village. On this road the little coal is twelve feet below it and is one foot thick. The black shale is present. Under the *Little Washington Coal* there is sandstone or sandy shale for thirty feet, containing many nodules of impure iron ore, but there is no evidence that Limestone I*b* is here. Above the village, on the Ten-Mile road, no exposures of the coal are found for nearly two miles, but the Middle Washington Limestone, IV, of the series, is well shown at a distance of little more than a mile, where it is of a bright buff color. At the Amity road crossing the coal is in the road, and at twenty feet above it is a thin coal underlying a handsome flaggy sandstone, which has been quarried. The same little coal is exposed half a mile farther down the creek. Between the creek and Amity the Middle and Upper Washington Limestones are exposed in the road. They are exposed also on the same road southward toward the Greene county line, and on the ridge between the two counties Limestone X is seen at the cross-roads.

At the township line a tributary enters Ten-Mile from the north and forms the boundary between Amwell and Morris. The crest of the Pinhook axis crosses the creek here, so that from this point northward the strata descend. Following up this branch we find the *Washington Coal* passing under it about one-third of a mile below the saw-mill, where it is in all nearly four feet thick. There is an old opening below this, but it has long been deserted and the coal is said to have been of inferior quality. At the mill Limestone II is exposed, about fifteen feet thick, and at twenty feet higher is a light blue limestone, ten inches. This is seen at the Baptist church farther up this stream, and is probably Limestone III. At a short distance above the railroad crossing, Limestone IV is seen eight feet thick, and near the saw-mill Limestone VI is shown with the following section:

1. Concealed	20′	0″
2. *Coal*	1′	4″
3. Clay	0′	8″
4. Shale	1′	0″
5. Nodular limestone	3′	0″
6. Ferruginous shale	6′	0″
7. Shaly sandstone	16′	0′
8. Dark shale	8′	0

The interval between Limestones IV and VI is not far from seventy feet. The latter remains in sight to the Franklin township line, where the next limestone, which is probably VII of the Greene county section, is exposed at the school house.

22. Morris Township, in Washington County.

This lies directly west from Amwell. It section extends upwards from the *Washington Coal* to two hundred feet above Limestone VI, a portion of the series possessing but little of interest, either scientific or economical. The synclinal northwest from the Pinhook axis crosses the township and passes in the immediate vicinity of the village of Prosperity.

From Van Buren, in Franklin township, a road leads to the Baptist church, in Amwell township, and passes through the north-west corner of Morris township. On this the Limestone VII (?) is seen at one-third of a mile from Van Buren, but no farther exposure occurs for a mile, when a rapid fall in the road brings it down to Limestone Vl, which is in the bed of the stream at the fork near Mr. J. Sander's house, where it has been burned for lime. At that place the dip hitherto southeast is reversed, and the rocks rise in that direction so that at the next fork the limestone is forty feet above the run. About one-fourth of a mile below, Limestone IV comes up, seven feet thick and near the last fork on this stream there is seen the

Little Limestone exposed at the Baptist church, and belonging thirty to forty feet above the *Washington Coal.*

At the junction of this fork of Ten-Mile with the creek the *Washington Coal* is shown in a bluff near the bridge. It is four or five feet thick, and is separated from the black shale below by ten feet of laminated sandstone, which contains the characteristic fragments of carbonized wood. The black shale seems to have a thin coal near the base. The exposure is not such as to admit of accurate measurements.

Along the creek for nearly half a mile the dip is very slight, as the course is with the strike, but beyond that the direction is changed, and the rocks fall rapidly towards the north-west. Just below where the axis crosses the stream the *Washington Coal* is mined by Mr. Johnson, at whose bank it shows three benches, eighteen, eight and fourteen inches respectively, while at another opening in the vicinity the *Little Washington* is two feet thick and twelve feet below the upper coal. On it rest three feet of black shale. At a short distance above these openings the coal and Limestone II pass under the creek. The little Limestone representing III first shows in a run coming in by the school house, and continues in sight for nearly one-fourth of a mile. Where it passes under the creek the following section was obtained:

1. Limestone VI	Fragments.
2. Shaly sandstones	80
3. Shale	15
4. Limestone IV	12′
5. Shale	18′
6. Calcareous sandstone	2′
7. Concretionary sandstone	7′
8. Clay shale	8′
9 Limestone III	10″

The Middle Washington Limestone is very ferruginous, and apparently concretionary. It weathers bright yellow to the depth of several inches, and the weathered portion scales off in concentric layers. From this point the creek flows south southwest, and the strata rise slightly about as fast as the bed of the stream, so that at Lindly's Mills Limestone III is in the creek and IV is in the bluff. The direction of the road is here changed to north-west, and the rocks again fall. At one-fourth of a mile above Lindly's Mills, Limestone IV is well exposed and shows a thickness of fifteen feet. It is dull, flesh-colored on fresh surface, but is ferruginous and weathers rusty yellow, mixed with dark patches, which give it an unmistakable appearance. Toward the base this stratum becomes less ferruginous and somewhat brecciated. Half a mile above Lindly's Mills its base is at the level of the creek, and it goes under altogether at one-third of a mile farther.

Above this there are no good exposures, but frequent fragments show that Limestone VI is coming down rapidly to the stream, and at the mouth of Short creek, one mile below Prosperity, it is level with the bridge. The upper layers are in the creek at the grist mill near the village, and at thirty feet above it in the hill is Limestone VII, the interval being filled with shaly sandstone.

At Prosperity a shaft was sunk for the *Pittsburg*, the impression being that that bed is only three hundred feet below the surface at that place. This impression was due, probably, to the fact that in a boring a mass of coal was reported as occurring at that depth. The shaft was begun at thirty feet above Limestone VI, and was continued to a distance of two hundred and thirteen feet from the surface and stopped in the *Waynesburg* "*b*," forty feet below the *Washington Coal*. The latter bed was found barely two feet thick, and of by no means good quality. From the bottom of the shaft to the *Waynesburg* or *Pinhook Coal* is probably little more than ninety feet, and to the *Pittsburg* it is about four hundred and ten feet.

Just above Prosperity Limestone VI comes up from the creek, and thence it rises north-westwardly, being sixty-five feet above the stream at the mouth of Craft's run. Near Mr. John Day's residence Limestone IV is seen in the road, its upper portion being six feet thick and very ferruginous, while the lower part is four feet thick and much less ferruginous. The two portions are separated by five feet of shale. On Craft's run, at somewhat more than a mile from its mouth, Limestone VI is one hundred feet above the stream, and half a mile farther down, near Mr. W. E. Craft's residence, Limestone IV is in the road. Limestone VI is seen for the last time on this run near School House No. 9.

Short creek enters Ten-Mile at one mile below Prosperity.

At its mouth the upper layers of Limestone VI are nearly level with the floor of the bridge, and they continue in sight for nearly half a mile. Limestone VII ? is exposed at the first cross-road, where it is nearly three feet thick. The westerly dip continues until within half a mile of Sparta, when it is reversed and the Upper Washington Limestone is again brought up, being exposed at both ends of that village. There the road turns south, and the dip being in that direction the limestone soon goes under. Back of Sparta in the hills a fine sandstone stands out in bluffs, at about one hundred and eighty feet above Limestone VI, and having at forty feet below it fragments of a blue limestone. About three-quarters of a mile above Sparta the road forks, one fork leading to Waynesburg and the other into East Finley township. At this fork in the road the massive sandstone is but eighty feet above the creek, the direction of the road from the village being south. Turning westward here one soon sees Limestone X, and above it the *Nineveh Coal.* The sandstone then is the one seen on Bates' Fork, above Hopkin's Mill, and on Brown's Fork, above Sargent's Mill, in Morris township, of Greene county. Between the village and this fork in the road much chert was found, but its horizon could not be ascertained.

From this place to the township line exposures are utterly wanting. At the township line near Mr. Samuel Day's residence the road is two hundred feet above Limestone X where last seen.

23. East Finley Township in Washington County.

This lies directly west from Morris, and is drained by Hunter's Fork of South Wheeling creek and its tributaries. The highest rocks of Washington county occur here, but the exposures of the upper strata are for the most part fragmentary and by no means always satisfactory. Though detailed sections are rarely found, yet one finds comparatively little difficulty in determining his position if in the vicinity of streams, since some of the well-marked limestones are almost constantly in sight. On nearly every stream Limestone VI is traceable for long distances. The section extends upwards from Limestone III to two hundred and forty feet above Limestone VI.

From the summit at the head of Hunter's Fork to the base of the hill at the first fork in the road the following succession is seen:

1.	Concealed.	
2.	Shale and sandstone	70′
3.	Limestone X seen	1′
4.	Concealed	70′
5.	Sandstone	15′

Limestone X is blue and brittle, and at fifteen feet above it is the blossom of the *Nineveh Coal.* This is in debris, and probably belongs somewhat higher. For nearly half a mile below this there are no exposures, but where the road crosses a run near Dr. Simpson's house Limestone VI is in the road, eight feet of it being exposed. The interval by barometer between this and No. 5 of the section is forty feet, but as the rocks are rising toward the west it is more likely to be sixty feet, thus making the total interval between the two limestones not far from one hundred and fifty feet. Many years ago a coal, eighteen inches thick and fifty feet above Limestone VI, was worked on Dr. Simpson's property, which is said to have been of very good quality. Its position could not be determined accurately, as the outcrop is not to be seen.

Below this the stream turns southward, and the rocks show no dip. One mile below Dr. Simpson's house Limestone VI is fifty feet above the road, and the hill side is covered with the white weathered fragments. Below it are sandy shales, which are occasionally compacted into sandstone. The overlying rocks are concealed here, but at a short distance farther on, at the second fork in the road below Dr. Simpson's, there is a very massive sandstone at fifty feet higher. It stands out in cliffs and resembles that seen above Limestone IX on Gray's Fork of Ten-Mile in Greene county. This massive character is purely local, having been observed nowhere else on this creek.

At a little way farther down the stream, and at seventy feet below Limestone VI, Limestone IV is exposed, much degraded. It consists of

1.	Limestone		4″
2.	Shale	3′	0″

3.	Limestone	5″
4.	Shale	3″
5.	Limestone	1′ 6″

The upper layers are blue and somewhat slaty, but the base is compact and ferruginous. Below it are sandy shales, which remain in sight to Hunt's Mill, where, at thirty-five feet below the last, Limestone III is seen, three or four feet above the dam. This rock is blue on the fresh surface, but for the most part is ferruginous, and weathers to a bright yellow. It continues above the creek to the township line, showing a slight rise in the direction of the stream. The Washington axis evidently crosses Hunter's Fork not far from the western boundary of this township.

On Gordy's Fork Limestone VI is exposed at the forks of the road, by Mr. Earnest's residence, near the line of Morris township. The stream flows almost south-west, so that the limestone is in sight all the way into West Finley, its line being often marked by the peculiar white fragments The altitude above the stream is from twenty to ninety feet, increasing with the descent of the creek. On Robinson's Fork the same limestone occurs at the township line, two miles above the village of Good Intent, and is in sight to the old saw mill on that stream, where it goes under. About one-fourth of a mile farther up, a limestone, probably VII, is seen twenty-five feet higher, with a little coal six feet above it. From this point, which is very near the head of the stream, the road rises rapidly to the ridge, where at the cross roads and two hundred and thirty feet above the Upper Washington Limestone there is a limestone eight feet thick in several thin layers.

Turning off here and taking the road leading by the steam saw-mill to Pleasant Grove, a limestone four feet thick is seen at the first fork in the road. Below this the exposures are incomplete, but give some information respecting the succession of the limestones as follows:

1.	Limestone	8′
2.	Red shale	50′
3.	Limestone	4′
4.	Concealed	70′
5.	Limestone	Fragments.
6.	Concealed	80′
7.	*Coal*	Blossom.

8.	Limestone VII?	Fragments.
9.	Sandy shale	30′
10.	Limestone VI seen	3′
11.	Concealed	18′

This brings the section down to the fork in the road below the saw-mill. No. 1 is the limestone referred to as two hundred and thirty feet above VI. In this it is two hundred and forty feet. The exposure of it is imperfect, only such as may be seen in a road cutting made many years ago. The rock is dark and very coarse. At the base of No. 2 there are dark shales resting on No. 3, which is a hard, flinty limestone, dark on the fresh surface but weathering dull yellowish white. Nothing could be ascertained respecting the coarse fragmentary Limestone No. 5, but it probably belongs higher up than the level at which the fragments were found, and may represent Limestone X, which if present should be somewhere in the interval four to six.

At the first fork below the saw-mill, the right hand fork leads to Pleasant Grove. On this, near the blacksmith's shop, and one hundred and eighty feet above VI there is a light blue limestone, but no exposures occur for sixty feet above or for one hundred feet below it. Rather more than half a mile beyond, at Pleasant Grove, No. 3 of the section has been exposed in a cut, and is a coarse, yellowish white limestone, about five feet thick. On the first road leading northward beyond Pleasant Grove and barely one-fourth of a mile beyond the last observation, a line of limestone fragments is seen at one hundred and twenty feet below the yellow limestone and at sixty feet lower VI is exposed. This road leads to the waters of Buffalo creek. At the fork near Mr. M'Clelland's house the following section is shown:

1.	Limestone VI	12′
2.	Sandy shale	28′
3.	*Jolleytown Coal bed*	1
4.	Impure limestone	1′ 2′′
5.	Sandstone	25′
6.	Limestone IV	3′
7.	Sandstone to run	9′

At the steam saw-mill No. 6 is in the banks of the creek. It evidently becomes much thicker northward, and is traceable along the road quite to the line of Buffalo township.

24. West Finley Township in Washington County.

This adjoins East Finley on the west. The section extends upwards from the *Waynesburg Coal* to the top of the series as exposed in Washington county.

On Hunter's Fork, Limestone III is almost constantly in sight from the line to the mouth of Owens' run. At the mouth of Templeton's run it is only a few feet above the stream; and the Upper Washington is exposed at Wallace's Mill, on the same stream, with a thin coal sixty feet above it. On the road leading from the mouth of Owens' run to West Finley village, the following section was obtained.

1. Concealed	180′
2. Limestone	Fragments.
3 Mostly sandstone	70′
4. Limestone	3′
5. *Coal*	Blossom.
6. Shaly sandstone	45′
7. Limestone VI seen	5′
8. Shale and shaly sandstone	50′
9. Dark limestone	4″
10. Shale	20′
11. Limestone IV	Fragments.
12. Shale	10′
13. Ferruginous limestone	6″
14. Sandstone	12′
15. Limestone	1′
16. Sandstone	17′
17. Limestone III	4′
18. Sandy shale to creek	6′

Thus making the interval between III and VI only about one hundred and ten feet, while in Franklin township, of Greene county, the same interval is two hundred and twenty feet. From this point the course of the stream is irregular, but the general direction is south-west, so that the dip is practically nothing. The creek falls rapidly and brings the lower rocks up, so that at Clouse's Mill, barely two-thirds of a mile south-west from the mouth of Owen's Run, the *Washington Coal* is exposed thirty-five feet below Limestone III, and only six inches thick with much black shale resting upon it. At one-third of a mile below the mill the general course of the creek having been south the following exposure was seen:

1. *Washington Coal Bed.*	
2. Concealed	10′
3. Drab shale	3′

4. *Little Washington Coal Bed*	1′
5. Laminated sandstone	8′
6. Limestone I, "b"	8′
7. Red shale	10′
8. Sandstone	15′
9. Shale, with iron ore	5′
10. Dark shale, (Waynesburg "b" ?)	3′
11. Limestone and shale	3′
12. Shale to creek	4′

At a short distance below this the Waynesburg sandstone is first shown. The upper portion for fifteen feet is compact, but below that the rock is little more than sandy shale. At probably a mile and a half due west south-west from Clouse's Mill the first openings in the *Waynesburg Coal* are found, but they are imperfectly drained, and cannot be measured. At M'Carrihan's Bank, a little way farther down, the bed has this structure:

1. Shale, seen	4 0″
2. *Coal*	1′ 5″
3. Clay	1′ 10″
4. *Coal*	1′ 5″
5. Clay	0′ 3″
6. *Coal*	1′ 7″

The roof shale is quite variable, for at one place in the entry, where it has fallen, the sandstone is exposed and the shale which is merely a badly slickensided clay is but one foot thick. The parting, No. 3, varies from ten inches to two feet, but rarely exceeds twelve inches. The coal is slaty and sulphurous, and is regarded as so much inferior to the *Pittsburg*, that farmers living in the interior of the township, choose to haul the latter or *Wheeling Coal* as it is termed here, from the Hempfield railroad, in preference to using the *Waynesburg*, which is more easily obtained. The Waynesburg limestone at twenty feet below the coal, is exposed in the creek near this opening. The rocks are dipping north-west here, but the rate is barely equal to the fall of the stream, so that the limestone is in sight to the mouth of Robinson's fork, and at the State line the coal is still fifteen feet above the creek. On Robinson's run, the coal disappears within less than half a mile. At the mouth of Croup's run, which enters the creek just below Robinson's fork, the coal is mined by Messrs. Ally & Sickel, whose openings give the following:

1.	*Coal*	1′ 6″	1′ 4″
2.	Clay	1′ 2″	1′ 5″
3.	*Coal*	0′ 4″	0′ 5″
4.	Clay	0′ 1½″	0′ 2″
5.	*Coal*	2′ 1″	2′ 1″
6.	Clay	0′ 1″	Not seen.
7.	*Coal*	0′ 3″	Not seen.

The coal does not differ in quality from that obtained at M'-Carrihan's.

On Croup's run from its mouth to school-house 10, a distance of three-fourths of a mile, the section is:—

1.	Limestone	6′
2.	Shale and sandstone	40′
3.	Limestone	Fragments.
4.	Shale	10′
5.	Limestone	6′
6.	Shale and sandstone	50′
7.	Limestone X	Fragments.
8.	Concealed	140′
9.	Limestone	2′
10.	Concealed	100′
11.	Limestone	5′
12.	Concealed	110′
13.	Sandstone	50′
14.	*Waynesburg Coal Bed*	5′ 6″

Though extremely imperfect, this section is of interest, as it shows the full extent of the Upper Barren Series as exposed in Washington county. Nos. 1 and 5, are the same with the limestones seen on top of the ridge at the head of Robinson's fork, in East Finley. No. 7, is found in place in the section, but its thickness cannot be determined. It is no doubt Limestone X, and the same with the fragmentary limestone seen near the steam saw-mill in East Finley, both having a dingy surface, but when freshly broken showing a mottled surface. Limestone VI is somewhere near the base of No. 8, and No. 11 is evidently Limestone II, the Lower Washington limestone.

On the ridge dividing the waters of Robinson's fork from those of Middle Wheeling creek, the top limestone of the section is frequently seen. It shows in the road near the head of Croup's run, at Mr. A. D. Hunt's; about a mile further on near Mr. W. M'Whorter's residence; again at Mr. S. Powers, near Mr. D. Hommeger's, a mile north-east from school-house No. 2, and near Mrs. Frazier's house, at the head of Middle Wheeling creek. Freshly fractured, it is very dark but the weathered surface is dull blue and very rough. Many of the farmers living on this ridge have burned it and found that it yields a good lime.

Returning to the main creek, we find the *Waynesburg Coal* mined at the mouth of Robinson's Fork, and passing under that stream at half a mile from the creek. The bed of the stream rises very rapidly, so that, at a mile from its mouth, the horizon of the *Waynesburg "b"* is reached, and at a little distance beyond, the *Washington Coal* is seen with its underlying shale and one foot thick. This remains in sight to M'Keag's Mill, above which it goes under the creek, having risen one hundred feet from its mouth. Above the mill, for two-thirds of a mile, only sandstones are exposed, and at the forks of the creek the *Washington Coal* is eighty feet under. One-fourth of a mile farther up, the Middle Washington limestone is exposed in the road, and at the fork leading to Wind Gap the following section is shown:

1.	Limestone VII (?)	2′ 6″
2.	Sandstone	20′ 0″
3.	Limestone VI	Fragments.
4.	Shale, dark at base	30′ 0″
5.	Concealed	12′ 0″
6.	Red shale to creek road	15′ 0″

Between this point and the saw-mill, No. 1 and the dark shale at the base of No. 4 are frequently seen. The dip continues northward to the mill, but there it is reversed, and at barely half a mile above, the Upper Washington limestone is twenty feet higher than the one over it is at one-quarter of a mile below the mill. Here in a drain the exposure is:

1.	Limestone	2′ 6″
2.	Sandy shale	30 0″
3.	Concealed	20′ 0″
4.	Sandy shale	25′ 0″
5.	Limestone	2′ 6″
6.	Flaggy sandstone	20′ 0″
7.	Limestone VI	8′ 0″
8.	Shale and sandstone	20′ 0″
9.	*Jolleytown Coal Bed.*	
10.	Concealed	15′ 0″
11.	Dark limestone	Fragments.
12.	Sandstone	30′ 0″
13.	Limestone IV	3′ 0″

No. 13 is reached in the creek just below the residence of Mr. J. Chase, and is Limestone IV, or the Middle Washington.

The rocks are rising very fast here, so that at Mr. B. Chase's house the lower limestone is in the road, and at the forks above that house Limestone VI is in the bluff. Another limestone is seen twenty-five feet higher, on which is a very handsomely ripple-marked flaggy sandstone. At the watering trough there is a thin limestone resting on calcareous sandstone and thirty feet below VI.

Near Mr. J. Patterson's residence, on the road turning north here, a section was obtained which fills some of the gaps in those previously given, thus:

1.	Limestone	6′
2.	Concealed	85′
3.	Limestone	5′
4.	Shale and sandstone	40′
5.	*Coal*	1′ 6″
6.	Shale	5′
7.	Limestone VII ?	2′
8.	Sandstone and shale	20′
9.	Limestone VI	10′
10.	Concealed	30′
11.	Limestone	2′
12.	Concealed to run	5′

No. 1 is the same with No. 7 of the Croup's Run section. The Coal No. 5 is quite persistent, and seems to be of good quality.

From Mr. Chase's house the direction of the road is north-east, and the rocks approach the creek so that at Good Intent Limestone IV is under the stream, and only Nos. 9 and 11 are exposed, the latter being on both sides of the creek. The dip is very sharp towards the south-east. At the first fork in the road above Good Intent Limestone VI is in the road, and twenty feet higher is the next limestone, with the coal at six or eight feet above it. From this point to the township line these rocks are constantly in sight, and at the line Limestone VI is fifteen feet higher than at the fork just referred to.

Descending from the dividing ridge to Middle Wheeling creek, the following section was obtained on the road crossing a little west from school-house No. 4, and leading to Mr. T. Frazier's house on the creek;—

1.	Limestone	Fragments.
2.	Concealed	50′
3.	Limestone	Fragments.

4.	Concealed	60′
5.	Limestone	6′
6.	Concealed	145′
7.	Limestone VI	Fragments.
8.	Concealed	30′
9.	Limestone, seen	2′

Here No. 5 is the same as No. 5 of the Croup's run section, and the second limestone of that obtained in East Finley, on the Pleasant Grove road. No. 3, is evidently the top limestone of those sections and the fragmentary rock No. 1, is the highest stratum of the series exposed in the county. Near the West Virginia line, Limestone VI is seen, and the interval between it and the *Washington Coal* seems to be not far from one hundred and sixty feet. The coal is only eighteen inches thick, and rests directly upon thirty feet of limestone and calcareous shale.

25. East Pike Run Township, Washington County.

This township is on the river border of the county and lies north from East Bethlehem. It is crossed by Pike run and its section extends upwards from the *Pittsburg* to one hundred and fifty feet above the *Washington Coal.*

On the hill known as Krepp's knob, one of the highest in the district, and almost directly west from West Brownsville, a section was obtained, which though imperfect, gives a general idea of the series as found in this township. It is:—

1.	Sandstone	40′
2.	Limestone	Fragments.
3.	Concealed	100′
4.	*Washington Coal bed*	Blossom.
5.	Concealed	130′
6.	Waynesburg sandstone	30′
7.	*Waynesburg Coal bed*	3′ 6″
8.	Shale and sandstone	45′
9.	Bituminous shale	1′ 6″
10.	Limestone	10′
11.	Shaly sandstone	40′
12.	*Uniontown Coal bed*	3′
13.	Limestone	8′
14.	Concealed	70′
15.	Limestone	50′
16.	*Sewickley Coal bed*	Blossom.
17.	Shaly sandstone	35′
18.	Limestone	25′
19.	Concealed to river	65′

No. 2 is not well exposed, but the profusion of fragments marks its position and show that it is not less than five or six feet thick. It is limestone IV, the Middle Washington. The blossom of the *Washington Coal* is large, but there is no opportunity to determine its thickness accurately. At an opening made many years ago by Mr. Krepp, but long deserted, the *Waynesburg Coal* was found to be three feet six inches thick and double. The *Uniontown* seems to be a very fair coal, but no openings have been made to determine its quality. The *Pittsburg* is at the base of the section, being exposed at low water, and is mined by Mr. J. S. Pringle, at West Brownsville, by means of a shaft forty feet deep. The opening was not measured, but Mr. Pringle gives the following as its section:—

Coal, 4 inches; shale, 6 inches; *coal*, 9 feet; total, 9 feet 10 inches.

He states that the lower division is a solid mass of coal and that the partings are not persistent.

At the head of Lilly's run, which enters the river a mile below West Brownsville, the *Waynesburg Coal* is mined by Mr. Lilly, at whose opening the structure is:—

Coal, 10 inches; clay, 3 inches; *coal*, 2 feet 6 inches; total, 3 feet 7 inches.

The coal does not differ from the *Waynesburg* elsewhere, being slaty and sulphurous. At one hundred and ten feet higher the smut of a coal is seen which may be that of the *Waynesburg "b."* For two and a half miles below Brownsville the *Pittsburg* is not exposed on the Washington county side of the river, being buried under the bottoms, and for the same distance exposures of the upper rocks are very unsatisfactory, but below that the following section is exposed in the river hills:

1.	*Waynesburg "a" Coal Bed*	Blossom.
2.	Sandstone	75′
3.	*Waynesburg Coal Bed*	4′ 4″
4.	Concealed	160′
5.	Limestone	65′
6.	*Sewickley Coal Bed*	Blossom.
7.	Sandstone	35′
8.	Limestone	20′
9.	Shale and sandstone	80′
10.	*Pittsburg Coal Bed* in river.	

The *Pittsburg* in mined by Mr. J. Imlay, about a mile above California, and by J. S. Neel and J. Reed & Co., at Greenfield.

At these banks the bed varies as follows:

Coal, 6 inches to 1 foot; clay, 10 inches; *coal*, 6 feet 6 inches to 7 feet 9 inches.

At all of these the partings occur in the lower division at about two feet six inches from the bottom, and are separated by two to four inches of coal. This is the bearing-in bench of the miner. After it is removed the coal above and below is easily brought out by wedges, and blasting is seldom necessary. For about fifteen inches below the partings, the coal is very good, but the bottom is so slaty and sulphurous that it is always left in the mine. At Mr. Reed's opening the *Pittsburg* is forty-five feet above the river. One-fourth of a mile below Greenfield the interval between the *Uniontown* and *Pittsburg* is finely exposed, the section being as follows:

1. *Uniontown Coal Bed*	3′
2. Limestone	12′
3. Shale and sandstone	28
4. Limestone and shale	88
5. Shaly sandstone	32
6. Limestone	30′
7. Sandstone and shale	85′
8. *Pittsburg Coal Bed*	8′
9. Concealed to river	48′

In this section neither the *Redstone* nor the *Sewickley* is seen. No. 6 is locally known as the "Soda-stone," and is said to have been used at Pittsburg many years ago in the manufacture of soda ash. One of its layers is especially pure, and is employed as a flux in the manufacture of glass at Belvernon. Other layers are used in iron smelting. No. 4 contains very little shale, and is almost wholly limestone.

Half a mile below Greenfield the *Pittsburg Coal* is mined by Messrs. Leadbeater and Co., who have the lower division, seven feet thick, and separated from the roof by ten inches of clay. The roof is in four benches, in all one foot ten inches thick, separated by thin partings of clay, each about two inches. At these works the section between the *Pittsburg* and *Uniontown* is again exposed. The *Uniontown* is three feet; the two portions of the great limestone are respectively ten and eighty-five feet, the lower containing, as in the other sections, very little shale; the *Sewickley* is absent, but at one hundred feet below it and twenty-five feet above the *Pittsburg*, the *Redstone* is found represented by one foot of bituminous shale. At the Dexter Coal Works, one mile below Greenfield, the *Pittsburg Coal* is sixty-five feet above the river, and has the same structure as at Leadbeater's Works.

On Pike Run the *Pittsburg* is in sight up to the township line, rising westward about as rapidly as the run does, and the openings are quite numerous. On a road leaving the run at one-third of a mile from the river the *Sewickley* is seen, and the horizon of the *Waynesburg* is reached on this road at half a mile north from the National pike.

Gorby's run enters Pike run at the village of Granville. On this stream the coal is in sight almost to the head, and rises from the river to where it disappears, nearly three hundred feet, the distance being hardly two miles. The openings are quite numerous on the west fork, and the coal goes under just above Mr. West's bank. On a branch of this fork the *Sewickley Coal* is found at Mr. M. Smith's place, where it is three feet six inches thick, and is separated from the Great Limestone by four feet of shale. It is mined by stripping for use in burning lime, and seems to be a very fair coal. It is at two hundred and twenty feet below the exposure of the *Waynesburg*, on the Nottingham road, near the western line of the township. On the eastern fork of Gorby's run the *Pittsburg Coal* is mined by Mr. Duvall, and farther up by Mr. J. White. At the latter opening it shows:

Coal, 1 foot; clay, 1 foot 1 inch; *coal*, 6 feet 11 inches.

The lower parting in the main division is two feet eight inches from the bottom of the bed. From Granville up Pike run the coal is seen almost constantly at about twenty-five feet above the stream, until near the township line it goes under, owing to a change in the course of the run. Nearly midway between Granville and the line, Mr. Jackson mines it quite extensively, and has the coal as follows:

Coal, 10 inches; clay, 8 inches; *coal*, 7 feet 3 inches.

The lower parting is two feet six inches from the base, and the bottom coal for one foot is worthless. The rest of the bed is of excellent quality.

26. West Pike Run Township, Washington County.

This lies west from East Pike Run and north from East Bethlehem. Pike run flows through it in a south of east direction. The section extends upwards from the *Pittsburg* to the *Washington Coal*, but there are no satisfactory exposure above the Waynesburg sandstone.

At the Eastern border of the township the *Pittsburg Coal* is about twenty-five feet under the run, but, as the stream turns and flows from the north, the coal is seen rising so that at the mouth of Gregg's run it is worked. From this point the openings are numerous up to Thompson's saw-mill, where the creek bed rises above it. At the opening belonging to Mr. A Jeffries, on Gregg's run, the coal is triple, showing thus:

Pittsburg Coal bed.	
1. *Coal*	0′ 2″
2. Clay	0′ 3″
3. *Coal*	1′ 0″
4. Clay	1′ 0″
5. *Coal*	7′ 3″

Overlying it is shale eighteen inches thick, which contains many impressions of *neuropteris* and *sphenophyllum*. At Mr. Moffat's opening on Little Pike run, this shale and the top coal are both absent, and there intervenes between the sandstone and No. 3 of this section, only five inches of clay containing thin streaks of coal. At Mr. S. Walker's bank below Taylor's saw-mill, the bed is again triple and the sandstone rests directly on it. Between this point and Thompson's mill, the bed was measured at openings belonging to J. L. Taylor, Geo. Sidwell and Adam Shaner, with the following results:—

Pittsburg Coal bed.				
1. *Coal*	0′ 3″	0′ 2′	0′ 4″	0′ 2″
2. Clay	0′ 2½″	0′ 4″	0′ 3″	0 2½″
3. *Coal*	1′ 1″	1′ 0″	1′ 1″	1′ 0″
4. Clay	0′ 11″	1′ 0″	1′ 2″	0′ 11″
5. *Coal*	6′ 10″	7′ 3″	7′ 2″	6′ 10″

In these the sandstone comes down upon the coal without any intervening shale, but is well exposed only at Mr. Shaner's opening, where it is forty feet thick. Immediately below the coal are shales containing iron ore in small quantity, and at fifteen feet below there is a dark blue limestone, two to three feet thick, having on it a bituminous shale, which may represent the *Little Pittsburg Coal.* At Thompson's mill, the coal is barely under the run, and is worked on Mrs. Thompson's property by a short inclined shaft. The overlying sandstone is forty-five feet thick, and the roof division is double, showing coal, three and thirteen inches, separated by clay, three inches. In the lower division the structure is:—

1.	*Coal*		8″
2.	Compact clay		½″ to 1″
3.	*Coal*	3′	3″
4.	Parting		½″
5.	*Coal*		2″
6.	Parting		½″
7.	*Coal*	1	7½″
8.	....*Coal*	1′	3″

No. 1 is probably a portion of the upper division and is quite bony. The clay parting No. 2 is most probably the main clay parting, much diminished in thickness. In this bank it is often so rich in bituminous matter as to resemble cannel coal. No. 3 is hard and bright, but contains many binders of pyrites. No. 7 contains laminæ of pyrites and leaves clinkers in the furnace. The bottom coal is quite clean and is used for smithing purposes. Much of the bed is open-burning, evidently too much so for coking, even were the sulphur present in smaller quantity. As a whole it is far from being an excellent coal, though it is superior to the *Waynesburg*, which is mined somewhat farther up the run.

The interval between the *Pittsburg* and *Waynesburg* is for the most part concealed. The *Redstone* does not seem to be present, at least it is absent at all localities where its horizon is exposed. The blossom of the *Sewickley* is seen at a few places directly underlying the great limestone, but is not mined, so that no opportunity exists for determining its thickness or quality. The most easterly exposure of the *Waynesburg* along the run is on the hill opposite an old still-house below Mr. R. Arnold's residence, where it is two hundred feet above the stream. It passes under the run immediately north from Beallsville. At the fork in the road below Mr. J. Hill's house this coal is seventy-five feet above the run, and the *Uniontown Coal* is seen at sixty-five feet under it, resting on a bright yellow limestone. Between this locality and the Pittsburg road, a distance of barely one mile, the *Waynesburg* is mined by J.

Hill, S. W. Rogers, C. Sellers and J. M. Miller. Measurements made at the openings are as follows:

Waynesburg Coal bed.				
1. *Coal*	1′ 0″	1′ 0″	0′ 9″	0′ 10″
2. Clay	1′ 1″	1′ 1″	1′ 0″	1′ 0″
3. *Coal*	3′ 0″	2′ 8″	2′ 4″	2′ 3″
4. Clay	0′ 2″	0′ 1½″	0 2″	0′ 4″
5. *"Brick" Coal*	0′ 3″	0′ 3″	0′ 3″	0′ 3″
6. Clay	0′ 2″	——	0′ 1″	——
7. *Coal*	1′ 6″	1′ 5″	1′ 3″	1′ 4″

The sandstone rests directly on the coal; the top bench is poor and is used only in burning lime. The ash and clinker render it unfit for anything else. The measurements were made in each case near the mouth of the entry and do not show the variations of the main parting. Within thirty feet this sometimes varies from one-half of an inch to the maximum as given above. The main bench is hard and very handsome when freshly mined, but is sulphurous. The "Brick" is free burning, leaves only a powdery ash and makes no clinkers, so that it is in high repute. The bottom is always poor and is rarely worked.

27. Allen Township, Washington County.

This little township is east from Fallowfield and East Pike run, and has a long front on the river, as it is situated in a bend of the stream. The section extends upwards from the *Pittsburg* to a considerable distance above the *Waynesburg*, but excepting along the river, exposures are rare and nothing of interest is shown.

In a ravine about a mile and a half below Greenfield, this section is found:—

1. Limestone	80′
2. Sandstone	40′
3. Limestone	25′
4. Shale	60′
5. *Redstone Coal bed*	0′ 8′
6. Shale	20′
7. *Pittsburg Coal bed*	10′
8. Shale	15′
9. Bituminous shale	1′
10. Limestone	4′
11. Sandstone	2′
12. Limestone	3′
13. Shale	5′

14.	Limestone	6′
15.	Concealed to river	25′

In this section the *Sewickley* is wholly absent, there being not even bituminous shale to represent it. The presence of limestone fragments above, shows that the whole of No. 1, which is the lower division of the Great Limestone is not fully exposed. This does not differ from its southern extension as seen in East Bethlehem and East Pike run. The *Pittsburg* is opened by Mr. Michael Wolf, at half a mile above Allenport, where it is sixty-five feet above the river. The roof division shows three benches, six, four and ten inches respectively, with partings of two and eight inches. The main parting is twelve inches, and the lower division is seven feet six inches. At Mr. S. Clark's opening, the lower division is seven feet eight inches and the roof is not exposed.

Here the river turns sharply northward, and the coal rises in that direction, so that on a little stream coming in half a mile or so above Belvernon, the *Pittsburg Coal* is one hundred and fifty feet above the river. The *Little Pittsburg* is represented by a bituminous shale twelve feet below the coal, the interval being occupied by limestone.

At Belvernon the coal is one hundred and seventy-five feet from the river. At Mr. Noah Speer's grist mill, which is a mile west from the river, there is an opening in the bed showing,

Coal, 2 inches; clay, 1 foot; *coal*, 6 feet 10 inches.

The disappearance of the roof division here is somewhat perplexing, since in the immediate neighborhood it is quite thick. The thinning cannot be accounted for by erosion, because the sandstone is separated from the coal by from seven to ten feet of clay shale. At this bank there are three partings in the middle of the lower division—a peculiarity of rare occurrence in this district, though it is characteristic of the bed in a large portion of Ohio. The coal is good throughout. On a road going due south from this mill the blossom of a thin coal was seen at one hundred and forty feet above the Pittsburg.

Opposite Belvernon a fine deposit of glass sand was seen at one hundred and fifty feet above the river. It is of excellent quality and is largely used. Reference has been made to it in a previous portion of this report.

28. Carroll Township, in Washington County.

This lies east and north from Fallowfield, and, being situated within a bend in the river, shows a long river front. The section extends upwards from one hundred and eighty feet below the *Pittsburg* to the *Waynesburg Coal.* Pigeon creek crosses the western portion of the township.

Along the river the *Pittsburg Coal* is extensively mined for shipment, and exposures of the overlying rocks are quite frequent. Three-fourths of a mile below Lock No. 4 this section was obtained:

1. Concealed	100′
2. Limestone	35′
3. Sandstone	30′
4. Shale	30′
5. *Redstone Coal Bed.*	
6. Shale and sandstone	48′
7. *Pittsburg Coal Bed*	9′ 10″
8. Shale and sandstone	40′
9. Limestone	12′
10. Shale and sandstone	65′
11. Limestone	4′
12. Shale to river	30′

The exposure seems to be complete, but the *Sewickley*, which should come directly under No. 2, was not seen, though it is present in the western part of the township. The *Redstone* shows an extensive blossom, and is probably three or four feet thick. The roof division of the *Pittsburg* is three feet three inches thick, and shows four benches of coal. The main parting is eleven inches, and the lower division is five feet eight inches. The exposure of the rocks below the coal is not complete, but there is no evidence that the *Little Pittsburg* is present. At an abandoned opening, one-fourth of a mile farther down the river, the roof is triple, with a total thickness of three feet eight inches. The bed is here one hundred and sixty feet above the river. Half a mile beyond it is one hundred and seventy feet, and at fifty feet higher the *Redstone* blossom shows itself. Two and a half miles below Lock No. 4 the coal is one hundred and eighty-three feet above low water mark, with a triple roof, and the lower division five feet ten inches thick.

Thus far the river has been flowing east, and the rocks have been rising in that direction, as the Waynesburg axis crosses at a little way above Lock No. 4. At Columbia the course is changed to north, and the coal rapidly approaches the river. Near that village the following section was seen on the property of Mr. Bradford Allen:

1. Limestone	?
2. Flaggy sandstone	35′
3. Shale	40′
4. *Redstone Coal Bed.*	
5. Shale	50
6. *Pittsburg Coal Bed*	8′
7. Concealed to river	175′

The *Sewickley* was not found, though it may be here, as the top of No. 2 is not well shown. That sandstone is an excellent flagging stone, and is quarried as such at this locality for shipment to Pittsburg. The flags can be obtained of any desirable size, and have a very smooth surface. Somewhat lower down the river, opposite Webster, the *Pittsburg Coal* is one hundred and twelve feet above the river, and is mined by Mr. Gilmer. The lower division is five feet, and the roof is quadruple, with a total thickness of five feet. Two and one-half miles from Monongahela city the coal is sixty-five feet above the river, and the lower division is five feet ten inches. At Robinson Bros.' Coal Works the altitude is only fifty feet.

Just above Robinson's works a little stream, Black Diamond run, enters the river. On this, at barely half a mile from its mouth, are several openings in the *Redstone* coal, which are rudely worked and in bad condition. The coal seems to vary from two feet eleven inches to three feet seven inches, and is apparently clean and of good quality. At forty feet above it is a fine flaggy sandstone, which is quarried. On the ridge, at the head of this run, the *Waynesburg* is reached, and at the fork in the road, about one-third of a mile north from the Baptist church, the Waynesburg sandstone is seen in the road. At one-fourth of a mile farther west the coal is said to have been worked, but as it was opened down the dip the water soon rendered mining impossible. The bed is reported to be four feet thick. Its blossom, which is frequently exposed along the ridge, indicates that the thickness is not far from that given.

Three-fourths of a mile above Monongahela City, the following section was obtained in a hollow coming down to the river:

1.	Limestone	75′ 0″
2.	Flaggy sandstone	35′ 0″
3.	*Sewickley Coal*, (bituminous shale)	1′ 0″
4.	Shale	40′ 0″
5.	*Redstone Coal Bed*	3′ 6″
6.	Limestone	4′ 0″
7.	Shale	50′ 0″
8.	*Pittsburg Coal Bed*	12′ 0″
9.	Concealed to river	40′ 0″

On Pigeon creek, which enters the river just above the city, there are numerous openings in the *Pittsburg Coal*, at all of which the roof division is concealed. The lower division exhibits little variation, and the following is the common structure:

Coal, 3 feet 3 inches; clay partings and *coal*, 3 inches; *Coal*, 2 feet 5 inches.

The lowest bench is usually divided nearly midway by a thin parting into the "Brick" and Lower Bottom. The latter is seldom of any value, being slaty and sulphurous, while the former, though sometimes poor, is ordinarily very good. Coal from the top bench is of excellent quality. This bench has some binders of pyrites, but they are not persistent.

Turning up Scott's hollow at this place, fifty feet of laminated sandstone are seen resting on the *Pittsburg Coal*, on which rests the *Redstone*. This is mined by Mr. Isaac Teeple, at whose bank it shows two benches eighteen and twenty-one inches thick. There is no clay parting, but the benches are distinct, and the portions in contact are slickensided. The upper bench is frequently cut away by clay-horsebacks from above. The coal, which is clean and somewhat free-burning, is in high repute for domestic use, and by many is preferred to that from the *Pittsburg*. The lower division of the Great Limestone is seen at fifty-five feet above the *Redstone Coal* at this opening, and rests on fifteen feet of laminated sandstone, which in turn rests on five feet of limestone. On the latter there is a thin bituminous shale. The interval between the *Redstone* and the *Waynesburg* is two hundred and forty-five feet; the latter bed having been seen on the ridge separating the waters of Black Diamond from those of Pigeon creek.

On Pigeon creek the coal goes under not far above the mouth of Scott's hollow. From the place of disappearance the stream comes from the south-west, so that at the Brownsville road, nearly two miles farther up, the *Pittsburg* is not more than forty-five feet under, and the *Redstone Coal* is exposed in the bluff. At a short distance up a little run coming in here from the north, the *Redstone* is mined, showing a thickness of two feet ten inches at the mouth of the entry, but becoming much thinner inside. The value of the bed is much diminished by clay veins and horsebacks, which cut it up very badly. Fifty feet higher in the hill, the interval being filled with sandstone, is an old opening on the *Sewickley*, in which the coal is said to be two feet thick. On the ridge at the head of this run the *Waynesburg* is seen three hundred feet above the *Pittsburg*, and apparently not more than two feet thick.

Southward from Pigeon creek on the Brownsville road, the *Waynesburg* is again seen, and on the summit at the boundary of the township, the *Washington Coal* occurs one hundred and forty feet above the *Waynesburg*.

At Monongahela City, there are several openings on the *Pittsburg Coal*. From this point to the township line, the coal constantly approaches the river, though for nearly a mile the dip is very slight. At the New Eagle colliery, about half a mile from the railroad depot, the coal is twenty-seven feet above the railroad. The roof division is concealed, but the following section of the lower division was obtained, which exhibits very fairly the structure of the bed in the vicinity:

Coal, 3 feet 2 inches; *coal* and partings, 9 inches; *coal*, 10 inches; parting, 2 inches; *coal*, 14 inches: Total, 6 feet.

Between the "Brick" coal and the upper bench, there are three partings all of which are persistent. In the upper bench a thin clay parting sometimes occurs, but it is irregular and not persistent. The lower bottom is not mined, being so impure as not to be marketable. Little pyrites is found in other portions of the bed. The fire-clay below the coal rests on a somewhat brecciated limestone which frequently disappears and is re-placed by the clay which then becomes five feet thick. No rocks above the coal are exposed here. At the Mingo colliery, somewhat farther down the river, an air shaft was sunk in which the following section was found. It was given from memory by the superintendent, Mr. T. S. Hutchinson, no record having been kept. The interval between *Pittsburg* and *Redstone* is evidently too small.

1. Sandy shale	12′	0′
2. *Coal* and shale	2′	0″
2. Clay	0′	10″
4. *Coal*	4′	0″
5. Limestone	1′	0″
6. Shale	6′	0″
7. Sandstone	4′	0″
8. Shale	7′	0″
9. Sandstone	20′	0″
10. *Pittsburg Coal bed*	—	—

It is very clear that Nos. 2, 3 and 4, represent the *Redstone Coal*, but the thickness is much greater than at any exposure in the vicinity. At these works the lower division of the *Pittsburg* is the same as at the New Eagle colliery. The roof division is double, and shows the benches ten and eighteen inches, separated by two and a half inches of clay. At a little distance up Mingo creek, near Ferree's distillery, the roof is triple, showing three, nine and twenty-two inches of coal, with a total thickness of four feet three inehes. The lower division is only four feet nine inches thick, and has but two partings in the middle. At both of these openings a thin bituminous shale rests directly on the coal. At fifty-five feet above Mr. Ferree's opening the *Redstone* is seen and appears to be not far from three feet thick.

29. Fallowfield Township, Washington County

This adjoins Carroll on the south and has a short line along the river between Allen and Carroll. Pigeon creek crosses the eastern portion. The section extends from one hundred and sixty feet below the *Pittsburg Coal* up to the *Washington.*

The only opening in the *Pittsburg* seen along the river is that belonging to Mr. H. M'Mahon, about a mile and a half below Belvernon. It shows —

1. *Bituminous shale*		10″
2. Ferriferous shale	1′	0″
3. *Coal*	1′	2″
4. Clay		6″
5. *Coal*	8′	9″
6. Concealed to river	160′	0″

The character of the coal obtained here does not differ from that in Carroll. On a fork of the run which forms the northern boundary of Allen township, the *Pittsburg* is mined by Mr. Thomas Redd, at whose bank the following exposure was seen:

1. Compact sandstone	35′	0″
2. *Bituminous shale*		10″
3. Roof division	3′	0′
4. Clay	1′	5″
5. Lower divsion	6′	1″

At this opening little pyrites occurs in the upper bench of the lower division, but is abundantly present in the lower bottom, which is not mined. The "brick" coal comes out in handsome blocks, but is much inferior to the upper bench. As a whole the coal is clean, and is said to be quite popular in the vicinity. At an opening on the adjoining farm the quality is the same. Just above Mr. Redd's house the *Redstone* is seen in the road, and at a short distance farther up the blossom of the *Sewickley* is exposed. The *Waynesburg* was seen on the ridge near Withrow's blacksmith shop.

The last bed is exposed at many localities along the ridge south from school-house No. 5; but coal can be obtained so easily from the *Pittsburg* openings, already referred to, that farmers choose to haul their coal rather than to bear the cost of keeping up an opening in the *Waynesburg*. Some years ago Mr. V. Colvin, residing near school house No. 5, opened this bed and found it about four feet thick, not including the upper bench, which he did not work. The *Washington Coal* is seen on his property at one hundred and forty feet above the *Waynesburg*, and between the two beds the blossom of the *Waynesburg "a"* is exposed.

In the greater portion of the township, that lying east from Pigeon creek, the exposures are very unsatisfactory. The *Pittsburg Coal* is available along the river border and the run separating Allen and Fallowfield, but on each of these lines openings are few. Elsewhere the land rises high and is so covered with debris that the *Waynesburg*, which is quite thin, is not often exposed even by its blossom. The strata dip rapidly toward the west, so that along the best line, that of the run referred to, the *Pittsburg* passes under at a little distance above Bailey's mill.

West from Pigeon creek, the *Waynesburg* is mined, as the distance to any opening in the *Pittsburg* is such as to render hauling somewhat expensive. At Mr. Grable's bank, the bed is in three benches, respectively ten, thirty and fourteen inches thick. On the south fork of Saw-mill run, openings were seen belonging to Messrs J. Warne and J. Hess, and on the north fork to Mr. Stacher. At the first and third of these, measurements were made as follows:—

Waynesburg Coal bed.

1.	*Coal*	0′ 11″	1′ 0″
2.	Clay	1′ 2″	1′ 3″
3.	*Coal*	2′ 10″	2′ 8′
4.	Parting	——	——
5.	*Coal*	1′ 1″	1′ 0″

At Mr. Warne's opening, three feet of bluish shale rests on the coal and contain great numbers of the plant impressions peculiar to this horizon. The top bench of the bed is very poor and slaty at all the banks, and the bottom bench is far from being a good coal. The middle bench, though somewhat sulphurous, must be much better than usual, if those who use it here are qualified to judge, for they assert positively that it burns to a fine ash and produces less ash and clinker than the *Pittsburg Coal.* At the same time it is difficult to believe that three bushels of this *Waynesburg Coal* is equal to five from the *Pittsburg;* such, however, is the statement made apparently in good faith.

On Saw-Mill run, from Mr. Stacher's opening, to Pigeon creek the following strata are exposed:

1.	*Waynesburg Coal Bed*	6
2.	Concealed	30′
3.	Limestone	Fragments.
4.	Concealed	40′
5.	*Massive sandstone*	25′
6.	Shale	10′
7.	Limestone and shale	75′
8.	Shaly sandstone	40′
9.	Limestone	4′
10.	Shale	30′
11.	*Redstone Coal Bed,* seen	1′

The Great Limestone shows rather more shale here than it does in the south-east portion of the county. The junction between it and the sandstone below is not fully exposed, but no evidence that the *Sewickley* is present was seen. The interval between the *Redstone* and *Waynesburg* is only two hundred and fifty feet, which is less than the interval between the *Waynesburg* and *Sewickley* in southern Greene county.

30. Somerset Township, Washington County.

This lies directly west from Fallowfield, and is drained by the North and South branches of Pigeon creek, which unite near the eastern line of the township. It is crossed by the Pinhook axis, and by the synclinal, between that and the Waynesburg axis. The section reaches from one hundred and fifty feet below the *Waynesburg Coal* to the Upper Washington limestone.

At Bentleysville the *Waynesburg* is one hundred and fifty feet above the stream. It is worked there by Mr. Buffington, at whose opening it shows three benches, respectively twelve, thirty-four and twelve inches thick. The top bench is worthless and is not removed, as it aids to support the somewhat insecure roof. The middle bench is said to be a very good coal, much better than at any other bank in the vicinity. It is free from slate, burns well, and leaves a bulky but not heavy ash. The bottom bench is pyritous and slaty. The Waynesburg sandstone is imperfectly exposed, but is evidently far less compact than at localities farther south. Between it and the coal is shale, eight feet. Three miles east from Bentleysville the same bed is mined by Mr. Richardson, and on the Beallsville road by Mr. Jones and others. Measurements at the banks named are as follows:

Waynesburg Coal Bed.

1. *Coal*	10″	11″
2. Clay	1′ 2″	1′ 1″
3. *Coal*	2′ 8″	2′ 6″
4. Clay	2″	3″
5. *Coal*	1′ 1″	1′ 3″

At Mr. Richardson's opening the shale is present over the coal, and is five feet thick, but it is absent at Mr. Jones'. At each locality the bottom bench is very impure, containing much slate and sulphur. The middle bench is a very fair coal.

At or very near Bentleysville the Synclinal crosses, and thence the rocks rise north-westward, so that the coal remains above the stream almost to its head. At a mile above Bentleysville Mr. Joseph Scott has opened this bed, and on a run coming in near Mr. Scott's Dr. Shaner mines it. These openings differ greatly in structure, as appears from a comparison of the measurements:

Waynesburg Coal Bed.		
1. *Coal*	2''—4''	1' 2''
2. Clay	1' 2''	1' 2''
3. *Coal*	3' 0''	1'10''
4. Clay	½''	2''
5. *Coal seen*	7''	1' 6''

At both localities the sandstone rests directly on the coal, and at Mr. Scott's bank it has cut out the bed quite seriously, the top bench being almost wholly removed. In the same opening the middle bench has an irregular parting of clay, which is not persistent. Near school-house No. 2, on this run, the *Waynesburg* shows a strange variation in thickness. At Dr. Shaner's opening the bed is more cut up than at that belonging to Mr. Scott, half a mile north, but on the properties of Mr. J. Van Voorhis and Mr. J. J. Hill, south from Dr. Shaner, the partings have either disappeared or have become so thin as not to be perceptible while the coal has thickened greatly, the exposure at Mr. Hill's opening being:

Sandstone, 15 feet; shale, 6 inches; *Coal*, 7 feet.

The opening belonging to Mr. Van Voorhis had fallen in at the time of examination, and no measurement could be made, but the owner says that in one part of his entry the coal shows a solid mass nearly nine feet thick, without any "horseback." The coal from these openings is said to be much better than that from most of the other banks, as it contains little slate and sulphur. At a short distance farther up the run the bed is double, the benches being twelve and forty-four inches, separated by twelve inches of clay.

On the south branch of Pigeon creek, above Mr. Scott's, openings were measured belonging to Mr. J. Huffman, near schoolhouse No. 3, Mr. J. J. Huffman, near the saw mill, and Mr. Burgen, near schoolhouse No. 4. The results are as follow:

Waynesburg Coal Bed.			
1. *Coal*	10 ''	1' 0''	1' 0''
2. Clay	1' 3 ''	1' 0''	1' 3''
3. *Coal*	3' 0 ''	3' 4''	3' 6''

4. Parting	½″	½-1″	2″
5. *Coal*	1′ 0″	6″	2′ 2″
Totals	6′ 1½″	5′ 11″	8′ 1″

At all of these openings the parting, No. 4, is so thin that it is overlooked by the miners, and Nos. 3 and 5 are regarded as forming but one bench. Certainly no such extreme development has been found outside of this township. Several deserted openings were seen near these, but their condition was such that no measurements could be made. The coal passes under the creek at Vanceville. No exposures of higher rocks are seen until the *Washington Coal* shows itself on the dividing ridge between the waters of Possum run and Pigeon creek, near school house No. 5. This coal is traceable by its blossom along the branch of Chartiers creek, which separates Somerset from South Strabane township. It has been opened by some of the farmers, but the works have been abandoned, as the thickness is only two feet six inches.

At the junction of the forks of Pigeon creek a boring for oil was made years ago to a depth of 700 feet. Gas was found in considerable quantity, but no oil was obtained. At one hundred and twenty feet below the surface a coal was reached which was supposed to be the *Pittsburg*, but which is most probably the *Sewickley*, as the well-curb is only sixty feet below the *Waynesburg*. No other information respecting this boring could be secured.

Along the north branch of Pigeon creek openings in the *Waynesburg* are quite numerous. For nearly two miles up the stream the bed shows characters similar to those observed on the other branch, as appears from the following measurements taken at banks belonging to A. Hetherington and H. Myers:

Waynesburg Coal Bed.		
1. *Coal*	11″	1′ 3″
2. Shale	1′ 1″	1′ 2″
3. *Coal*	2′ 7″	2′ 10″
4. Clay	2″	5″
5. *Coal*	1′ 2″	1′ 4″

The lower parting is so thin as to cause no annoyance in mining.

The Waynesburg sandstone forms an insecure roof, and is much degenerated, being a miserable shaly sandstone. A similar section is seen at a little way south from the creek, where the coal is mined by Mr. R. Graham; but at Mr. A. Gamble's opening in the same neighborhood, the coal shows the structure commonly found in the eastern portion of the county, the clay parting, No. 4, being eighteen inches thick, while the middle bench is four feet. Mr. Graham removes the whole bed, as it is all marketable. The coal is evidently popular, for the annual yield of the bank is about 20,000 bushels.

In the vicinity of school-house No. 7 there are several banks, in one of which, belonging to Mr. Hiram Myer, the bed is said to show eight feet of solid coal, exclusive of the roof bench. The entry had fallen in at the mouth and could not be examined, but in the same vicinity, Mr. J. Myer's opening shows only two feet eight inches in the middle bench and ten inches in the upper, so that if the statement respecting Mr. Hiram Myer's bank be true, the variation is extreme even for this bed.

About a mile farther up the creek is Mr. Couch's bank, which closely resembles that belonging to Dr. B. Emery, near the church, half a mile beyond. The measurements at these banks are as follows:—

Waynesburg Coal bed.

1. *Coal*	1′ 2″	10″
2. Clay	1′ 0″	1′ 2″
3. *Coal*	2′ 10″	2′ 8″
4. Clay	6″	8″
5. *Coal*	1′ 3″	1′ 4″

The roof is shale, of which ten feet are exposed at Mr. Couch's. The Waynesburg sandstone seems to have lost itself here and to have become only a loose sandy shale. For this reason the top bench is not removed, though it is a fair fuel. The middle bench is quite poor throughout, and for ten inches at the base is very bad. The lower bench is so impure that it is not mined. The coal goes under the creek near the cemetery. About half a mile farther up the stream, Mr. Samuel Gamble has an opening in the *Waynesburg "a."* The character of the coal is not such as to encourage much expenditure upon development of the bed in this vicinity. The section at his opening is as follows:—

1. Sandstone	10′ 0″
2. *Coal*	10″
3. Clay	4″

4.	*Coal*, slaty	5″
5.	Clay	2″
6.	*Coal* and shale	1′ 4″

Half a mile farther up, the *Washington Coal* is seen in the road near Mr. J. Barr's residence, on the Monongahela pike. West from this, near Mr. R. Barr's house, the Upper Washington limestone is exposed at one hundred and sixty feet above the coal and it is extensively quarried for use on the road.

31. Union Township, Washington County.

This is the most northerly township along the river. It is crossed by Peters creek, and Mingo creek flows along its southern border. The synclinal between the Pinhook and Waynesburg axes reaches the river at Huston's station. The section extends from one hundred feet below the *Pittsburg* to the limestone underlying the *Washington Coal.*

From the mouth of Mingo creek to Limetown, at the mouth of Huston's run, the *Pittsburg Coal* is seen constantly approaching the river, so that at a short distance above that station it is at the railroad level. It is mined there at the Cincinnati works very extensively for shipment. The available portion of the bed is barely five feet, and one foot at the bottom is left in the mine as unfit for market.* Two hundred yards below these works, a deserted opening shows the intimate structure of the roof division, which is :—

1.	*Coal*	4″
2.	Shale and *coal*	5″
3.	*Coal*	2″
4.	Clay	1″
5.	*Coal*	2″
6.	Clay	1′ 0″
7.	*Coal*	10″
8.	Clay and *coal*	1′ 0″
9.	*Coal*	10″
	Total	4′ 10″

At the mouth of Huston's run, the blossom of the *Redstone Coal* is seen at fifty-two feet above the *Pittsburg*, the interval being filled with shale and sandstone, with one foot of limestone at six feet below the upper coal. Above this, exposures are very imperfect. Midway between Limetown and Coal Bluff, the *Redstone* is exposed at fifty feet above the *Pittsburg* and is four feet six inches thick. In two openings, which are worked to supply the owners of the land, the coal is evidently very good and seems to be preferred to that from the *Pittsburg*, for at the roadside immediately below these, are two deserted openings in that bed. At seventy feet above the *Redstone*, the interval being concealed, the Great Limestone is exposed for nearly sixty feet, but it is losing its purity and contains nearly thirty feet of calcareous shale. The lower portion of this limestone, which is about eighteen feet thick, shows layers which will no doubt yield a hydraulic lime. The roof division of the *Pittsburg* is double here, the benches being fourteen and twenty-eight inches, separated by ten inches of clay. At Coal Bluff, the coal is mined extensively for shipment. The rise of the bed here is at the rate of about sixty feet per mile, as shown by the relative levels of the two entries at this place. The Pittsburg sandstone is massive for fifteen feet at the base. Near the county line, an opening in the *Pittsburg Coal* was seen at the roadside, in which the roof is single and two feet five inches thick, but worthless, as it contains numerous thin streaks of clay. The lower division was not measured, as the entry was filled with water.

On Peters' creek the *Redstone Coal* has been opened on the property of Mr. Morrison, where the following exposure occurs:

1. Shale	15′
2. *Redstone Coal*	3′ 6″
3. Concealed	95′
4. Limestone	2′
5. Sandstone and shale	40′
6. Dark shale to creek	4′

The *Redstone Coal* is hard, but brittle, breaking up into small lumps. It answers well for burning in grates, but has too much sulphur to admit of using it in the blacksmith's fire. The ash is white and powdery. The Limestone No. 4 is exposed at several localities farther up the creek, where it is seen to be about ten feet thick. The shale at the base of the section contains many impressions of plants, most of which are well preserved.

At Curryville the *Pittsburg* is mined by several persons, and at Mr. J. Lytle's opening shows:

Roof, 5 feet 3 inches; clay, 1 foot 3 inches; lower division, 6 feet 3 inches.

Resting on the roof division is a layer of bituminous shale, one foot thick, which sometimes is so rich as to have a cannel-like fracture, while at others it is merely a black fissile shale. Underlying it is a very slaty coal, eight inches below which the roof shows three benches, six, five and thirty-nine inches respectively. None of these is worked, as they are all bony. The overlying rock is shale, fifteen feet, on which is the Pittsburg sandstone, of which twenty feet are exposed. This bank is worked quite extensively, and the coal has a good reputation.

Somewhat more than a mile south from Curryville, near the residence of Mr. J. Figley, the following section shows the highest rocks observed in this township:

1. Limestone I, "a"	Fragments.
2. Shale and sandstone	25′
3. *Waynesburg "b" Coal Bed*	Blossom.
4. Concealed	70′
5. *Waynesburg Coal Bed*	Blossom.
6. Concealed	60′
7. *Uniontown Coal Bed*	3′

The *Waynesburg* is here two hundred feet above the *Pittsburg* at Curryville, so that the interval cannot be more than two hundred and fifty or three hundred feet. The rocks are dipping south-east, but the rate of dip is decreasing, as the synclinal crosses not more than two-thirds of a mile south from the locality where this section was obtained. The *Uniontown Coal* is no longer exposed, but a number of years ago it was worked by Mr. Figley, who says that it is three feet thick, and a good coal for use in grates. The *Waynesburg* is exposed here and at several other localities in the vicinity, but only its blossom is seen. On Mr. S. Kennedy's place, somewhat more than a mile south-west from Mr. Figley's, it was found in a well, and only two feet thick. It was opened near by, but proved too thin to pay for working. The Little Coal No. 3 of the section is most probably the *Waynesburg "b,"* as the buff limestone at the top is evidently the stratum underlying the *Washington Coal* in a large portion of the county. On Mingo creek the *Pittsburg* passes under the stream at a little distance above Wiley's mill. A section of the bed has been given in the description of Carroll township.

Above Curryville, along Peters' creek and its vicinity, the *Pittsburg* is accessible to the township line, but most of the openings are deserted, and the business is done now by only two or three. The coal shows little variation, as appears from the following measurements taken at the banks of Mr. Finley, at Finleyville, and Mr. Bell, near the township line:

1.	*Coal*	3′ 2″	3′ 1″
2.	*Coal* and partings	4″	3½″
3.	"*Brick*" *Coal*	1′ 5″	1′ 6″
4.	Parting.		
5.	*Lower Bottom Coal*	1′ 2″	1′ 2″

The lower bottom is not removed, as it is too impure to be used as fuel for any purpose. Nos. 1 and 3 are excellent coal. The roof division is well exposed at Mr. Finley's, where it has a total thickness of five feet five inches, and shows four benches of coal, ten, five, four and thirty inches respectively. The bottom is not compact coal, but is made up of alternating layers of coal and clay.

32. Nottingham Township, in Washington Cotnty.

This lies west from Carroll and Union. Peter's creek forms its northern boundary and Mingo creek flows across its southern portion. The Pinhook axis passes through the centre and the synclinal south-east from it crosses the extreme south-east corner. The section extends from the *Pittsburg Coal* to the upper Washington limestone.

Along Mingo creek the *Pittsburg* disappears near Kennedy's mill, about half a mile east from the township line, but as the synclinal crosses in that vicinity, the following section is exposed below M'Millan's tannery:

1.	Limestone	50′ 0″
2.	Shale and sandstone	35′ 0″
3.	Dark shale	2′ 0″
4.	Shale	4′ 0″
5.	Limestone	5′ 0″
6.	Shale	25′ 0″
7.	*Redstone coal*	3′ 6″
8.	Limestone	3′ 0″
9.	Sandstone to creek	50′ 0″

The *Redstone* was formerly worked here, as well as at some

other localities farther up Mingo creek, but the openings have all been deserted. The coal is said to be very fair.

About three miles farther up the creek the *Waynesburg Coal* is mined by Messrs. M'Clain and Kammerer, and at a short distance above by Mr. Devore. The measurements at these banks are:—

Waynesburg Coal Bed.

1.	*Coal*	1' 0"	1' 2"
2.	Clay	1' 3"	1' 2"
3.	*Coal*	3' 1"	3' 0"
4.	Clay	2"	3"
5.	*Coal*	9"	8"

The Waynesburg Sandstone has become quite shaly, and is rather an arenaceous shale. It affords a very insecure roof, so that the top bench cannot be taken out. No. 3 is slaty and is decidedly inferior coal, leaving a large quantity of ash and cinder. The bottom bench is similar to the middle one, and is used. At the creek near M'Clain & Kammerer's opening the *Uniontown Coal* is seen, resting on fifteen feet of limestone and only fifty-five feet below the *Waynesburg*. The last bank in the *Waynesburg* seen on the waters of Mingo creek belongs to Mr. H. Warne, but was not measured, as the mouth of the entry was choked with debris and contained much water.

At the extreme south-west corner of the township, near Mr. J. B. Mosier's residence, an imperfect exposure occurs on a hill above the cut on the Hempfield railroad extension, reaching from the Upper Washington Limestone to Limestone II.

1.	Upper Washington Limestone VI	20'
2.	Concealed	30'
3.	*Jolleytown Coal Bed*	Blossom.
4.	Concealed	65'
5.	Sandstone	4'
6.	Shale	8'
7.	Calcareous shale	3'
8.	Limestone III	8'
9.	Sandy shale	20'
10.	*Bituminous shale*	2'
11.	Shale	3'
12.	Limestone II	10'

No. 1 is seen in the summits of the hills, and many of the farmers have burned it for lime. The *Jolleytown Coal* is not more than one foot thick. The exposure in the cut is a very favorable one, showing Limestones II and III distinctly. At most of the road exposures the debris of the upper stratum is continuous with that of the lower, so that one usually finds difficulty in separating them. III is buff colored, and shows many bryozoans on its weathered surface. The black, somewhat calcareous shale resting on it has many lamellibranchiates, together with occasional specimens of teeth and scales of fish, the latter belonging to *Rhizodus*. Bivalve crustaceans are not rare in this stratum. The carbonaceous shale No. 10 is rich in bivalve crustaceans and remains of fish, with now and then a leaf, which gives evidence of long maceration before final burial. Limestone II is fossiliferous here, and contains innumerable specimens of minute univalves. It is not fully exposed, and is certainly much thicker than is given in the section.

On Peters' creek, about half a mile from the Union township line, the blossom of the *Redstone Coal* is seen at fifty-five feet above the *Pittsburg Coal*, and resting upon an impure limestone. At thirty feet higher is a dark shale, separated from the great limestone above by thirty feet of sandy shale. Fragments of the limestone occur through a vertical space of fifty feet, but there is no exposure of the rock in place. On a little run coming in at the Methodist church the *Pittsburg* is worked by Mr. L. Phillips, and at a short distance farther up by Mr. Cassber, opposite Thomas' saw mill. These banks show as follows:

Roof division	Concealed.	Concealed.
Clay	1′ 0″	1′ 1″
Coal, Lower Division	5′ 8½″	6′ 0″

The partings associated with the bearing-in bench are each one-half inch thick. The quality of the coal differs at these banks. In the first the top and "brick" benches are so clean as to be superior smithing coal, whereas in the second the top bench contains many persistent "binders" of pyrites. The "lower bottom" is not removed at either. Not far from Thomas' mill the *Pittsburg* goes under, and no openings are seen in any other bed for nearly two miles. At a short distance above the forks of the stream, near Mr. J. Rainey's house, the *Uniontown* coal is seen in the road resting on its characteristic yellow limestone. Near by the *Waynesburg Coal* is mined by Mr. J. Patterson, and at a mile south by Mr. Atkinson. Both of these openings are new, and have been worked but little. The coal is poor, being slaty and sulphurous. The middle bench is three feet six inches at the former, and two feet eight inches at the latter.

Coming down from the hill back of Mr. Patterson's to the fork in the road at Mr. Rainey's the following section was seen:

1. *Washington Coal Bed*	Blossom.
2. Concealed	50′ 0″
3. *Waynesburg "b" Coal Bed*	2′ 0″
4. Concealed	90′ 0″
5. *Waynesburg Coal Bed*	4′ 6″
6. Concealed	50′ 0″
7. *Uniontown Coal Bed*	Blossom.
8. Limestone	4′ 0″

The blossom of the *Washington Coal* is very large and covers a vertical space of about eight feet. The *Waynesburg "b"* is a dull-brownish coal and evidently inferior. The feature of interest in the section is the smallness of the intervals as compared with the same in Greene county, or even in the south-eastern portion of this county.

At Munntown, Limestone II is in the road. At school house No. 3, a coal three feet thick is shown at eighty feet below the *Waynesburg*, which is possibly the *Sewickley*.

33. Peters Township, in Washington County.

This lies west from Union and north from Nottingham and North Strabane. Peters' creek separates it from Nottingham and a fork of Chartiers creek from North Strabane. The Pinhook axis passes through the south-east corner and the synclinal between that and the Washington axis crosses almost the centre of the township. The section extends from the *Pittsburg* to the *Washington Coal*.

At the mouth of Brush run the great limestone is well exposed, as follows:

1. Concealed	117′ 0″
2. Limestone	1′ 0″
3. Sandstone	5′ 0″
4. Limestone, brecciated	2′ 6″
5. Concealed	15′ 0″
6. Limestone	4′ 6″

7. Sandstone	14′	0″
8. Concealed	15′	0″
9. Shale	5′	0″
10. Limestone	12′	0″
11. Shale	12′	0″
12. Limestone	50′	0″
13. Sandy shale to creek	15	0″

No. 6 is coarsely brecciated near the top, and contains great numbers of minute univalves, which can be seen only on the weathered surface.

No. 12 contains every variety of rock from clean ringing limestone to the dull impure material which breaks up on exposure. The former will yield good lime and the latter will give a hydraulic lime or possibly cement. The junction between 12 and 13 is not well exposed, but some coaly matter is there which represents the *Sewickley Coal.*

The *Uniontown Coal* should occur in the exposed portion of the section, but no evidence of its presence was observed. About half a mile south-west from the village of Thompsonville the *Washington Coal* is seen in the road near Mr. W. Hawes' residence, and appears to be not far from four feet thick. Somewhat more than a mile south from this the blossom of the *Waynesburg* is exposed in the road near the house of Mr. J. P. Skile, where it is about three feet thick and rests on twelve feet of shale, below which is a limestone.

At a mile and one-half above Thompsonville, in a field belonging to the M'Loney heirs, the *Waynesburg* blossom is shown. Near Bowerhill this coal was once worked by Mr. Hixon, but was found so poor and thin that the working was abandoned. A similar condition exists on the property of Mr. Bell, one mile west from Bowerhill. In the south-western portion of the township the following succession was observed on the road leading north from Mr. S. Speer's residence.

1. Limestone II	25′
2. *Washington Coal Bed*	Blossom.
3. Concealed	120′
4. *Waynesburg Coal Bed*	Blossom.
5. Concealed	45′
6. *Uniontown Coal Bed*	Blossom.
7. Limestone	20′

The *Uniontown* appears to be not less than three feet thick, but it has not been opened in the vicinity. The feature of especial interest in this section is the greatly diminished intervals between the coals. Throughout the western portion of this township the *Washington* and *Waynesburg Coals* are too thin to be worked, even if the coal were good for anything. Near the Presbyterian church, in the north-east part, the former was once mined by stripping. About one-fourth of a mile north-west from that church, on a hill known as Rocky Ridge, is seen the remnant of a massive sandstone. The fragments on this outlier look like huge bowlders, and many of them are of great size, weighing several tons. The nearest continuous bed of the rock is nearly half a mile away. This, like the broken stratum at Hillsborough, is an interesting example of atmospheric erosion. The rock is two hundred and seventy feet above the *Waynesburg Coal*, and therefore belongs near the horizon of Limestone VI.

About two miles south from the church, on a tributary of Peters' creek, the *Waynesburg* was once opened by Mr. Joseph Townsend, who found it in all not far from six feet thick. The bed lay high up on the hill, and as it had but little cover the coal was of exceptionally bad quality, and the working was abandoned. From this old opening to the run is one hundred and fifty feet. At three hundred yards farther down the stream the *Redstone Coal* is seen, so that the extreme interval between the two coals is not more than one hundred and seventy-five feet, thus making the interval between the *Waynesburg* and *Pittsburg* not more than two hundred and twenty-five feet; whereas in Marshall county of West Virginia, just beyond the south-west corner of the district, the interval between the same coals is four hundred feet.

At half a mile farther down this stream the *Redstone* was mined many years ago by Mr. E. R. Townsend, who states that it is four feet thick, and yields an excellent coal for grates, that burns freely and leaves a fine white ash. It rests on limestone and has a limestone roof. The chief drawback to successful mining is the frequency of clay-veins and horsebacks. The same bed was opened on Mr. T. M'Comb's land nearer Peters' creek, where it showed the same characteristics.

The *Pittsburg* is mined on Peters' creek, near Thomas' mill, by Mr. R. M'Comb, and at a mile south-east on a tributary of that stream, by Mr. A. Buckingham. The measurements of the lower division at these openings, are as follows:—

1. *Coal*	3′ 2″	2′ 11″
2. *Coal* and partings	2½″	3½″
3. *Brick coal*	1′ 6″	1′ 8″
4. Lower Bottom *coal*	10″	1′ 0″

At Mr. M'Comb's opening, the roof division is exposed and shows three benches, five, four and eighteen inches thick, all of them very impure. The coal differs in quality at these banks. At the former it contains so much pyrites in binders and lumps that no portion of the bed can be used by blacksmiths, whereas at the latter, the upper bench is good smithing coal throughout. In both, the lower bottom is slaty and sulphurous and is not removed. The coal rests on a sandy clay, below which is an irregular limestone.

34. Cecil Township, Washington County.

This adjoins Peters on the west and is separated from it by Chartiers creek. Miller's run flows eastwardly through its centre and the Washington axis passes through its eastern portion. The section extends from the base of the Great Limestone to the *Washington Coal.*

The exposures, aside from those in the cuts on the Chartiers railway, are extremely poor. The *Washington* and *Waynesburg Coals* are so thin, that for many years no openings have been in operation, and their blossoms are so frequently concealed that one's position is often doubtful. The Great Limestone is exposed along most of the streams, but for the most part the exposures are so indefinite that the relative horizon cannot be determined. The work is rendered the more difficult by the fact that the synclinal between the Washington and Claysville axes crosses the township, and cannot be definitely located, as it is very shallow and the western anticlinal is becoming ill-defined as it goes northward.

Descending the hill from the cemetery south from Venice, the following intervals were found in the road:

1. Limestone II	Fragments.
2. *Washington Coal*	Blossom.
3. Concealed	50′ 0″
4. Limestone I "*a*"	Fragments.
5. Concealed	40′ 0″

6. *Waynesburg Coal*		Blossom.
7. Concealed to creek at Venice		140′ 0″

There is a constant diminution in the intervals between the coals. There is no evidence of coal associated with the limestone No. 4, and it is quite possible that *Waynesburg* "*a*" and "*b*" are absent. The place of the *Uniontown Coal* is concealed, but the bed is so persistent that there can hardly be any doubt of its presence. It is, however, of no economical value, being exceedingly thin wherever seen in this portion of the county.

In the extreme northern portion of the township at the cross-roads half a mile south from the railroad, the blossom of the *Washington* (?) is seen three hundred feet above the Pittsburg, as exposed at M'Donald's station.

Limestone II is seen at the cross-roads near J. J. Johnson's hotel and store, about half a mile from the point where Miller's run crosses the county line. About two miles farther southeast on a tributary to Chartiers creek, Mr. White found a little coal near the saw-mill, in the following section:—

1. Shaly sandstone		10′ 0″
2. *Coal.*		
a Coal	′8	1′ 10″
b Clay	4″	
c Coal	10″	
3. Shale		15′ 0″
4. Limestone seen		6′ 0″

This is evidently the *Waynesburg Coal*, but owing to the lack of exposures in the vicinity its relations cannot be absolutely determined. Along the Chartiers Valley railway the exposures are quite satisfactory from the base of the Great Limestone up to the *Waynesburg Coal.* About half a mile below Cannonsburg, the base of this limestone is shown to the thickness of twenty-seven feet. For the most part it is dull, almost cream colored, has a smooth fracture, and when long exposed to the weather it breaks down into irregular fragments. It does not yield good lime, but if burned with care might give a fair cement. The bottom layer contains vast numbers of minute uni- and bivalves, which are very distinct on the weathered surface, but being imperfectly silicified they are usually defective, so that identification would be difficult. On the surface of fresh fracture they appear only as points, giving the rock a "birdseye" appearance. Directly under the limestone is a mass of coaly matter eighteen inches thick, which represents the *Sewickley Coal*, and rests on a thinly laminated sandstone, of which only ten feet are exposed.

The lower division of the Great Limestone continues in sight to where the railroad crosses the creek, and is nearly fifty feet thick. Two layers in the upper portion, one of them at the very top and quite ferruginous, the other at ten feet lower, yellowish to almost white and somewhat brecciated, are fossiliferous. The latter contains only minute univalves, while the other shows only obscure bryozoans. In each case the fossils are not distinguishable on the surface of fresh fracture. Twenty feet above the top is another fossiliferous limestone, containing only minute univalves.

Between Vaneman' station and Grier's, the upper division of the Great Limestone is seen, including a thin coal from three to six inches thick, which may possibly prove to be the *Uniontown*. At Grier's, about seventy feet above the track, there is an old opening in the *Waynesburg*, which shows as follows:

Waynesburg Coal Bed.

1. *Coal*	6″
2. Clay	2″
3. *Coal*	1′ 4″
4. Clay	2″
5. *Coal*	4″

This coal is said to be very good, but the opening has been deserted for many years. Pyrites cannot be present in large quantity, for a most excellent spring issues from the bank. A section obtained here shows the upper part of the great limestone as follows:

1. Sandstone	10′
2. *Waynesburg Coal Bed*	2′ 6″
3. Concealed	8′
4. Sandstone	35′
5. *Uniontown Coal Bed*	Blossom.
6. Limestone and shale	8′
7. Shale	6′
8. Limestone and shale	8′
9. Sandstone	9′
10. Limestone	6′
11. Shale	2′ 6″
12. Limestone seen	8′

The rest of this section as obtained at the mouth of Brush run has already been given in the description of Peters township.

35. Chartiers Township, in Washington County.

This adjoins Cecil on the south, and has Chartiers creek for its eastern boundary. The west fork of that creek flows eastward across it. The surface has been worn away by erosion, so that the section reaches from the *Pittsburg* only to the *Waynesburg Coal.* The Washington anticlinal skirts along the eastern border, and the synclinal between that and the Claysville axis runs through the central portion of the township. There are few exposures aside from those of the *Pittsburg Coal,* as the whole surface is covered with a thick mass of debris, which, though quite inconvenient to the geologist, is an excellent soil, and renders this one of the finest agricultural townships in the county.

Along Chartiers creek, the openings are quite numerous, beginning above the Pittsburg pike and continuing to just below Cannonsburg, where the coal goes under the creek. The first opening seen is that operated by Mr. Peter Ashurst, who mines by an inclined shaft, reaching the coal at a vertical distance of fourteen feet. At one-third of a mile north, the coal is opened by Mr. James Arthur. Measurements made at these openings show a somewhat wider range of variation than is commonly found within so short a distance.

Pittsburg Coal.

1.	*Coal*	8″	2″
2.	Clay	1′ 6″ to	5″ 4″
3.	*Coal*	2′ 6″	8″
4.	Main clay parting	4″	1′ 2″
5.	*Coal*	2′ 10″	2′ 7″
6.	*Coal* and partings	3″	3″
7.	"Brick" *coal*	11″	10″
8.	"Lower Bottom" *coal*	1′ 0″	1′ 4″

The main clay parting is not often so thin as at the place where the measurement was made in Mr. Ashurst's entry, though in this township it is sometimes very thin. The roof division is utterly worthless for fuel and is not removed; but it is of service in strengthening the roof, which is not always secure. Some shale occasionally intervenes between the coal and sandstone above, but it is not persistent.

At both of these openings, No. 5 is an excellent coal. For the most part it seems to be free from pyritous binders, though in Mr. Ashurst's entry a rather persistent streak of pyrites occurs at one foot from the top. This bench contains much semi-cannel and is said to be in high repute as fuel for domestic use. The "Brick" coal is said to be equally good. A parting occurs between it and the lower bottom, but commonly it is so thin as to be imperceptible on the face of the coal. The benches however are always distinct and are separated by the miner. The lower bottom is exceedingly bad and is invariably left in the bank.

On the Pittsburg pike the coal is extensively mined by Hon. J. H. Ewing and by Mr. J. Allison. At the former opening it is still below the surface, but at the latter it is above the road and is worked by a tunnel. The coal is thicker here than at the banks just referred to, as appears from the following measurement of the lower division:—

1. *Coal*	3′ 0″	3′ 4″
2. *Coal* and partings	3″–5″	3″–5″
3. "Brick" *coal*	10″	1′ 0″
4. "Lower Bottom" *coal*	1′ 6″	1′ 0″

At Mr. Ewing's shaft the roof division appears to have but one bench of coal which is from one foot to one foot six inches thick. At Mr. Allison's opening, the roof is not well exposed, only a confused mass of coal and shale two feet thick having been seen. The coal is mined very largely by these gentlemen, not only for the supply of Washington, but also for shipment. It has a high reputation and is undeniably a very fine coal. Here, as at the other banks, the lower bottom is not removed.

Between these and the mouth of the West fork of Chartiers creek, no other openings were seen, although the coal is available on all the farms; but just below Houstonville, the bed has been opened by Mr. D. C. Houston, as well as by Mrs. Foley, on the adjoining property. At the former, the bed shows:

1. Shale	8′ 0″
2. Roof division	2′ 0″
3. Clay	1′ 0″
4. Lower division	5′ 6″

At Mrs. Foley's opening the roof is double, the benches being two and nine inches, separated by only two inches of clay. The main-parting varies from nine inches to less than half an inch. The coal seen on the dump at these banks is exceedingly clean and shows little pyrites. It is soft, caking and no doubt is a good gas coal. Thin streaks of cannel occur in the top bench, and at ten inches from the top, there is a thin but persistent band of pyrites. In the vicinity of Cannonsburg, the openings are quite numerous. On a little stream coming in at the upper end of the village, the following section was obtained in the entry to Mr. Banfield's works:

1. Laminated sandy shale	8′ 0″
2. Clay shale	1′ 0″
3. Roof division	0′ 10″
4. Clay	0′ 2″
5. Lower division seen	4′ 6″

The coal of the lower division is good and resembles that from the works already mentioned. The diminution of the roof and the main clay parting is a little curious. Following up the road past these openings, no good exposure is seen though a vast amount of limestone in fragments is strewn over the road. Where this intersects the Ridge road, a coal blossom is shown immediately overlying a yellow limestone and two hundred and thirty feet above the Pittsburg coal where last seen. This blossom is that of the *Uniontown Coal*. At a little way north from this fork in the road, a knob rises nearly one hundred feet higher and should catch the *Waynesburg*, but no evidence of any coal was found on it. North-west from this point along the ridge for nearly two miles, the *Uniontown Coal* and its associated limestone are frequently seen.

On Plum run, which enters the west fork of Chartiers near Houstonville, the *Pittsburg Coal* is worked by Mr. S. Skile and Mr. T. Thompson, at whose openings the lower division of the bed shows as follows:

1. *Coal*	2′ 10″	2′ 9″
2. *Coal* and partings	3″	3″
3. *Coal*	1′ 9″	2′ 0″
Total	4′ 10″	5′ 0″

Just above the last is another, belonging to Mr. Hamilton, which was not measured. In Mr. Thompson's bank a two inch streak of cannel occurs at three to six inches from the top. In Mr. Skile's, at six inches from the clay there is occasionally a clay binder, and at ten inches there is a very persistent streak of pyrites, one-fourth of an inch thick. At these openings the lower bottom is taken out, and is little inferior to the rest of the bed. At Mr. Thompson's bank the roof and the clay are wanting, and the sandstone rests directly on the coal; but at Mr. Skile's they are both present, though the roof is but two inches thick. The coal passes under the run at a short distance above Mr. Hamilton's opening, and the *Redstone* comes down to the road, where it is seen resting on a thin limestone. It is too thin to be of any value.

On the west fork of Chartiers creek no opening is in operation below Connellsville; there the coal is mined by John Fee. From the mouth of the stream up to this point the *Pittsburg* is available on every farm, and at many places the limestone below it is exposed in the road. At Mr. Fee's opening the section is—

1. Sandstone		Not measured.
2. *Pittsburg Coal.*		
Coal	1′ 0″	
Clay	9′	6′ 11″
Coal	5′ 2″	
3. Clay and shale		8′ 0″
4. Limestone		4′ 0″

The clay and shale immediately below the coal contain a good deal of nodular iron ore, but the quantity is not sufficient to be of economical value. Above Connellsville the coal rises for some distance at nearly the same rate as does the bed of the stream, but goes under finally about half a mile below the crossing of the Hickory road.

On that road, between the west fork of Chartiers and the poor-house, there are few exposures. The Great Limestone is shown near each stream, and on the ridge portion of the road the blossom of the *Uniontown Coal* is frequently exposed, accompanied by its yellow limestone. This coal has been opened at a little distance off the road by Mr. Fergus.

36. North Strabane Township, in Washington County.

This lies east from Chartiers and south from Cecil and Peters. The north fork of Chartiers crosses it from south to north, and Chartiers creek itself is the western boundary of the township. The section extends from the *Pittsburg Coal* to the Upper Washington Limestone. The horizon of the *Pittsburg Coal* is reached only on Chartiers creek, and the rest of the township lies chiefly above the *Waynesburg Coal.*

Near Houstonville station, on the Chartiers Valley railroad, Mr. T. Boon mines the *Pittsburg*, which shows:

Roof division	0′ 9½″	7′ 9½″
Clay	1′ 0″	
Lower division	6′ 0″	

A band of pyrites is persistent in the top bench of the lower division and is frequently double, the two portions being one and two inches respectively. The lower bottom was not exposed and its thickness, two feet, is given on the authority of the pit-boss. This part is very poor and is not mined. The rest of the coal is very good.

Opposite Cannonsburg, another opening is seen just above the creek and belonging to Mr. Myers. The section here is:

1. Sandstone		25′ 0″
2. Sandy shale		13′ 0″
3. *Redstone Coal bed*		0′ 3″
4. Sandy shale		12′ 0″
5. Carbonaceous shale		2′ 0″
6. *Pittsburg Coal bed.*		
Roof division	0′ 7″	5′ 6″
Clay	1′ 0″	
Lower division	3′ 11″	
7. Shale and sandstone		20′ 0″
8. Limestone in the creek		2′ 0″

The little coal above the *Pittsburg* is found also at Mr. Banfield's works on the opposite side of the creek, and is without doubt the *Redstone*. The carbonaceous shale just over the coal shows indistinct impressions of plants with many minute scales and teeth of fish, and occasional lamellibranchiates. The sandy shale No. 4, is thinly laminated and contains much broken vegetable matter, so that it resembles the shale under the *Washington Coal*. The *Pittsburg Coal* here is said to be very good but somewhat lighter than that obtained on the other side of the creek. The thinness of the lower division is due to the upper bench which is only eighteen inches, while the other benches retain their thickness. The Limestone No. 8, is crowded with minute univalves. At a short distance below this, the coal passes under the creek.

On the old road to Pittsburg, about one-fourth of a mile north from the township line, the Upper Washington limestone is exposed at the fork in the road. A section was obtained here which gives the succession down to the Middle Washington limestone:

1.	Upper Washington limestone, VI	8′
2.	Sandstone and shale	12′
3.	*Jolleytown Coal bed*	—
4.	Concealed	75′
5.	Limestone	5′
6.	Sandstone	30′
7.	Middle Washington limestone, IV, seen	3′

Limestone VI exhibits its usual character, weathering almost snow-white with a slight tinge of blue. The surface of fresh fracture varies in different portions of the bed, from black to nearly white. Limestone No. 4 is very coarse, somewhat brecciated and ferruginous, and the weathered surface is lamellar. The *Jolleytown Coal* is thin. At the fork in the road, nearly a mile from the township line, Limestone IV is in the roadside, showing its characteristic bright yellow color. Turning off southward here, the following succession is seen in the road and ends at the township line below Mr. D. Quail's house:

1.	Middle Washington limestone IV	——
2.	Concealed	70′
3.	*Washington Coal bed*	Blossom.
4.	Concealed	50′
5.	*Waynesburg "b" Coal bed*	Blossom.
6.	Concealed	60′
7.	*Waynesburg Coal bed*	Blossom.
8.	Concealed	43′
9.	*Uniontown Coal bed*	Blossom.
10.	Concealed	35′
11.	Limestone	3′

Joining this and the preceding section together the interval between the *Washington Coal* and Limestone VI is found to be one hundred and forty feet. The blossoms indicate that the coals in the section are all thin. No. 5 is referred to the *Waynesburg "b,"* because the blossom evidently belongs higher up in the hill than its place in the section shows. It is clearly a slip.

Returning to the old Pittsburg road, which here is simply the Cannonsburg road, one finds, near the residence of Mr. J. Haines, both the *Washington* and the *Waynesburg "a"* in the road and seventy-five feet apart. Near the same place Limestone I "*b*" and II are quite thick, each being not far from twenty feet. From this to the Presbyterian church, Limestone II is almost constantly in sight along the road and has its characteristic black shale resting on it On the road to Cannonsburg, turning off at the Presbyterian church, there is a stratum of calcareous shale, containing coprolites and teeth of fish in great numbers, many of which were collected years ago by Prof. Jones, of Washington and Jefferson College, in whose possession they now are. The specimens are very fine, and some of the coprolites are two inches long. The stratum is now concealed, but is, as nearly as I can determine from the description given, about three hundred feet above the *Pittsburg Coal.* About a mile and a half north-east from the church the *Uniontown Coal* is seen near Mr. E. M'Clelland's residence, and its limestone is directly under it.

In the north-east portion of the township the *Uniontown Coal* is exposed at the cross-roads near Mr. J. Scott's residence, where it seems to be about two feet six inches thick. Two-thirds of a mile south from this, near the head of a little tributary to the north fork of Chartiers creek, the same coal appears in the road near Mr. Thomas' house. On this stream Mr. Lyons many years ago opened the *Waynesburg*, which is three feet thick and only forty feet above the *Uniontown.* The same bed was once opened also by Mr. Bell and Mr. Black on this run, but all the openings were long ago deserted, and no direct information can be obtained respecting the coal. All agree in stating that the quality was inferior. This bed is exposed at the cross-roads on the creek near Mr. R. M'Clelland's house and near Mr. S. Pollock's residence, two-thirds of a mile west from the last. At Mr. M'Clelland's the *Waynesburg* "*a*" is fifty feet above the *Waynesburg* and is associated with Limestone I "*a*."

At the cross-roads near school-house No. 2, the *Waynesburg* "*a*" and Limestone I "*a*" are exposed, and in the immediate vicinity the *Waynesburg* is worked by Messrs. W. & T. Rees, at whose openings the following measurements were made:

1. Sandstone	20′ 0″	
2. Shale	6′ 0″	10′ 0″
3. Bituminous shale	3″	4″
4. Clay	1′ 0″	1′ 2″
5. *Coal*	2′ 2″	2′ 6″
6. Clay	5″	6″
7. *Coal*	7″	5″

The Waynesburg sandstone is more compact here than is usual in this portion of the county, being massive and resisting the weather, so that it forms bluffs. Some quarrying has been done above the mill, and the stone obtained proves excellent and durable for building purposes. The shale overlying the coal seems to be utterly barren of vegetable impressions, as diligent search was not rewarded by the discovery of any specimens. The main bench, No. 5, fully maintains the evil reputation of the bed. It contains so much slaty and pyritous matter that the ash is said to equal the combustible matter. The bottom bench is still worse and is not used. Below it is a bituminous shale, which is not persistent.

On a little tributary to the North fork of Chartiers creek flowing south-east along the southern border of the township, the *Waynesburg* "*a*" is seen near the steam saw-mill. One-third of a mile north the Upper Washington limestone is exposed at the forks in the road, the total interval between it and the coal being two hundred and thirty feet. The blossom of the *Washington Coal* is seen between the two points at one hundred and sixty feet below the limestone.

37. South Strabane Township, Washington County.

This adjoins Chartiers and North Strabane on the south, and the National Road is, for the most part, its southern boundary. The synclinal between the Washington and Pinhook axes crosses very nearly through the central portion. Chartiers creek is the north-west boundary, and the North fork of that stream flows through the eastern part. The section extends from the *Pittsburg Coal* to about two hundred feet above the Upper Washington limestone. The higher rocks are not well exposed. Washington is situated in the south-west corner of the township.

On the road leading from Washington to Hickory the *Washington Coal* is shown in a cutting made to alter the grade of the road within the corporate limits of the former town. It is here quite thick, and at ten or twelve feet below it is the *Little Washington Coal* about eight inches thick. Under this is the Limestone I "*b*," which is associated with the *Washington Coal* throughout the western portion of the county, but is absent in the eastern part. It is ferruginous on top, but light-colored below. The same rocks are exposed at a little distance beyond on the opposite side of the hollow. From this point to the immediate vicinity of Chartiers creek there are no exposures, but as one approaches that stream he sees many fragments of the Great Limestone in the road.

Where the Pittsburg road passes out of Washington the *Washington Coal* and Limestone II are seen in the roadway. They continue in sight for a short distance along the pike, but the road soon descends, and at the toll-gate the *Waynesburg Coal* is at only a few feet above the level of the road. Many years ago it was mined here, but the opening is deserted, and no direct information is within reach. The bed is said to be about two feet six inches thick, but this may refer to the main bench alone. At a short distance farther on a coal blossom is in the road, which may be the *Waynesburg "a,"* but its relations cannot be determined positively, as the rocks are rising northward quite rapidly, and there are no exposures in the vicinity. Where the road descends to Chartiers creek the Great Limestone is seen in the road.

At the toll-gate a road turns off northward from the pike. On this, at half a mile from the fork, Messrs. Harding and Warrick have sunk a shaft upon the *Pittsburg Coal*, in which, according to the superintendent of the works, the following section was obtained:

1. Soil			4′ 0″
2. Limestone			45′ 0″
3. *Sewickley Coal Bed*			4″
4. Limestone			30′ 0″
5. Shale			45′ 0″
6. *Redstone Coal Bed*			3 0″
7. Sandstone			20 ′0″
8. Shale			1′ 0′
9. *Pittsburg Coal Bed.*			
Roof division	2′	0″	8′ 10″
Clay	1′	0″	
Lower division	5′	10″	

The Little Coal, No. 3, is evidently the *Sewickley*, but the sandstone which underlies it in the greater part of the district has disappeared, and the limestone below it has become more compact than at any locality in this neighborhood where the rock is at the surface. The *Redstone Coal* is of fair quality.

In the original shaft it is two feet six inches thick, but in the escape shaft, one hundred yards from the other, it is four feet. The *Pittsburg* at these works yields a coal of excellent quality, which is mined extensively to supply Washington.

Along the Monongahela pike there are no exposures for several miles, excepting one of the Upper Washington limestone at the first southerly fork of the road beyond the toll-gate.

On the old Pittsburg road, the *Washington Coal* continues in view to the first fork in the road. Beyond that, about half-way to Ewing's station, the road crosses a high hill on whose northern side the following section occurs:—

1. Middle Washington limestone IV..	——
2. Concealed with limestone II at base	70′
3. *Washington Coal bed*	4′
4. Concealed	50′
5. *Coal*	1′
6. Imperfectly exposed, mostly sandstone	60′
7. *Waynesburg Coal bed*	Blossom.
8. Concealed	70′

As the road approaches Ewingsville, the Great Limestone is seen but the exposures are incomplete.

Returning to the fork of the road, we go north north-east, and at the cross-roads near Mr. J. Munce's residence, the *Washington Coal* is seventy feet above us. On the hill beyond, the thick limestone underlying that coal is well exposed and has a thin coal, probably *Waynesburg* "*b*" directly under it. The *Washington Coal* is shown just beyond where the Cannonsburg road turns off. This road reaches the *Waynesburg Coal* at the cross-roads near Mr. M'Nary's house, and the *Uniontown* at the township line. On the other fork, the Upper Washington limestone is first seen where the road to Clokeysville turns off, and remains in sight on the higher portions of the road quite to the township line. The Middle Washington limestone is seen on the cross-road leading to the old Pittsburg road and on the same road the *Washington Coal* is exposed with the Limestone II, resting almost directly upon it.

On the proposed extension of the Hempfield railroad, which for a mile or more runs almost alongside of the National pike, the exposures are very good. In a low cut just east from Washington the Limestone IV is well shown to a thickness of eight feet. In the inside it is dull flesh-colored and extremely hard, but being ferruginous it weathers, after long exposure, to the depth of several inches. The weathered surface gradually exfoliates and the mass finally breaks down into a coarse powder. This character is finely exhibited in the long fill west from the cut.

The upper layer of this bed contains numerous fossils, but few of them are silicified. For the most part they are replaced by crystalline calcium carbonate and appear as glistening points on the surface of fresh fracture. On the weathered surface, the larger specimens are simply agglomerations of spar, showing no structure. The most of these seem to be branching bryozoans. At least those are the only ones of considerable size that can be made out. Some very minute univalves, among which are *Euomphalus* and *Bellerophon*, are silicified and occur in vast numbers. Occasionally one finds what appears to be the section of a brachiopod.

Resting on this limestone is one foot of bituminous shale, and at twenty-five feet higher is a thin coal. Above this to the Upper Washington Limestone there are only shaly sandstones, except about midway, where a thin limestone is seen. The upper limestone is shown at the tunnel about one mile east from Washington, where the following section was obtained:

1. Debris	10′	0″
2. Limestone	1′	0″
3. Shale	3′	0″
4. Limestone	2′	6″
5. Shale	2′	0″
6. *Coal*		8″
7. Clay	1′	0″
8. Dark shale	10′	0″
9. Limestone VI	20′	0″

The Limestones 2 and 4 are drab-colored, quite fetid, and contain much pyrites. Some minute ostracoids were seen in them. In the dark shale, No. 8, are thin layers, crowded with impressions of leaves and bark, which appear to be closely related to *Sigillaria Menardi*. Ferns are extremely rare, but they certainly occur here, for I saw one leaflet of a *Neuropteris*. The Upper Washington Limestone is light colored at the base, but the layers next ascending are dark and somewhat mottled.

These are fetid when struck, and contain fragments of vegetables, fishes and minute molluscan shells. The greater part of the bed is dark, almost black on the surface of fresh fracture, brittle, and contains vast numbers of ostracoids. The top layers are slaty.

On the National road, near the tunnel, this limestone has been quarried for use on the road and for burning into lime, for both of which uses it answers admirably. The road here rises above the limestone nearly two hundred feet, but without affording any exposures. A shaft was sunk from the top of the hill to the tunnel, but no record of the section is for the present accessible. From this point to the eastern line of the township there are no exposures along the road.

At the United Presbyterian Church, in the south-east corner of the township, about half a mile north from the National road, there occurs a limestone, which at another exposure is found to be sixty-five feet above Limestone VI. It is light blue, and in its mode of weathering bears much resemblance to that rock. The same stratum is seen on the Amity and Dunningsville road, somewhat more than half a mile farther west. From this point it is easily followed along the road northward for nearly a mile to the cross roads near Mr. R. D. Henry's house. Turning off here to the west and descending to the north fork of Chartiers creek, this exposure is found, beginning with the limestone just mentioned:

1.	Limestone	3′
2.	Concealed	25′
3.	Limestone	Fragments.
4.	Concealed	40′
5.	Limestone VI, seen	15′
6.	Concealed	80′
7.	Limestone IV, seen	6′

No. 7 is exceedingly ferruginous and somewhat nodular. The larger nodules are frequently coated with hematite. The base of the section is in the creek below Mr. J. Zedeker's residence. The roads form a triangle here, of which the base runs east and west. Limestone IV is exposed on all the roads. From this point to Washington, rather more than two miles, there are no good exposures. The Upper Washington Limestone is seen on the road leading from the school-house to the National pike.

The synclinal between the Washington and Pinhook axes crosses near the school-house.

38. Canton Township, Washington County.

This lies west from Chartiers and South Strabane, and has the National road for its southern boundary. The Hempfield railroad passes through its southern portion. The section extends from the Great Limestone to the Upper Washington Limestone.

On the railroad the exposures are quite satisfactory. At the line between this and Buffalo township is a tunnel exposing a section of the rocks above the Middle Washington Limestone, which has been given in the description of Buffalo township. Between the tunnel and Chartiers creek, the strata between that limestone and the *Washington Coal* are very handsomely shown, and the following section gives the succession as there exhibited:

1.	Limestone IV	20′ 0″
2.	Shale	2′ 0″
3.	*Coal*	Blossom.
4.	Shale	10′ 0″
5.	*Coal*	Blossom.
6.	Limestone and calcareous shale III, (?)	7′ 0″
7.	Shale	15′ 0″
8.	Calcareous sandstone	4′ 0″
9.	Sandy shale	20′ 0″
10.	Limestone and shale II	23′ 0″
11.	Shale, with nodular iron ore	4′ 0″
12.	*Washington Coal Bed*	6′ 11″
13.	Laminated sandstone	10′-12′ 0″
14.	*Little Washington Coal Bed*	8″
15.	Dark argillaceous shale	5′ 0″

Thus far the rocks have been rising eastward quite rapidly, but in the next cut they are falling in that direction, and the following section is exposed:

13.	Laminated sandstone	8′-12′
13*a*.	Lead-colored shale	0′- 9′
14.	*Little Washington Coal Bed*	1′
15.	Sandy shale	2′-10′
16.	Limestone I "*b*"	10′

No. 10 consists about equally of limestone and shale. The *Washington Coal*, as exposed here, is in eight divisions of coal and shale, and shows in all four feet nine inches of coal, the bottom bench being two feet six inches. Throughout it seems to be poor and slaty, but the amount of pyrites is certainly small, for in the long cut, where the coal is exposed for nearly one-fourth of a mile, it shows no signs of yielding to the weather.

No. 13 is the laminated sandstone, which seems to accompany the coal everywhere in this district, except, perhaps, along the extreme western border. It contains the characteristic fragments of carbonized vegetable matter, and, as at Waynesburg and other localities, is gashed vertically, the gashes extending from the upper to the lower coal. These fissures vary in width from one to fourteen inches and are filled with a lead-colored argillaceous shale, which is vertically laminated. They have no definite course, some bearing north-east and south-west, while others have a contrary direction.

The *Little Washington Coal* varies from eight to twelve inches and seems to be rather good coal. The interval between it and the *Washington Coal* is quite irregular. In the cut where the first section was obtained, the lower bed describes a number of short and shallow curves, all of them at the expense of the sandstone and without affecting the coal above. Where the second section is seen, this variation is very finely shown. The top of the cut is covered with debris, so that the upper coal is not exposed, and the laminated sandstone is the highest rock seen. Between this and the little coal is a lead colored shale, the same with that filling the fissures in the sandstone. At the east end of the cut this shale is three feet thick, toward the middle of the cut it is nine feet, while at the west end the shale has entirely disappeared and the coal, so to speak, has worked upwards into the sandstone fully three feet, so that the interval between the coals is fully twelve feet less than it is midway in the cut, though the distance is but a few yards. Unfortunately the cut ends abruptly and the hill is cut off, so that the final result of this approximation cannot be ascertained; but if the lower coal continued in its course for ten yards farther, the two coals certainly came together. As the *Little Washington Coal* approaches the sandstone above or recedes from it, so it recedes from the limestone below or approaches it. Midway in the cut the underlying shale is only a few inches thick, whereas at the west end it is nearly twelve feet. The Limestone I "*b*," is ferruginous, weathering with a very smooth surface. It is the lowest rock exposed along the railroad within the township. This cut opens out on the National pike and the railroad enters Franklin township.

Away from the railroad, exposures are for the most part poor. The Claysville axis enters the township near the head of Brush run, and the trough east from it crosses near Mr. M'-Clay's residence, on the road leading from that run to Washington. The Upper Washington Limestone is occasionally seen on the higher hills in the western portion, though it is nowhere well exposed. The peculiar color of its fragments enables one to identify it without difficulty. In the same localities, fragments of the Middle Washington Limestone are seen, but this too is not well exposed in place. The upper limestone is fairly well shown near Mrs. Slemon's residence, in the north-west near Mrs. M'Kee's, at a little distance south from the last and near Mr. E. Wolf's, near the central part of the township.

The blossom of the *Washington Coal* is found at from one hundred and forty to one hundred and sixty feet below Limestone VI at many places, but the coal seems to be worked only on Mr. Taggart's property at the extreme north-west. There it shows a total thickness of five feet six inches and consists of seven divisions of coal and clay. Of coal there are in all four feet, but the bottom bench is not fully exposed and the thickness is most probably greater by six inches, which would make the amount of coal very nearly the same with that in the cut on the railroad. Of Limestone II, only two feet can be seen here, and three feet of shale come between it and the coal bed. The coal from this opening is slaty and sulphurous, though the bottom bench is said to be of moderately fair quality. It is used to but a slight extent, chiefly toward spring, when the supply of coal runs short and the roads are so bad that none can be hauled from the mines on the Pittsburg. In the south-eastern part of the township, this coal is exposed by its blossom in the road near Mr. S K. Weirick's residence, and at the cross-roads near Mr. J. Kelly's. At the latter locality, the blossom of the *Waynesburg* is seen at one hundred and five feet below the *Washington*. The bed is evidently very thin.

On the road leading from Washington to the head of Brush

run the *Waynesburg Coal* has been opened by Mr. D. M'Clay, who has it as follows:

Coal	6″
Clay and *Coal*	10″
Coal	1′ 0″
Clay	3″
Coal	4″

In all, two feet eleven inches thick. The coal is very poor throughout, and is no longer worked. The opening extends for nearly one hundred and fifty feet, having been pushed so far in the hope that the bed would thicken up under the hill.

On a road reaching this from the south, near Mr. J. Boon's residence, the following section was obtained:

1. Upper Washington Limestone	Fragments.
2. Sandy shale	20′
3. *Jolleytown Coal Bed*	Blossom.
4. Concealed	100′
5. Limestone II	15′
6. *Washington Coal Bed*	Blossom.
7. Concealed	50′
8. Sandstone and sandy shale	45′
9. Concealed to creek to level of *Waynesburg Coal Bed*	20′

This section was obtained very nearly on the line of the synclinal between the Washington and Claysville axes. The intervals show some interesting variations from those observed in Buffalo township, on Brush run.

39. Franklin Township, Washington County.

This adjoins Amwell on the west and Morris on the north. Chartiers creek flows through it from south to north, and tributaries to Ten-mile flow from each side of its southern half. The Washington axis passes across the northern corner, and the synclinal between that and the Pinhook axis barely touches the extreme south-east point. A slight local axis occurs in the north central part. The exposed section extends upwards from one hundred feet below the *Washington Coal*, to about one hundred and fifty feet above Limestone VI, but a shaft at Washington, in the north-east corner of the township carries it down to the *Pittsburg Coal*. That portion of the section above the Upper Washington Limestone is very imperfectly shown.

In the immediate vicinity of Washington, the following section was obtained on Cemetery Hill:—

1.	Limestone, seen	10′
2.	Argillaceous shale	5′
3.	*Coal*	1′
4.	Sandstone	10′
5.	Dark shale	8′
6.	Upper Washington Limestone VI	30′
7.	Concealed	50′
8.	*Coal*	Blossom.
9.	Imperfectly exposed	80′
10.	Limestone II, seen	12′
11.	*Washington Coal bed*	7′
12.	Clay	4′
13.	Sandstone	9′
14.	Concealed	10′
15.	Limestone I "*b*," seen	2′

The Little coal, No. 3, is the same with that seen at the tunnel east from Washington, and referred to in the description of South Strabane township. The detailed section of the Upper Washington Limestone has been given in another connection. It is a mass of limestone and shale and is quarried almost directly opposite the cemetery. The highest limestone of the section is only partially exposed with the little coal below in the cut on top of the hill, but is fully shown in a quarry on the other side. Fragments of the Middle Washington Limestone, IV, occur in a field below the toll-gate, but they were not found in place. In the lower portion of the interval No. 9, is a fine sandstone which is quarried. The *Washington Coal* is well exposed in the brickyard, where it consists of eleven divisions of coal and shale, the bottom coal being two feet ten inches thick. Its quality seems to be very inferior as the *Pittsburg Coal* is used in burning the brick. The laminated sandstone underlying the coal is blue, and crowded with fragments of carbonized wood.

A shaft was sunk upon the *Pittsburg Coal* near the depot of the Chartiers Valley railroad at Washington. It begins at ninety-five feet below the *Washington Coal*, which is exposed in the Hempfield railroad cut overlooking the depot. The section as given by Dr. Creigh, of Washington, is as follows:

1.	Soil and clay	4′	0″
2.	Gravel	5′	0″
3.	*Black slate*	1′	6″
4.	Limestone	4′	0″
5.	Blue shale	15′	0″
6.	*Coal*	0′	8″

7. Gray shale	6′	0″
8. Sandstone	5′	0″
9. Limestone and shale	170′	0″
10. *Black slate*	12′	0″
11. Gray limestone	13′	0″
12. Blue shale	50′	0″
13. Sandstone	15′	0″
14. Shale	3′	0″
15. *Pittsburg Coal bed*	5′	6″

Thus making the interval between the *Pittsburg* and *Washington Coals* four hundred feet. In south-eastern Greene county it is five hundred and forty feet. The Little Coal, No. 6, which is about one hundred and twenty-five feet below the *Washington*, is the *Waynesburg*, which in the western portion of the county is very thin and of no economical value. The great mass of limestone and shale No. 9 is utterly unlike anything obtained in the sections above ground. It is described as "Gray Limestone, interstratified with shales, varying in thickness from six inches to three feet." The sections along Chartiers creek below this indicate much limestone in this interval, but except at the base, where there are fifty feet of limestone, the shale and shaly sandstone predominate. I am inclined to think that the record had been condensed before it came into the hands of Dr. Creigh, as it seems hardly possible that so great a variation in conditions should occur within two or three miles. Near the middle of this mass, salt water was found in a white limestone. The absence of the *Redstone Coal* is not strange, although that bed is found in the Enterprise Shaft less than two miles away. It is so variable in this vicinity that it may be barely an inch thick here, and for that reason it may have escaped observation. No. 10 is at the horizon of the *Sewickley Coal*, and No. 13 contains vegetable remains as at Cannonsburg.

The coal obtained from this shaft was of excellent quality, and was mined to the extent of one thousand bushels per diem; but the works have been abandoned for the present, at least, as the proprietors do not feel justified in incurring the expense ot putting down the escape shaft, which is required by the mining law of the State. The estimated cost of the additional shaft is not far from twenty thousand dollars.

On the Hempfield railroad the *Washington Coal* is exposed in the first cut immediately beyond the trestle, and Limestone I "'b' is imperfectly shown at twenty-five feet below it. In the long cut, beginning near the National road and directly opposite the Chartiers depot, the coal and its overlying limestone are seen thus:

1.	Limestone and Shale II	33' 0''
2.	Shale	5' 0''
3.	*Washington Coal Bed*	6' 4''
4.	Laminated sandy shale	15' 0''

Limestone II consists of sixteen feet of limestone in numerous layers, interstratified with shale, which is more or less calcareous, and the shale between it and the coal contains many calcareous and ferruginous nodules. In the cemetery section the *Washington Coal* is seven feet four inches thick, while in the cut it is six feet four inches, but in both sections the amount of coal is precisely the same, being four feet eight inches. The proportion of pyrites cannot be very large, as at each of these localities the coal, though long exposed to the weather, shows little tendency to break up and is still compact. The chief drawback is the great quantity of ash, which renders it an inconvenient and by no means economical fuel.

This cut opens out on the National road about half a mile west from Washington. On a high hill on this road, Limestone VI has been quarried. It is exposed on the property of Mr. Moses Little, just south from the road, where also Limestone IV is seen in a ravine, with a fine flagging stone resting upon it. The latter limestone is exposed in the National road, near the township line, opposite the first tunnel west from Washington.

On the Waynesburg road, leading south from Washington, one soon comes down to the *Washington Coal* after crossing the Cemetery hill, and finds it alongside Chartiers creek, where the road forks. On the easterly fork, which leads past Point Lookout, the coal does not go under until within one-eighth of a mile from Mr. S. Wilson's old house, where Limestone II is seen fully eighteen feet thick. Near Mr. Wilson's residence Limestone IV is exposed to the thickness of six feet. It is quite ferruginous, and there is the blossom of a thin coal ten or fifteen feet below it. From this locality the road rises rapidly, and Limestone VI comes in sight, with a thick limestone above it. Neither of these is well exposed, and the road is covered with fragments for a vertical distance of nearly forty feet. About midway in this there is a coal blossom, the same with that seen on the Cemetery hill, near Washington. Ninety feet higher is another limestone, which has been quarried near Point Lookout. It is light-colored, apparently good and seems to be used for burning into lime. The interval between this and the mass below is concealed, nothing being exposed except a coal blossom at sixty feet below. Thirty-five feet above this limestone is another, and in the forks of the road at Point Lookout, ten feet above the last, fragments of a dark blue limestone, probably Limestone X, are seen. This is the divide between the waters of Ten-Mile and Chartiers.

Returning to Chartiers creek and following it from the point where the road forks, the *Washington Coal* is traceable to opposite Mr. J. Brownlee's house, where it passes under the stream. For a short distance it rises southward under the influence of the local axis already referred to, but at a little above the toll-gate the dip is reversed and the coal falls rapidly towards the south. On a road, turning east at three-quarters of a mile above the toll-gate, it was opened many years ago, but the opening is now deserted. The bed is said to be of very inferior quality and two feet thick, this referring no doubt to the available thickness of the coal. Limestone II occurs here in loose fragments.

About two-thirds of a mile above Mr. Brownlee's, Limestone IV is quarried for use on the road, and at twenty feet above it is the blossom of a coal, which is possibly that of the *Jolleytown*. Beyond this to the summit there are no exposures, but at the forks on the hill there are fragments of Limestone VI, and at Van Buren the next limestone is in the road. Turning northward on the divide, and taking down the first road leading eastward, a little tributary to Ten-Mile is reached. There are no satisfactory exposures here until at the tannery, the limestone above the Upper Washington is seen under a heavy mass of sandstone. The dip is quite sharp to the south-east, and this stratum continues in sight to the junction with the stream, which forms the boundary between Amwell and Morris townships. Here, at almost the extreme south-east corner, this limestone, including the several thin layers of shale, is ten feet thick. Under it is the little coal resting on a sandstone.

On the road following up this stream to Point Lookout, this limestone continues in sight to within a mile of Point Lookout school-house. Above it there are no exposures. The little coal below it was once worked by stripping on the property of Mr. W. D. Andrew, and at a cross-road about one-third of a mile farther down the stream, it is exposed in a cut where it seems to be two feet six inches thick.

In the western portion of the township, near the head of Ten-Mile creek, Limestone VI is in the forks of the road near Mr. J. M'Millan's house, and at twenty feet above it is the blossom of a coal. It is one hundred and seventy feet, by barometer, above the *Washington Coal.* Thirty feet below it is the blossom of the *Jolleytown Coal*, which indicates a thickness of not less than two feet. Near Mr. S. Walter's place Limestone II is in the bed of Ten-Mile creek, and at school-house No. 5 the greater part of it is exposed. Near Mr. J. Cooper's house it is altogether above the creek, and the blossom of the *Washington Coal* is seen under it. This coal was formerly mined by Mr. Cooper, but the opening is in such condition that only the following imperfect measurement could be made:

1. *Bituminous shale*	2′ 0″
2. Shale	3″
3. *Coal*	6″
4. Clay	2″
5. *Coal*	1′ 0″

The coal is very slaty and evidently poor fuel. On the little stream coming in here, the following section was obtained, beginning at the cross-roads just east from Mr. B. Craig's residence:—

1. Concealed	60′
2. Upper Washington Limestone VI	25′
3. Concealed	30′
4. *Jolleytown Coal bed*	Blossom.
5. Concealed	40′
6. Middle Washington Limestone IV	8′
7. Concealed	30′
8. Sandstone	20′
9. Limestone II	20′
10. *Washington Coal bed*, seen	3′

The *Jolleytown Coal* is double here, and shows a large blossom. Limestone IV has the ordinary character, being buff in the upper layers and light colored, somewhat brecciated in its lower portion. Limestone II occurs in unusually thick layers, and many of the blocks which have separated are of very large size.

The *Washington Coal* is exposed at the next cross-roads below and near school-house No. 6. The Washington axis crosses this road somewhat less than a mile from the line of Buffalo township.

40. Buffalo Township, Washington County.

This lies west from Canton and Franklin. Buffalo creek flows through it in a north-westerly direction and receives many tributaries from the north-east. The section extends upwards from the *Uniontown Coal* to eighty feet above the Upper Washington Limestone. The Claysville axis crosses Buffalo creek just below Taylorstown, and the synclinal south-east from it crosses the Hempfield railroad just west from school-house No. 2. Near the line between Buffalo and Franklin and three-fourths of a mile from the National road, Limestone VI is seen in the Ridge road. It is about twenty-five feet thick and yields good lime, for which it has been burned by Mr. J. Mountz. Nearly a mile farther south, at the head of the East fork of Buffalo creek, there is another limestone sixty feet above the last. Its upper portion is somewhat conglomerate, but the base is light colored and fine, breaking with a regular fracture. About half a mile farther west, Limestone II is a little way above the stream, fully twenty feet thick and showing some very massive layers at the base. Blocks of it lie scattered around which measure five feet in every direction. Under it is the blossom of the *Washington Coal*, and the dark shale above seems to be a coal at this locality.

From this point westward the creek falls rapidly, but the dip of the rocks is equally rapid, so that at the forks of the road, near Mrs. Gantz', only the top of Limestone II can be seen, while resting on it is the little coal. Above this is a massive sandstone, which has been quarried for building purposes.

Below this the dip slackens, so that where the National road crosses, the bridge is on the level of the *Washington Coal*, and at a short distance farther down the stream the *Little Washington Coal* is exposed, with the laminated sandstone over it. Both of these are much contorted. The underlying Limestone I "*b*" is soon seen, and continues in sight almost to the railroad, where the little coal is level with the track, and the debris of the *Washington Coal* shows itself at the top of the cut, while fragments of Limestone II are scattered on the hillside. The rocks are now rising very fast northward, and at the junction of the two forks of Buffalo creek a thin limestone is seen in the road. Its exact relation to the *Washington Coal* could not be determined, as there are no exposures in the vicinity, but it is probably the same with one seen opposite Taylorstown.

Turning off eastward where the railroad crosses the creek, a road follows up a branch of this fork to the line of Canton township. Its direction for more than two thirds of a mile is east of north, so that the *Washington Coal* is constantly in sight for that distance, rising somewhat more rapidly than the road. At the first fork the coal is in the road as follows:

1. *Washington Coal Bed* seen	2′ 0″
2. Shale and laminated sandstone	14′ 0″
3. *Little Washiugton Coal*	1′ 6″
4. Shale and shaly sandstone	16′ 0″
5. Limestone I "b" seen	6′ 0″

Two or three hundred yards farther east the coal is mined on property belonging to Mr. D. Hagerty, where it has a total thickness of eight feet nine inches, with this section:

Coal and shale	0′ 10″
Clay	2′ 0″
Coal and clay	3′ 1′
Clay	0′ 10″
Coal	2′ 0″

In No. 1 there are four layers of coal and clay, the two being of equal thickness; in No. 3 there are seven layers, of which four are coal, with an aggregate thickness of one foot ten inches, so that the total of coal in the bed is four feet three inches, very nearly the same as at Washington. The coal is of wretched quality, containing much sulphur and a great quantity of ash. From this place the direction of the road is eastward, and the rocks fall to the distillery, near school-house No. 2, where the dip is reversed. There Limestone II is in the road. No exposures occur until the tunnel on the township line is reached, where the section shows that the road has been rising more rapidly than the coal. It is,

1.	Sandy shale and sandstone	17′ 0″
2.	Limestone	2′ 0″
3.	Shale, dark at base	8′ 0″
4.	Limestone	1′ 6″
5.	Shale, with streaks of *Coal*	3′ 0″
6.	*Coal*	1′ 0″
7.	Laminated sandstone	14′ 0″
8.	Middle Washington Limestone IV	4′ 0″

At the east end of the tunnel No. 8, it is fifteen feet thick, and of a peculiar yellowish color. The section along the railroad, from the *Washington Coal* to this limestone, has already been given in the description of Canton township. The coal is nearly one hundred feet below the limestone, so that from the distillery to this point the rise in the strata is about fifty-five feet. South from the railroad the Upper Washington Limestone is seen before the National pike is reached.

On the West fork of Buffalo creek the synclinal crosses near the United Presbyterian church, where the Middle Washington Limestone appears in the road. At the National pike Limestone II is shown, and at a little distance beyond, the *Washington Coal* is exposed with Limestone I, "*b*" twenty-five feet thick, at a few feet below it. From this to the junction with the East fork there are no exposures. The West fork passes under the railroad at the east end of a very deep cut, and thence along the railroad to the south-western boundary of the township where cuts are numerous. In the deep cut the following section is exposed:

1.	Middle Washington Limestone IV	8′ 0″
2.	*Coal*	Blossom.
3.	Shale, sandy and argillaceous	33′ 0″
4.	Sandstone	1′ 0″
5.	*Coaly shale*	4′ 0″
6.	Limestone II	17′ 0″
7.	Shale and sandstone	14′ 0″
8.	*Washington Coal Bed*	10′ 2″
9.	Sandy shale to track	5′ 0″

The *Washington Coal* is in two main benches, separated by shale, with thin streaks of coal, four feet, and sandstone one foot. The upper bench is three feet ten inches, and consists of seven layers of coal and clay, containing in all one foot two inches of coal. The lower bench shows no clay, and is one foot four inches thick. About midway in the cut the lower part of No. 3, as well as the whole of Nos. 4 and 5, suddenly thins out toward the south, and the overlying rocks show an abrupt dip of nearly three degrees for a distance of somewhat more than one hundred and fifty yards. The little sandstone No. 3 contains much comminuted vegetable matter. The rocks are dipping southwardly quite rapidly. At the southern end of the cut a little stream passes under the track and exposes the *Washington Coal.* Between this and the *Little Washington Coal* below, the interval is only about six or eight feet, and the lower coal is eighteen inches thick. At barely one-eighth of a mile south from this cut the track enters tunnel No. 5, where the section is a short one, embracing only Nos. 1 to 5 of that just given. At the north end of the next tunnel Limestone IV is at the track level, and the following section is exposed above it:

1. Sandstone and sandy shale		14′ 0″
2. *Jolleytown Coal Bed.*		
1. *Coal*	4″	
2. Shale	6′ 0″	7′ 0″
3. *Coal* and black shale	8″	
3. Limestone		1′ 6″
4. Sandstone		15′ 0″
5. Shale		15′ 0′
6. Limestone		1′ 6″
7. Shale		5′ 0″
8. Limestone IV		8′ 0″

The National road passes over this tunnel. The Upper Washington Limestone cannot be more than a few feet above the top of the section, but it is not exposed. At a short distance from this point the direction of the road changes to west, and the strata rise, so that at the next tunnel, on the township line, Limestone IV is again at the level of the track.

Following down Buffalo creek from the junction of its forks no exposure occurs until Taylortown is reached, where a run comes in from the north-east. Two miles up this run the *Washington Coal* is worked by Mr. A. Henderson and Mr. I. Carson. The sections here are as follows:

1. Sandstone seen	10′ 0″	
2. Bituminous shale	1′ 0″	
3. *Coal*	5″	2″
4. Clay	8″	1′ 0″

5.	*Coal*	3″	4″
6.	Clay	6″	1′ 4″
7.	*Coal*	2′ 9″	2′ 4″
8.	*Coaly* shale	1′ 3″	1 3″

No. 7 is variable, sometimes reaching three feet in Mr. Henderson's bank. In another opening, a little way east from that, it is said to be three feet six inches, but no measurement was made because the entry had fallen in. The coal from the main bench is rather handsome and bright when freshly mined, and shows but little pyrites. The proportion of ash is evidently very great, and in bulk is said to be almost equal to the coal before burning. Still the coal is liked for use in grates, as, being free-burning, it is more easily managed than that from the *Pittsburg*. The best coal is about nine inches thick and comes from the bottom of No. 7. The coaly shale at the base has been used for burning lime and answers very well. Trials of it as fuel in grates have not resulted so satisfactorily. The upper benches are exceedingly poor, but are mined.

Descending the run to Taylortown, one sees at the first cross-road below these openings Limestone I "*b*," which shows the same characters here as at the exposure on the west fork near the railroad, being very bright buff above, but light-gray at the base. The run falls faster than the rocks, so that in a bluff near Taylortown the following section is exposed, beginning about forty feet below Limestone I "*b*":

1.	Fragments of limestone	——
2.	Shaly sandstone	25′ 0″
3.	*Coal*	Blossom.
4.	Limestone	3′ 6″
5.	Sandstone	15′ 0″
6.	*Coal*	10″
7.	Limestone	4′ 0″

No. 1, is the limestone seen in the road near the junction of the forks of the creek. The *Waynesburg Coal* is certainly not more than forty feet below the road at Taylorstown. On a road passing out westward from that village, the following succession is shown:—

1.	Limestone IV	10′
2.	Concealed	30′
3.	Dark shale	5′
4.	Limestone II	9′
5.	Concealed	24′

6. *Washington Coal bed*	6′
7. Laminated sandstone	12′
8. *Little Washington Coal*	1
9. Concealed	85

The coals Nos. 6 and 8, are exposed only by their blossoms, which are so deep that the coal cannot be reached for measurement. No. 9, is given by estimate. The true interval by barometer is one hundred and twenty-five feet, but the rocks are falling eastward, and the horizontal distance covered by that interval is nearly one mile. About half a mile below Taylorstown, a road comes in from the north-east, on which the *Waynesburg Coal* is exposed at forty-five feet above the creek. At twelve feet below it, there is a layer of bituminous shale, eighteen inches thick, and at thirteen feet lower is the limestone, three feet. Between this point and the summit, the following section is exposed:—

1. *Washington Coal bed*	Blossom.
2. Limestone I "*b*"	10′ 0″
3. Concealed	70′ 0″
4. *Waynesburg "a," Coal bed*	0′ 3″
5. Shale	2′ 0″
6. Sandstone	40′ 0″
7. Shale with thin limestone	2′ 0″
8. Concealed	3′ 0″
9. *Waynesburg main Coal bed*	2′ 5″

In the imperfectly exposed interval No. 8, there are indications of an upper bench to the *Waynesburg*. In No. 7, is the most easterly exposure of the limestone which, from this line westward into Ohio, accompanies the *Waynesburg Coal*. On the road next west from this, the interval between the *Washington* and *Waynesburg Coals* is one hundred and thirty feet. The latter is clearly double, but the exposure is not such as to admit of accurate measurement. Still farther down the creek, near Mr. W. Donahy's residence, the coal is in the road and shows:—

Coal, 4 inches; clay, 6 inches; *coal*, 2 feet 4 inches: Total, 3 feet 2 inches.

It is exposed where the road again comes down to the creek and shows a similar section. On the other side beyond the bridge, it is concealed. At half a mile below this bridge which which is the second below Taylorstown, a by-road turns off southward, on which the following succession is exposed:—

1.	Limestone in fragments	——
2.	Shale and sandstone	40′ 0″
3.	*Coal*	Blossom.
	3. *a* Shale	10′ 0″
4.	Limestone	6′ 0″
5.	Concealed	20′ 0″
6.	Limestone VI	12′ 0″
7.	Sandstone	15′ 0″
8.	*Jolleytown Coal bed*	0′ 8″
9.	Concealed	40′ 0″
10.	Limestone IV	15′ 0″
11.	Concealed	45′ 0″
12.	Limestone III	10′ 0″
13.	Concealed	5′ 0″
14.	*Coal* (?)	——
15.	Concealed, including Limestone II	25′ 0″
16.	*Washington Coal bed*	6′ 0″
17.	Concealed, including Limestone I "*b*"	35′ 0″
18.	*Waynesburg "b" Coal Bed*	Blossom.
19.	Concealed	40′ 0″
20.	*Waynesburg "a" Coal bed*	Blossom.
21.	Waynesburg sandstone, seen	10′ 0″

The section ends at the junction with the creek road. At the bridge beyond, the *Waynesburg Coal* is found fifty feet below No 21, so that the intervals are, from Limestone VI to the *Washington Coal*, one hundred and fifty-five feet, and from the latter to the *Waynesburg Coal*, one hundred and twenty-five feet. Nodules representing Limestone I "*a*" are seen above the *Waynesburg* "*a*," and fragments of Limestone I "*b*" and II occur in the intervals eighteen and sixteen. From this point to two-thirds of a mile beyond Noble's mill there are no exposures, but below that to Atchison P. O. the sandstones above and below the *Waynesburg Coal* are constantly in sight, and at forty feet below the place of that coal there is a bright yellow limestone, with a thin coal resting upon it. At Atchison the coal and limestone are under the creek, but on the other side of the store they come up, and the coal has been mined by stripping. This is the *Uniontown*, and the limestone is the upper division of the *Great Limestone.*

At Atchison Brush run enters the creek. It is the northern boundary of Buffalo township. Near its mouth are several deserted openings in the *Waynesburg.* For nearly a mile the stream flows from the north-east, and that coal is carried under by the rise of the bed, but at the boundary between Hopewell and Independence townships the course is changed to east and west, so that within a short distance the *Uniontown Coal* and twelve feet of the underlying limestone are above the creek. Somewhat less than two miles from Atchison openings in both the *Washington* and *Waynesburg* are seen in the same hill, the former belonging to Mr. Samuel Donaldson, and the latter to Mr. Charles Denning. The relations here are as follows:

1. *Washington Coal Bed*	5′	4″
2. Concealed	110′	0″
3. *Waynesburg Coal Bed*	6′	10″
4. Sandstone	40′	0″
5. *Uniontown Coal Bed*	1′	0″
6. Limestone to run	12′	0″

The *Washington Coal Bed* shows three benches, eight, six and three inches respectively. The coal is a very poor article, except for one foot at the base, which is said to be quite good. Little work has been done here, and the opening seems to be deserted. The *Waynesburg* is somewhat extensively mined by Mr. Denning to supply the adjoining country. It is in two well-marked divisions, which, however, have no clay parting, but show slickensided surfaces at the line of junction. The upper division varies from three to four feet in thickness, and contains at the bottom two benches of coal, separated by a thin clay parting. The upper part is merely a bituminous shale, with coaly streaks. The coal from this division is very poor, and can be burned only in grates. The lower division varies from three to four feet, and consists of two benches of coal thirty to thirty-six inches, and two to four inches respectively, separated by a clay parting two to eight inches thick. The upper bench contains much ash and sulphur, and cannot be used by blacksmiths. It is said to make an extremely hot fire, and the engineer of a mill on this stream regards it as the best steam producing coal in the county. It is used both in stoves and grates throughout this neighborhood. The dip of the bed in this opening is very irregular. Not far from the mouth of the entry a roll was found, which proved so serious that it had to be tunneled, and at the time of examination another roll had just been reached.

Half a mile farther up the run the *Waynesburg* was formerly worked by Mr. R. Barr, and on the road leading by that gentleman's house the *Washington* is exposed. The same section is shown also on a road coming down to the run at the old mill, about a mile and a half farther east. The two sections are as follows:

1.	*Washington Coal Bed*	Blossom.	Blossom.
2.	Concealed	12′	35′
3.	*Little Washington Coal*	Blossom.	
4.	Concealed	37′	
5.	*Waynesburg "a" Coal Bed*	2′	Blossom.
6.	Shale and limestone	6′	55′
7.	Sandy shale	30′	
8.	*Waynesburg Coal Bed*	2′	2′
9.	Sandstone	40′	40′
10.	*Uniontown Coal Bed*	Blossom.	Blossom.

The *Waynesburg* thins out rapidly eastward, and no other workings were seen in the township, though the bed continues in sight above the stream almost to the line of Canton township.

41. Donegal Township, in Washington County.

This lies directly west from Buffalo, and is crossed in its southern portion by the Hempfield railroad. The Dutch Fork of Buffalo creek flows northward through the township, and the creek itself forms the northern boundary. An obscure synclinal crosses the railroad at Coon's Island station and the Claysville anticlinal crosses it near the village of that name. The section extends from the *Waynesburg Coal* up to about seventy-five feet above the Upper Washington Limestone.

On the line between this and Buffalo townships there is a railroad tunnel, at which the following section is exposed:

1.	Debris containing *coal*	5′	0″
2.	Sandy shale	4′	0″
3.	Dark shale	4′	0″
4.	Limestone	7′	0″
5.	Dark shale	10′	0″
6.	*Coal*		4″
7.	Limestone IV	14′	0″

The Middle Washington Limestone IV, is more or less ferruginous, and for the most part exceedingly tough. The more compact layers weather to a dirty cream color and some of them are quite brittle. The slaty portions are almost drab, and in some instances contain ostracoids with fragments of what seem to be fish remains. The dip at the extreme west end of the approach to the tunnel is sharply eastward, but in the tunnel the rate diminishes, and at the east end is flexuous. At a short distance west from the tunnel, there is an old cut now overgrown with vegetation, but showing near the track a coal blossom, which is clearly the one seen under Limestone IV in Canton township, near the tunnel. The same coal is seen in a cut just east from Claysville, alongside of the National road. Beyond this there are no exposures until nearly a mile west from Claysville, where the railroad crosses the National road. There the *Washington Coal* is imperfectly exposed above the road. On the first road turning north from the pike, west from this crossing, the *Little Washington Coal* is seen with Limestone I "*a*," at ten feet below, but the *Washington Coal* and all else are concealed. On the next road, which is about half a mile east from Coon's Island station, the section is :—

1. Sandstone			10′ 0″
2. Concealed			30′ 0″
3. Limestone III, seen			2′ 0″
4. Concealed, with dark shale at base			10′ 0″
5. Sandstone			6′ 0″
6. Concealed			14′ 0″
7. Limestone II, seen			0′ 6″
8. Concealed			4′ 0″
9. *Washington Coal bed.*			
	Bituminous shale	1′ 0″	14′ 8″
	Sandstone and shale	3′ 0″	
	Bituminous shale	0′ 2″	
	Laminated sandstone	9′ 0″	
	Coal	1′ 6″	
10. Clay			0′ 4″
11. Laminated sandstone			9′ 0″
12. Argillaceous shale			8′ 0″
13. *Coal, Little Washington Bed*			1 1″
14. Shale to bridge			4′ 0″

No. 3 is Limestone III, and No. 7 is Limestone II. The extraordinary degradation of the *Washington Coal* is the peculiar feature of the section. There is, of coal, only the thin lower division, while all the upper portion is represented by only fourteen inches of bituminous shale. The sandstone above the coal is the same in character and position with that seen in the deep cut on the railroad in Buffalo township. But there it is only one foot thick. The two laminated sandstones of the section are precisely similar, being almost shale, and containing much carbonized vegetable matter in minute fragments. The thickness of Limestone II cannot be determined, as, with the exception of the six inches, it is altogether concealed. The presence of the dark shale at the base of No. 4 leads me to suppose that possibly the whole interval may be occupied by a limestone or calcareous shale representing it. The full thickness of Limestone III is exposed.

At Coon's Island station the *Washington Coal* is exposed, by its blossom, in the road above the mill. A small excavation, east from the mill, shows the blossom of the *Little Washington Coal* separated by ten feet of slate-colored shale from Limestone I "*b*," which, as far as exposed, is of a bright yellow color. The *Washington Coal* is again seen on the railroad about one-third of a mile beyond the station with eighteen feet of shale above it. Fragments of Limestone II are numerous on the hill-side, but cannot be found in place.

Two-thirds of a mile north-west from Coon's Island station the *Washington Coal* is mined by Mr. John O'Connor, on a branch of Dutch Fork, where it shows the following structure:

1. Shale	1′ 0″
2. *Coal*	8″
3. Clay	9″
4. *Coal*	8″
5. Clay	8″
6. *Coal*	9″
7. Clay	6″
8. *Coal*	2′ 6″
Total	6′ 6″

At nearly two hundred and fifty yards east from this opening, the *Little Washington* is seen at the saw-mill resting on ten feet of argillaceous shale, beneath which one sees seven feet of Limestone I "*b*." The limestone is thirty feet lower than at the exposure at Coon's Island.

The *Washington Coal*, at this locality, bears a striking resemblance to the *Waynesburg* as seen in Greene county, both in general structure and in the character of the coal. So exact is the resemblance, that it has been mistaken for that bed by excellent geologists, and was referred to the *Waynesburg* in the report on the First Geological Survey of the State. This error was excusable and even natural under the circumstances. Limestone I "*b*" is seen below, holding the same relation as does the underlying limestone to the *Waynesburg*, while here Limestone II is at an unusual distance above the coal, is thin and much like the one over the Waynesburg in the Panhandle of West Virginia. In the Panhandle, only six miles away, the *Washington Coal* is very thin, and the limestones associated with it attain their full thickness, so that the bed has little resemblance to itself as exposed on Dutch Fork. In the absence of connected sections, the mistake was almost unavoidable, and it reflects no discredit upon the geologists who made it, for they labored under many and serious disadvantages, from which geologists of the present day are happily free.

The outcrop of the *Washington* at the O'Connor opening looks like dark shale, and the coal contains a very large proportion of ash. The slaty layers are so numerous as to give the coal the appearance of semi-cannel, and of course to render it open-burning. Besides being slaty, it is sulphurous, so that it is not in high repute, though used to a slight extent in the immediate neighborhood. Most of the inhabitants prefer to purchase the *Pittsburg Coal*, which is brought from Wheeling by the railroad company. Directly opposite this opening a section is shown on the railroad, the base of which is fifteen feet above the coal, the interval being concealed. It is,

1. Limestone III	2′
2. Sandy shale	15′
3. Clay shale, dark at base	5′
4. Limestone and Shale II	6′
5. Shale to track	7′

No. 1 is clearly the same with that which occurs in the road east from Coon's Island, and which in that section I have called Limestone III. It is quite compact, dark-colored within, but weathering pinkish white. When struck, it splinters and yields angular fragments. Over it is a dark shale. The grade of the road rises rapidly from this point to West Alexander, and in the next two cuts, which are close together, the following succession is seen:

1. *Coal*	0′ 6″
2. Shale	2′ 0″
3. Limestone	7′ 0″
4. Red sandy shale	15 to 22′ 0″

5.	Limestone IV	6′ 6″
6.	Red shale	40′ 0″
7.	Limestone III	2′ 0″

At the end of this cut, West Alexander is in sight, but owing to want of exposures anywhere, there is difficulty in joining the section just given with that obtained in the tunnel under the village. The interval between the two can scarcely be more than ten feet, and is evidently occupied by sandstone. The exposure in the tunnel approaches is:—

1.	Sandstone seen	3′ 0″
2.	Shale	8′ 0″
3.	Limestone	6′ 0″
4.	Shale	8′ 0″
5.	Limestone	3′ 0″
6.	Dark shale	6′ 0″
7.	Sandstone	12′ 0″
8.	Limestone and shale	7′ 4″
9.	Sandstone seen	15′ 0″

This is the only locality in the whole district where any uncertainty exists respecting the Upper Washington Limestone. No. 3, of the section before the last, has the characteristics of that rock. It is dark within, weathers white and has the little coal above it, but the interval between it and the Washington coal is only one hundred and twenty feet, which is much smaller than at five miles east, whereas locally it ought to be somewhat larger, as the intervals have shown a tendency to increase toward the central lines of the troughs. The next limestone above No. 8 of the last section is at more nearly the calculated interval, but does not exhibit any of the characters of the bed. No. 5 of the former section is beyond doubt the Middle Washington Limestone, although much degraded in character, being very earthy. In the tunnel section, No. 7 bears a striking resemblance to the laminated sandstone always found underlying the Washington Coal. It has the peculiar grayish color, as well as the fragments of carbonized wood.

Along the line beginning just east from Coon's Island station, and continuing to a little beyond West Alexander, all the limestones of the series, as far as exposed, are thin and earthy. Limestones III and V of the tunnel section cannot be identified with any observed elsewhere. Just north from the railroad at West Alexander a little coal was found in sinking a well. It is the one overlying the Upper Washington Limestone in the section before the last.

In order, if possible, to ascertain the relations of these rocks the section was carried several miles into West Virginia, and the following succession was found in passing from the tunnel at West Alexandria to the station of Valley Grove, two or three miles west from the State line:

1.	Concealed	40′	0″
2.	Limestone IV	20′	0″
3.	Sandstone and shale	64′	0″
4.	Limestone II	20′	0″
5.	Shale	5′	0″
6.	*Washington Coal Bed*	1′	3″
7.	Laminated sandstone	14′	0″
8.	*Little Washington Coal Bed*		10″
9.	Shale	8′	0″
10.	Limestone I "a"	12′	0″

The section begins a little below the level of No. 8 of the tunnel section, so that the interval between it and the coal is about one hundred and forty feet. This is the same as the interval obtained on the other side in Pennsylvania. Limestone IV shows the usual character, being a ferruginous rock, and therein differs greatly from its nature as shown in the cuts between Coon's Island and West Alexander. Limestone III is absent.

The *Waynesburg Coal* is reached at the Point Mills, in West Virginia, and the interval between it and the *Washington* is ninety-six feet. Farther south, toward Wheeling, the interval between the *Pittsburg* and *Waynesburg* is two hundred and sixty-five feet, as ascertained by me in 1872. The details of these sections are given elsewhere in this report and need not be repeated here.

The distance of the *Pittsburg Coal* beneath the railroad station at West Alexander is not far from five hundred feet. The *Waynesburg* can be reached at two hundred and sixty-five feet less, but is not worth seeking, since, as exposed in the Ohio Panhandle of West Virginia, it is seldom more than three feet thick and of by no means superior quality.

North from the railroad in Donegal township exposures are few. Both the *Waynesburg* and *Washington* appear to be so thin as to be of no economical importance, and no openings were seen in either coal. The Upper Washington Limestone occurs on all the hills, and the horizon of the *Waynesburg Coal* is reached only on Buffalo creek and its Dutch Fork.

On the latter stream two and a half miles north from the railroad, a little run comes in by the Baptist church. Up this, say half a mile, the *Waynesburg* is seen in the bed of the run, associated as follows:

1. Dark blue limestone	4′ 0′		
2. Sandstone	6′ 0″		
3. Shale	1′ 0″		
4. Limestone	1′ 6″		
5. *Waynesburg coal bed.*			
Shale, with streaks of coal	3′ 0″	}	4′ 0″
Coal seen	1′ 0″	}	

At one hundred and twenty feet measured up the dip above No. 1, the blossom of the *Washington Coal* is seen near the head of the run. Neither of the coals is mined even by stripping. Below the mouth of Dutch Fork the north-west dip of the rocks is reversed, and they rise in that direction along Buffalo creek.

At the mouth of Dutch Fork the *Washington Coal* is one hundred and thirty feet above the creek, but at the West Virginia line there are exposed sixty feet of the great limestone, and one mile and a quarter farther down the stream the *Pittsburg Coal* shows itself in the bed of the creek.

On a road coming down by Mr. Ralston's residence to a stream entering Buffalo creek, near the north-east corner of the township, a section was obtained almost precisely similar to that found in Buffalo township, about two and a half miles farther east. The *Washington Coal* is fairly exposed here, but is not worked.

42. Mount Pleasant Township, in Washington County.

This lies west from Cecil, and north-west from Chartiers and Canton. The dividing ridge between the waters of Chartiers creek on the east, and those of Raccoon and Cross creeks on the west, passes irregularly north and south through the central portion. For the most part the rocks are concealed by debris, and such exposures as do occur are usually far from satisfactory. The section extends upwards from the *Pittsburg Coal* to Limestone VI, but the latter is found only in fragments on top of the highest hills.

The *Pittsburg Coal* is mined near the Chartiers line, on the Hickory and Cannonsburg road, by John Maretta and W. S. White. The measurements at these banks are as follows:

1. Shale and *coal*	0″	4′ 0″
2. *Coal*	8″	1′ 6″
3. Clay	8″	1′ 11′
4. *Coal*	3′ 6″	2′ 10′
5. *Coal* and partings	5″	4″
6. "Brick" *Coal*	9″	1′ 0″
7. "Lower Bottom" *Coal*	1′ 6″	1′ 1″
Total	7′ 6″	8′ 8′

As these openings are barely half a mile apart, the variation is extreme. At both banks the coal is of very fair quality, containing little pyrites and not much ash. The proprietors claim that it is free burning in Nos. 4 and 6. This no doubt is true respecting the upper portion of No. 4, but the rest contains some thick layers of bituminous coal, so that, while it can hardly be called caking, it cannot be regarded as open-burning. The lower bottom is usually quite free from sulphur, but being too brittle to bear handling, it is not mined. At both of these openings clay veins varying in thickness from one inch to two feet, are numerous and annoying. Less than one-fifth of a mile above Mr. White's bank the coal disappears under the run, and the *Redstone Coal* is seen in the road at a short distance beyond, twenty feet above the *Pittsburg*. On this road, within half a mile of Hickory, a finely brecciated, somewhat fossiliferous limestone is exposed, having a bituminous shale at five or ten feet above it, which may represent the *Sewickley Coal*. It is seventy-five feet above the *Pittsburg*. From this point to Hickory everything is concealed, but in that village, near the summit of the hill the *Waynesburg* is seen, two to three feet thick, and two hundred and thirty-five feet above the *Pittsburg* where last exposed. This is not far from the true interval, for the Claysville anticlinal crosses a little way east from Hickory, and at that village the rocks are falling westward quite rapidly. On the road leading from Hickory to Washington there are no exposures until near the Chartiers line, where the Great Limestone is seen. The *Sewickley* is reached just beyond the township line, and is represented by a bituminous shale.

Just west from Hickory, where the road forks, the *Waynesburg* is shown. At the first fork in the West Middletown road, south-east from this point, part of the Great Limestone is exposed, but what portion of it cannot be determined, as the rest of the section is concealed. There are no exposures along this road to Cross creek, nor along the creek to the township line. The road turning northward from the creek, just above the township line, shows nothing until it meets the road leading to Cross Creek village, where the *Washington Coal* makes a large blossom. At the Presbyterian church, about one-third of a mile from this fork, the same coal shows a great blossom, and cannot be less than six feet thick. On the lane leading from the church to the Burgettstown road the following section is seen:

1. *Washington Coal Bed*	5′
2. Shale	10′
3. *Little Washington Coal Bed*	Blossom.
4. Shale	12′
5. Limestone I "a"	15′
6. *Coal*	——
7. Shale imperfectly exposed	48′
8. *Waynesburg Coal Bed*	Blossom.
9. Sandstone	10′

On the Burgettstown road there are no exposures, but the stream along which it leads—the West Fork of Raccoon creek—evidently falls less rapidly than the rocks, for on the road leading from this creek to Rankin's schoolhouse, the following succession was observed:

1. Concealed	50′
2. Limestone IV	20′
3. Concealed with fragments of Limestone II	30′
4. *Washington Coal bed*	5′
5. Concealed with fragments of Limestone I "b,"	40′
6. *Coal*	—
7. Concealed to the road	30′

The Waynesburg is but a few feet under the road at the base of the section. Crossing the summit one finds Limestone IV, in the road at the school house. The exposure is imperfect, but it shows that the amount of limestone must be very great, since for a distance of nearly forty feet that rock seems to be continuous. The *Washington Coal* is exposed at a little way below the school house, and from that place down to the East fork of Raccoon creek the following section is in the road:

1.	*Washington Coal bed*	Blossom.
2.	Imperfectly exposed, but much limestone	30′
3.	*Coal* (?)	——
4.	Limestone and shale not well exposed	8′
5.	*Waynesburg Coal "a"* (?)	Blossom.
6.	Concealed	30′
7.	Limestone blue	4′
8.	Shale	8′
9.	*Waynesburg Coal bed*	3′
10.	Sandy shale	20′
11.	*Uniontown Coal Bed*	1′
12.	Clay	2′
13.	Limestone, bright yellow, seen	6′

I have given these sections in detail, imperfect as they are, because they exhibit very clearly the relations of the coals, and show the extraordinary thinning out of the intervening rocks. The *Waynesburg Coal* was formerly worked here by J. H. Miller, but the bank has been deserted for two years, as the bed is thin and mining is difficult. The section, as far as could be determined, is

Coal, 4 inches; clay, 2 inches; *coal*, 16 inches; clay, 2 inches; *coal* seen, 6 inches.

No. 5 is said to be one foot thick, and the coal resembles that usually obtained from this bed.

On this creek there are few exposures down to the township line. The stream runs through the Great Limestone all the way, and the *Sewickley* is seen at Cherry Valley Postoffice, on the line. At this point a branch of the creek comes in, on which the Great Limestone is exposed for more than two miles.

At a little distance west from Primrose station, on the Panhandle railroad, the *Pittsburg Coal* is mined by the Robbins Block Coal Company. The roof division is not exposed and the lower is as follows:

1.	*Coal*	2′ 9″
2.	*Coal* and partings	4½″
3.	"Brick" *coal*	8″ to 1′ 3″
4.	"Lower bottom" *coal*	9″ to 1′ 6″

The whole bed is mined here for shipping. The lower bottom is somewhat variable in quality, but the poorer portions are easily separated. At the mouth of the main entry the coal was a block, whence the name given to the works, but at a short distance the character changed and it became bench coal. At the same time the structure is unusual, as the partings are not persistent and the coal as a whole bears marked resemblance to semi-cannel. For about fifteen inches on top the coal is handsomely open-burning, but the lower portions contain much bituminous coal, and cake. Films of calcium carbonate are found on the faces of the coal. Clay-veins and horse-backs are frequent and annoying. The dip is irregular and swamps cause much trouble to the miners.

43. Robeson Township, in Washington County.

This is the most northern township on the Allegheny county border. Robeson run, along which the Panhandle railroad passes, is its southern boundary. The geology is uninteresting excepting in this, that the northern line of outcrop of the *Pittsburg Coal* crosses the township. The section reaches nearly three hundred feet down into the lower barren series and extends upwards to the Great Limestone. On the railroad, the *Pittsburg Coal* is mined near M'Donald's station. At the Briar Hill Coal Works the bed consisted of first rate block for a short distance from the mouth of the main entry, but soon changed into the ordinary bench coal. The only block now found in the pit is a layer about fifteen inches thick and in the upper bench. Even this contains much bituminous coal. At the Laurel coal mines, in the same vicinity, the general structure of the bed is similar. The total thickness of the lower division varies from four feet to four feet six inches. The partings are not regular or persistent, and clay-seams and horsebacks are very troublesome.

For two or three miles north from the railroad the coal is not opened, but is frequently exposed at the roadside and makes a large blossom. It is well shown near North Star P. O., and is worked in Allegheny county about a mile north from that place.

About a mile north-east from Bavington the *Pittsburg* is mined by Mr. Wilson, near the head of Karr's run, where the roof division is six feet thick. It shows five benches of coal, in all five feet, the thickest clay parting being only three inches. The main clay parting averages nearly two feet. The lower division is only partially exposed, but it shows a feature which is seldom seen, except in the northern portion of the district. At about one foot below the clay, there is a thin but constant parting, sometimes two inches thick. This was observed at but one exposure in Greene county and at none in this county along the river, but it occurs not infrequently along the Pan-handle railroad. The great thickness of the roof division shows that this exposure, though so near the extreme northern outcrop, is very far from the northern line of the bed, as it existed before it had been subjected to erosion. The *Pittsburg Coal* must have reached quite to Lake Erie. At this bank clay-veins cut the coal very badly, and some of them extend from the under-clay through the coal into the overlying shale.

The most northern exposure of the *Pittsburg Coal* is on the property of Mr. Bigger, near Raccoon creek, and nearly one mile south from the Beaver county line. It was worked here at one time, but being at the top of a high hill and without sufficient cover, the quality was poor and the bank was abandoned. No measurement of the coal could be made, but the blossom shows that the roof is very thick. Mr. White gives the following as the section below the coal:

1. *Pittsburg Coal Bed*	10′ 0″
2. Concealed	160′ 0″
3. Sandstone	50′ 0′
4. Shaly sandstone	40′ 0″
5. Bituminous shale	10′ 0″
6. *Coal*	3′ 7″
7. Sandstone and shale	35′ 0″
8. Crinoidal Limestone	4′ 0″
9. *Coal*	1′ 4″
10. Concealed to the creek	25′ 0″

No. 3 is somewhat coarse-grained, but is easily dressed, and is quarried for building purposes. The coal, No. 6, has been opened on Bigger's run by Mr. Coventry, who mines it by stripping as well as by drifting. It is divided by a clay parting, one inch thick and one foot from the top. The coal is good and is preferred to that from the *Pittsburg* by many of the farmers. The crinoidal limestone has a greenish color, and is literally crowded with fragments of crinoids and brachiopod shells. It burns into a strong but rather dark lime.

44. Hanover Township, in Washington County.

This occupies the extreme north-west corner of the county, and lies north from Smith and Jefferson. The dividing ridge between the waters of Raccoon and Wheeling creeks passes almost north and south through the middle of the township. The section extends from about three hundred feet below the *Pittsburg Coal* to the Great Limestone. Exposures of the lower barren series are in all cases unsatisfactory.

The extreme north-western limit of the *Pittsburg Coal* is the Ridge road leading from Frankfort to Paris, but the limit of available coal may be regarded as marked by the road leading north from Florence to Frankfort as the west line, and the Pittsburg and Steubenville pike west from Florence as the north line. In the triangular space bounded by these three roads the coal is caught by the higher hills, and there are several small patches which will be of local service for years to come. In the western portion of the township the coal is found only on the ridge, as south from the Steubenville pike to the railroad erosion has been so effective as to remove all the upper beds. The general south-east dip brings the rocks down so that in the eastern portion the coal is available over a more extensive area.

Between Florence and Frankfort the road follows the ridge, and the *Pittsburg Coal* is frequently exposed. The only opening seen is one-fourth of a mile north from the former village and seventy feet below the pike at that place. It is deserted. The village of Frankfort is in Beaver county, its southern end being, perhaps, two hundred yards beyond the Washington county line. On a road leading south-east from the village the *Pittsburg Coal* is mined by several persons to supply the neighborhood. At Mr. John Frazier's opening the following section was obtained:

1. Roof division	2′	0″	5′ 11″
2. *Coal*		9″	
3. Clay		2″	
4. *Coal*	1′	1″	
5. *Coal* and partings		3″	
6. *Coal*	1′	8′	

The bed is under a thin but sufficiently sound cover, and the coal is said to be very good, or at least as good as any obtained from Florence, where the cover is much thicker. It is quite free burning, contains comparatively little ash, and shows pyrites only near the top. On the other side of the hill there is at seven feet above the *Pittsburg*, a bed of cannel four feet thick, containing much ash, but only a small proportion of sulphur. At one time it was mined on Mr. Frazier's property, but the bulky ash rendered the stuff unsalable, and the opening was abandoned. Owing to the utter lack of exposures in the vicinity, it is impossible to fix the place of this coal with any degree of accuracy, but I am inclined to think that it is the *Sewickley*.

On the ridge road leading from Frankfort to Paris the coal is occasionally exposed, and at about two miles from the former village it is worked. It is possible that the coal may be found in the tops of two hills lying north from this road, but if so, it will be utterly worthless, as the rocks are rising very rapidly to the north-west.

At Jackson's mill, on the South fork of King's creek, the road is three hundred feet below the *Pittsburg Coal*. No connected exposures can be found between the coal and this place, but in the interval, there are limestones at sixty and two hundred feet, and a coal, fifteen inches thick, at two hundred and fifty feet below the *Pittsburg*. Coming up from the creek to the Steubenville pike, the blossom of the *Pittsburg Coal* is seen near School-house No. 5, and on the pike the coal is worked at the first fork west from Florence. There the following section is exposed:

1. Sandstone.		
2. *Coal*	1′ 10″	3′ 10½″
3. *Coal* and partings	4½″	
4. "Brick" *coal*	10″	
5. "Lower Bottom" *coal*	10′	

There is no roof here, and the sandstone comes down directly upon the coal. Twenty feet higher is a nodular limestone, evidently very thin, and at twenty feet above that is another of similar character. At Florence is a high knob rising far above anything for miles around, which shows near its top a well-marked bench. This may be the horizon of the *Waynesburg*

Coal, but no evidence can be obtained bearing on this matter. There is no other point in the township where that coal is reached.

About half a mile east from Florence, on the pike, the *Pittsburg Coal* is mined to supply the village. One of the banks shows the following section:

1. Roof division	1′ 8″	
2. *Coal*	1′ 3″	
3. *Coal* and partings	5¾″	
4. "Brick" *coal*	1′ 1″	3′ 6″
5. Parting	2″	
6. "Lower bottom" *coal*	6″	

The roof is simply clay containing thin streaks of coal, which sometimes combine to form a layer several inches thick. The coal from the lower division at these openings is of very fair quality, and is rather open-burning. There is also an opening just South from Florence, but it shows no material difference from these.

Nearly a mile farther east, a road turns off to the north and crosses a branch of Raccoon creek. On this a deserted opening was seen at seventy feet below he pike. A bank in similar condition was found on the other side of the creek at a mile north from the last, but neither of these was in condition to afford any information. No exposures occur on the creek, but on a road, coming down to it from the north near the saw-mill, the *Pittsburg Coal* was formerly worked, several deserted openings having been seen on it.

On the Steubenville pike east from Florence a thick, bituminous shale was observed at many localities. It is seventy feet above the *Pittsburg Coal*, and is associated with sandy shales above and below.

45. Smith Township, Washington County.

This lies south from Hanover, and south-west from Robeson. The two forks of Raccoon creek flow through its southern portion, and unite at the railroad which crosses the northern part of the township. The Bulger anticlinal is seen in the north-east, and the synclinal west from it crosses the railroad at Burgettstown. The section extends from one hundred feet below the *Pittsburg Coal* to the Upper Washington Limestone, and some of the local sections obtained here are of very considerable interest.

Along the Panhandle railroad the *Pittsburg Coal* is extensively mined for shipment, and the largest works are at Midway station, near the line of Robeson township. At the Midway Block Coal Works the workable portion of the bed varies from three feet six inches to four feet, and rests on the "bottom" which is from ten to fifteen inches thick, and of such poor quality that it is not mined. The workable portion shows no partings and is a true block coal, yielding masses of any desired size, which show the full thickness of the bed. Many blocks, measuring two feet each way, were lying on the dump when the works were examined. The coal is rather more than semi-cannel, but contains a good deal of bituminous coal in streaks from one-third to one-half inch thick. It is extremely clean and is commonly regarded as the purest coal obtained along the road. The raw coal has been tried at Sharon Furnace in the proportion of two-thirds coke and one-third Midway coal, and the experiment is said to have been thoroughly successful. It is supposed that a less proportion of coke would yield equally good results, but the amount of bituminous coal present is sufficient to render the success of such an experiment somewhat doubtful. The percentage of sulphur is small, and the coal has proved well fitted for use in the manufacture of illuminating gas. The coke is quite brittle, as might be expected from the character of the coal. The roof division is not exposed. Clay-veins and horse-backs are numerous, the latter coming in both from above and from below. On the north side of the track the bed shows the partings and contains less of the block coal. Only three feet of it there are regarded by the owners as marketable coal.

On the other side of the station, at the Walnut Hill Coal Works, the coal is a block, not much if at all inferior to that from the works just described. The variations are the same, for where the coal is a block it shows no partings, and is from three to four feet thick. There is no bottom, this being the full thickness from roof clay to bottom clay. The ordinary bench coal comes in "cat-a-cornered," as the miners express it, and when it does come in it shows the following structure:

Coal, 2 feet 10 inches; *coal* and partings, 3 inches; *coal*, 1 foot 6 inches; total, 4 feet 7 inches.

The roof division is not exposed, and the "brick" cannot be distinguished from the "lower bottom." The slack is coked in ovens, of which there are eight, and the coke is surprisingly compact, considering the character of the coal from which it is made.

In a cut near these works the following short section is shown:

1. Limestone, yellow, and shale	9′
2. Dark shale	12′
3. Sandy shale	3′
4. Concealed to coal	45′

In the next cut west, twenty-five feet of dark shale are seen resting on the limestone. The limestone is constantly in sight until Bulger station is passed, and is the limestone exposed in the approach to the tunnel just west from that station. At the tunnel a light colored limestone rests on the dark shale above the last, while at seventy feet higher there is another limestone, the interval being concealed. From this tunnel the track falls very rapidly, so that at a mile and a half west from Bulger the *Pittsburg Coal* is again seen at a few feet below the track, at Mr. Russell's banks. Here the bed shows a structure so remarkable that I give it in detail as follows:

1. Limestone, yellow		Fragments.
2. Shale, with nodular limestone		8′
3. Sandy shale		35′
4. Roof division.		
A. *Coal*	2′ 3″	8′ 5″
B. Clay	10″	
C. *Coal*	2″	
D. Clay	4″	
E. *Coal*	3″	
F. Clay	10″	
G. *Coal*	3″	
H. Clay	4″	
I. *Coal*	1′ 1″	
J. Clay	2″	
K. *Coal*	1′ 1″	
5. Main clay parting		0′ 10″
6. Lower division.		
A. *Coal*	1′ 2″	5′ 0″
B. Clay	2″	
C. *Coal*	2′ 0″	
D. Clay	½″	
E. *Coal*	2″	
F. Clay	⅓″	
G. "Brick" *Coal*	5″—8″	
H. "Lower B." *Coal*	1′ 0″	

The parting B. in the lower division is well defined and persistent throughout the entry. The top of this division is slickensided. The coal was once worked in the hollows adjoining, but the banks have been deserted. At Raccoon station the coal comes to the level of the track, and in a cut beyond I find:

1. *Coal*	5′
2. Concealed	10′
3. Limestone at track	2′
4. Sandstone	15′
5. Limestone	5′
6. Shale and sandstone to creek	40′

No. 1 is part of the roof division and is unbroken save by very thin partings, which are barely perceptible, and in its structure bears much resemblance to the ordinary appearance of the lower division. Respecting the latter, one cannot speak positively, as it is concealed by debris, but the total thickness of the whole bed cannot be less than fifteen feet. About one-fourth of a mile west from this station the coal is mined by Dr. Dunning, at whose bank the lower division varies little from five feet, shows the usual partings and is rather more bituminous than the coal from the openings already described. Just west from Burgettstown station the coal is worked by the Whittaker heirs, at whose opening the structure of the bed is quite as peculiar as at that belonging to Mr. Russell. I give it in detail:

Pittsburg Coal Bed.

A. *Coal*		8″	7′ 9″
B. Clay		4″	
C. *Coal*	1′	0″	
D. Clay		4″	
E. *Coal*	1′	0″	
F. Clay		4″	
G. *Coal*		3″	
H. Clay		2″	
I. *Coal*	1′	9″	
J. Clay and *Coal*		5″	
K. *Coal*	1′	6″	

It is difficult here to determine the exact line of division between the roof and lower portion of the bed, but I am inclined to regard it as occurring at No. H. The coal contains much slaty matter and mineral charcoal, but is said to be good, and is in fair repute in the vicinity. Near the mouth of the pit there is a clay vein, which exhibits the usual conditions very well. The vein is not more than six inches wide, but its influence is perceptible for two feet and a half on each side. The coal is thrown up sharply and without any snapping at the angles, but the upturned portion is badly shattered.

From Bulger to this place the dip has been westward, but here the direction changes. At the same time the grade of the road changes, so that before reaching the Dinsmore tunnel the track is above the coal, but at the tunnel the road again descends westward, and at the west end the coal is two feet above the track. The former rises and the latter falls quickly, so that at the end of the tunnel cut the coal is thirty feet above. The following section was obtained at the tunnel:

1 Concealed	90′
2. Sandstone	40′
3. Dark shale	2′
4. Variegated shale	10′
5. Limestone	4′
6. Sandy shale	10′
7. *Black shale*	2′
8. Clay, with nodular limestone	43′
9. *Pittsburg Coal Bed*	—
10. Clay	4′-5′
11. Limestone	0′-2′
12. Sandstone	20′
13. Limestone	6′ to 6″
14. Calcareous clay	4′

The clay No. 8 is not laminated. The black shale above it is all that is left of the *Sewickley Coal*. The limestones Nos. 11 and 13, vary abruptly; No. 11 thins out and runs into shale, while No 13 is cut out by the overlying sandstone. The coal is singularly variable. Near the west mouth of the tunnel it consists of thirteen layers of coal and clay, so arranged that it is impossible to determine where the line falls between the roof and the lower division, while at a little distance the whole section of the bed is:—

Coal, 1 foot; clay, ½ inch; *coal*, 11 inches. Total, 1 foot, 11½ inches.

The bed is badly cut up by horsebacks from below. One of these has removed the lower portions for a horizontal distance of thirty-five feet, and is associated with a clay-vein which breaks through the coal above and unites with the overlying clay. Sections of several of these horsebacks show them to be depressed-conical in shape, and to be frequently associated with clay-veins. These clay-veins must be of much later origin than the coal, and clearly date from some time after its consolidation. It may be that they mark lines where the coal yielded during the folding of the strata. Beyond this tunnel the *Pittsburg Coal* soon rises well up in the hills, and the openings are at a considerable distance from the railroad.

In the northern part of the township there are few exposures. The *Pittsburg Coal* was seen on the road from Burgettstown to Candor, about a mile from the former village; but aside from this, everything seems to be concealed. South from the railroad exposures are numerous and satisfactory.

On the east fork of Raccoon creek the *Sewickley Coal* is seen at the township line near Cherry Valley Postoffice, and at a mile and a half farther down the stream the *Pittsburg Coal* is reached by the road. Just above where the Burgettstown road leaves the stream, there is an opening in this coal which appears to be occasionally worked. It shows three benches of coal nine, seven and twenty-seven inches thick, separated by thin layers of shale. The lower bench has a persistent parting almost midway. The exposure here is clearly imperfect, and the poverty of the roof as compared with other localities in the neighborhood leads me to suppose that here is seen only a portion of the roof division, and that the lower division is not reached. A sandy shale, of which eighteen feet are seen, rests directly on the coal.

On the Burgettstown road the following section is shown between the creek and the summit:

1. Sandstone	35′
2. Limestone, yellow on top	40′
3. Shale	6′
4. Limestone and shale	25′
5. Shale	20
6. *Sewickley Coal bed*	Blossom.
7. Shale	10′
8. Limestone and shale	15′
9. Sandy shale	20′
10. *Pittsburg Coal bed*	10′
11. Concealed	12′
12. Limestone	5′
13. Sandstone	12′

The two divisions of the Great Limestone are imperfectly exposed, and it is more than possible that the thickness assigned to them in the section is exaggerated. The *Sewickley* is represented only by a bituminous shale which contains some coaly matter. The *Pittsburg* shows such thickness that the opening just referred to must be in the roof division.

On the opposite side of the stream the road leads to Midway. On it, between the creek and the summit, a section was obtained similar to that just given, but the intervals are larger, as they were measured up the dip.

On the road leading from Bulger station to the east fork of Raccoon creek, and coming to it near Cherry Valley Postoffice, there are no exposures. On the first road turning west below the township line, and leading to the west fork of the creek, there are no exposures until the summit is reached, where Limestone IV, the Middle Washington, is seen. Descending the other side the following succession is shown:

1.	*Washington Coal bed*	Blossom.
2.	Concealed	65′
3.	*Waynesburg Coal bed*	Blossom.
4.	Sandy shale	15′
5.	Limestone, yellow	——

The *Uniontown Coal* is not exposed, and may not be present.

Beyond the west fork of Raccoon creek, on the way to the State road, the two coals are again seen, and are seventy feet apart with a limestone almost midway between them. Fragments of the Upper Washington Limestone occur at the junction with the State road, and the stratum itself is ten feet above, while the *Jolleytown Coal* is exposed at a few feet above the junction of the two roads. The Upper Washington Limestone is only six feet thick here, and has a thin coal resting upon it. Though so much thinner, the limestone retains its peculiar characteristics of color and purity, and contains the little ostracoids in great numbers. It is only one hundred and ten feet above the *Washington Coal.* At all these exposures the *Waynesburg* is very thin and the *Washington* is little better.

The *Waynesburg* is again seen on Raccoon creek, west fork, where the State road reaches the stream; but below this to Burgettstown there are no exposures. At that village the

Pittsburg Coal is worked by a number of persons, but the openings for the most part closely resemble those on the railroad which have been described. On the road from Burgettstown to Eldersville there are no exposures until the summit at the township line is reached, where the *Waynesburg Coal* is seen associated as follows:

1. Concealed	30′
2. Limestone	8′
3. *Washington Coal Bed*	Blossom.
4. Shale, not well exposed	20′
5. Limestone, I	10′
6. Concealed	20′
7. *Waynesburg Coal bed*	Blossom.

The base of this section was obtained in Jefferson township, where its relations are better shown.

46. Jefferson Township, in Washington County.

This adjoins Smith on the west, and is south from Hanover. Harman's creek is its northern boundary, and Cross creek flows through its southern portion. The section extends from about two hundred feet below the *Pittsburg Coal* to the Middle Washington Limestone. Possibly Limestone VI is caught in some of the hills, but it is not exposed. Openings in the *Pittsburg Coal* are very numerous, but owing to the season in which this township was examined, few of them were measured. They are worked chiefly for domestic use, so that during the summer they are usually in bad shape and cannot be measured.

On the road leading from Burgettstown to Eldersville, the *Waynesburg* is seen at the school-house by the first cross-roads west from the township line, and appears to be about four feet thick and double. At one-eighth of a mile farther west it is again seen in the road, well exposed and resting on fifteen feet of sandy shale. The hill above it is very high, and catches much of the upper barren series, but there are no exposures. At the next cross-roads the coal is again shown, and judging from its blossom, is not less than four feet thick. Fifty feet higher is the blossom of the *Washington Coal*, and Limestone IV is seen near the summit of the road just beyond. The hill rises fully one hundred feet above the *Washington Coal*, and must certainly catch the Upper Washington Limestone.

These upper limestones are thinning out very fast. The *Washington Coal* continues in sight to Eldersville, where it is seen on top of the high hill on which the village is built.

The *Pittsburg Coal* is mined at one-third of a mile north from Eldersville to supply the vicinity. Between that and the *Washington Coal* the following imperfect section was obtained:

1. *Washington Coal Bed*	5′
2. Clay	6′
3. Limestone I "b"	4′
4. Concealed, mostly sandstone	75′
5. Limestone	1′
6. Concealed	100′
7. Sandstone	30′
8. *Pittsburg Coal Bed*	5′
9. Clay	3′
10. Concealed	7′
11. Limestone	6′
12. Sandstone	20′

Thus making the interval between the *Washington* and *Pittsburg Coals* only two hundred and sixteen feet, so that calculating the interval between the *Washington* and *Waynesburg* as fifty feet, according to the last measurement made on the road, we have as the interval between the *Waynesburg* and *Pittsburg* only one hundred and sixty-six feet. This is the most northern exposure of the *Washington Coal;* indeed it is the most northern point at which the coal occurs. The *Waynesburg* is not seen at any point north from this, and the only locality where it can occur is in the vicinity of Florence, in Hanover township.

The *Pittsburg Coal* was examined at several openings in the ravine north from Eldersville, and measurements were made at two as follows:

1. Sandstone	4′ 0″	3′ 0″
2. Clay	1′ 0″	5′ 0″
3. *Coal*	0″ to 0′ 5″	0′ 0″
4. Clay	0″ to 1′ 0″	0′ 0″
5. *Coal*	2′ 9″	2′ 8″
6. *Coal* and partings	3¾″	5½″
7. *Coal*	1′ 0″	1′ 1″
8. Parting	0″	½″
9. *Coal*	1′ 2″	1′ 2′

The roof division is very irregular, and is frequently cut out by sandstone horsebacks. which, however, rarely affect the lower division. The coal at these openings is very bituminous, showing layers sometimes three inches thick, without any lamination. Some cannel occurs occasionally, but the quantity is very small. The coal is said to be exceedingly good for fuel. On the road from Eldersville to Hanlin's Station, an opening was seen at one hundred and thirty-five feet above the railroad, which resembles those already referred to. The dip south-east is quite sharp, so that when the coal is again reached on the north side of the railroad it is two hundred and twenty-five feet above the track.

On the Eldersville and Cross Creek road the *Waynesburg Coal* is seen soon after leaving the former village, and is frequently exposed at the roadside quite to the township line. Along this road it appears to be about three feet thick. The dip is so sharp southward that at Bethel Church this coal is at the roadside. About midway between the church and Gillespie's mill, on a branch of Cross creek, the *Pittsburg Coal* is exposed, with a white limestone almost directly under it. Below that, for nearly one hundred feet, is a sandstone, varying from shaly to massive, soft, and of irregular structure, some parts weathering into cavities. It is seen in cliffs along the stream.

On the east fork of this stream, at about a mile and a half above the mill, a deserted opening in the *Pittsburg* was seen. The coal is reached on the divide between the forks, but there are no openings until well up on each fork. On the west fork the first openings are within two miles of Eldersville, and the coal is in sight along the stream for nearly a mile. The banks were all in bad condition, and could not be measured. Near Eldersville the Great limestone is in the road, and shows a serious decrease in its greatness, being represented by three layers, which in all contain only fourteen feet of limestone. The *Waynesburg Coal* rests directly on the upper layer, the whole interval having disappeared.

47. Cross Creek Township, in Washington County.

This lies south-east from Jefferson and south-west from Smith. Cross creek flows along its southern border, and two important branches of that stream flow across the township. A synclinal passes through its south-eastern corner, and the section extends from the *Pittsburg Coal* to the Upper Washington Limestone.

On Cross creek the *Pittsburg Coal* is worked by Mr. Plummer near the south-west corner of the township, and by Mr. Bushfield a mile farther up the stream. The measurements at these banks are as follows:

1. Roof division	Concealed.	Concealed.
2. *Coal*	2′ 7″	2′ 5″
3. *Coal* and partings	3″	4″
4. *Coal*	1′ 10″	2′ 2″
Total	4′ 8″	4′ 11″

No. 2 is the best coal and contains much less sulphur than No. 4. At Mr. Jones' bank, near the last, the roof division is altogether wanting, having been removed during the deposition of the overlying sandstone, for fragments of coal are scattered through that rock. Here the coal goes under. The dip is rapid to the east, so that at the mouth of the next tributary above, the *Uniontown Coal* is seen sixteen inches thick and resting on the upper division of the Great Limestone. The *Waynesburg Coal* is exposed in a little gulch at the cross-roads below Mr. J. Thompson's residence, where it shows the following association:

1. Shaly sandstone	25′ 0″
2. Limestone	3′ 0″
3. Shale	2′ 0″
4. Limestone	6″
5. Shale	4′ 0″
6. *Waynesburg Coal Bed*	2′ 0″
7. Sandy shale to creek	10 0″

At Wilson's Mill the coal is three feet six inches thick, and is mined to supply fuel to the mill. The limestone is exposed in the roof. Above this point the bed rises very fast, and at the township line is seventy feet above the creek. On the road leading from Bushrock school-house to this stream the following succession is exposed:

1. Sandstone	25′
2. Concealed, containing fragments of limestone	120′
3. *Washington Coal Bed*	Blossom.
4. Concealed	100′
5. *Waynesburg Coal Bed*	Blossom.
6. Concealed	55′
7. *Uniontown Coal Bed*	Blossom.
8. Limestone	8′
9. Concealed to creek	7′

No. 1 of this section is a massive sandstone, soft and liable to weather into fantastic forms. Only a little of the bed remains, barely enough to cover one-fourth of an acre, but on the hillsides there are enormous fragments, many of which contain three thousand cubic feet. At several localities farther east this rock is seen underlying Limestone VI.

On the other branch of the South Fork the *Waynesburg* is exposed sixty feet above the stream at Plumpsock, and at a little way lower down the limestone underlying the *Uniontown* is shown. Just below the forks of this creek Mr. W. Rea has an opening in the *Washington,* which yields the following measurement:

1. Concealed, said to contain two thin *Coals.*		
2. Shale		8″
3. *Coal*	1′	6″
4. Clay		1″
5. *Coal*	2′	3″
6. Clay		4″
7. *Coal*	1′	2″

For the most part this is quite bad, but No. 5 contains about ten inches of very good coal, called the "Brick," and the whole of No. 7 is of fair quality. Mr. Rea seems to think it fully equal to that from the *Pittsburg.* The rest of the bed is very poor and can be burned only when large quantities of it are thrown together. At one time the *Waynesburg* was worked here, but it was found so thin that the working was unprofitable. On Mr. Rea's property the massive sandstone, already referred to, is seen in fragments, as is the case on another hill about a mile north-west. At each locality the top of the highest fragment is about one hundred and twenty-five feet above the *Washington Coal.* The *Waynesburg* was formerly worked by Mr. Lawton about a half mile farther down the stream, but the opening has fallen into decay. On the steep bluff above this old bank there is a deserted opening in the *Washington,* and the distance between the two coals is ninety-five feet. There are no exposures in this interval, but midway there is a row of limestone fragments

Crossing the divide between this stream and the North Fork of Cross creek, the following section was obtained in coming down to Patterson's Mill, on the North branch:

1.	Shaly sandstone	30′
2.	Limestone	Fragments.
3.	*Waynesburg Coal Bed*	Blossom.
4.	Concealed	50′
5.	*Uniontown Coal Bed*	Blossom.
6.	Limestone	Fragments.
7.	Concealed	45′
8.	Limestone	10′
9.	Concealed	40′
10.	Shale and limestone	55′
11.	Massive sandstone	25′
12.	*Pittsburg Coal Bed*	——

Thus making the interval between the *Pittsburg* and *Waynesburg Coals* about two hundred and twenty-five feet, measured with the dip. The former coal is worked by W. Patterson below the mill and by L. Patterson at half a mile farther up the creek. The sections at both the openings are similar and vary little from the following:

Coal, two feet six inches; *coal* and partings, four inches; *coal*, two feet; total, four feet ten inches.

At the lower bank the roof division is absent, but it occasionally occurs at the upper. That it has been cut out by the overlying sandstone is well shown at the upper opening, where the sandstone has a very irregular bottom, resting at times directly on the lower division, while again it rises and lets in the roof. The changes are abrupt. The sandstone is massive and stands out in cliffs along the creek. The upper bench of the coal is quite good, but the lower one is inferior and its outcrop is covered with copperas.

Just above Mr. Patterson's opening the coal goes under the stream. From that point to a fork in the road, immediately west from Cross Creek village, the outcrop of the Great Limestone is conspicuous in the hills. East from that village the blossom of the *Washington Coal* is shown in the road, and above it there seems to be an enormous mass of limestone and shale, extending upwards for nearly eighty feet. Fragments of that rock lie thickly strewn everywhere in the vicinity of the village. Half a mile south the *Waynesburg* was once worked on the property of Rev. J. Stockton, but the bed proved to be only two feet thick and the opening was abandoned. Just above it is the limestone which seems to be persistently associated with it in this portion of the county, though absent in the eastern part. Near Mr. J. Allen's residence, a mile south from the village, the following section was obtained:

1.	Limestone IV	30′
2.	Concealed	30′
3.	*Washington Coal Bed*	Blossom.
4.	Concealed	40
5.	Limestone Ia	10′
6.	*Waynesburg "a" Coal Bed*	Blossom.
7.	Shale and sandstone	40′
8.	*Waynesburg Coal Bed*	Blossom.

This is one of the few localities in the north-western part of the county where the *Waynesburg* "*a*" is satisfactorily identified. It underlies the limestone which in this region has been found so often between the *Washington* and *Waynesburg Coals.* About a mile farther south the same series is exposed near Mr. R. Buxton's residence, as follows:

1.	Concealed	80′
2.	*Washington Coal Bed*	Blossom.
3.	Concealed	40′
4.	Limestone Ia	10′
5.	*Waynesburg "a" Coal Bed*	Blossom.
6.	Sandstone and shale	45′
7.	*Waynesburg Coal Bed*	Blossom.

At a mile east from the last locality, near Mr. J. M. K. Reed's, the two coals are seen ninety-five feet apart. Just north from this the *Waynesburg* was once worked by Mr. Simpson, who found it four feet thick. On land of Mr. Jeffrey, one mile north-west from Woodrow Postoffice, there is an old opening in the *Waynesburg*, which shows:

Coal 7 inches; clay, 2 inches; *coal*, 18 inches; clay, 3 inches; *coal*, 17 inches; total, 3 feet 11 inches.

At ninety-five feet above this the blossom of the *Washington Coal* is seen.

Hopewell Township.

This lies south from Cross creek and Mount Pleasant. Its northern boundary is Cross creek, and its southern, Brush run, a tributary to Buffalo creek. The section extends from the Upper Washington Limestone downward almost to the *Pittsburg Coal*, which appears on Cross creek at a short distance below the north-west corner of the township. The available coals are the *Washington* and *Waynesburg*, both of which are worked. The upper limestones attain great thickness.

On Brush run the *Waynesburg* is first seen at two-thirds of a mile below the Canton township line, near Mr. Caldwell's residence, where it is sixty-five feet below a massive limestone exposed at the cross-roads, half a mile from the creek. The strata fall westwardly, but not so fast as the run, for at Mr. S. Work's the coal is sixty feet above the stream, having fallen fifty feet from where it first appears. It was once opened by Mr. Work, but was so thin that mining proved unprofitable, and the opening was abandoned. At this locality the associated rocks are exposed as follows:

1.	*Washington Coal Bed*	Blossom.
2.	Concealed	40′
3.	*Waynesburg "a" Coal Bed*	Blossom.
4.	Shale and sandstone	50′
5.	*Waynesburg Coal Bed*	2′ 6″
6.	Shale and sandstone	25′
7.	Nodular Limestone	3′
8.	Sandstone and shale	37′
9.	*Uniontown Coal Bed*	Blossom.
10.	Great Limestone seen	5′

The *Waynesburg "a"* seems to be not less than two feet thick. At a little way farther down the run the limestone underlying the *Uniontown Coal* is exposed to the thickness of fifteen feet. At the head of a little run coming in about a mile from the Independence line, the *Washington Coal* was once worked by Mr. A. Farrar, but the opening is no longer in operation, and no information could be obtained respecting the coal. Half a mile farther west Mr. Bartilson mines the *Waynesburg Coal*, which is roofed by one foot six inches of shale, on which rests a slaty limestone. It shows four benches of coal, six, eleven, thirty-six and six inches respectively, all quite inferior in quality, and much the same as that shown by the bed at almost every exposure, whether in this or Greene county.

At the township line, on Brush Run, the *Uniontown Coal* is below the bed of the stream. Following up a branch, which comes in here, this section is found at half a mile from the run:

1.	Limestone and shale II	35′
2.	*Washington Coal Bed*	Blossom.
3.	Concealed	100′
4.	Shale	5′
5.	Limestone	4′

6.	Shale, with coaly matter	2
7.	*Waynesburg Coal Bed*	3′
8.	Concealed to stream	40′

The interval between the two coals is one hundred and eleven feet. No opening in the *Washington* was found in this vicinity, but it makes a broad smut in the road. The *Waynesburg* was once opened here by Mr. Hemphill, and, at half a mile farther up the stream, Mr. J. James' opening on the same coal is still operated. There the bed shows a total thickness of six feet eight inches, and is roofed by one foot of shale, on which rests five feet of limestone. The coal is in three benches, the bottom being three feet five inches, and is somewhat superior to the average from this bed. The bottom bench is quite good and has much less sulphur than usual, as is shown by the diminished tendency to clinker. The slaty matter is unexpectedly small, and the owner states that the coal burns down to a powdery white ash. Here the bed passes under the stream. The main, or eastern, branch of this run flows from the east, and the strata rise so rapidly in that direction that at the residence of Mr. J. M. Rush, near the head of the run, the *Washington Coal* is still eighty feet above the stream.

Crossing the divide to a tributary of Brush run, the following succession is seen near school-house No. 5:

1.	*Coal?*	Blossom.
2.	Limestone IV	35′
3.	Concealed	40′
4.	*Coal?*	——
5.	Limestone III	10′
6.	Shale	20′
7.	*Coal?*	——
8.	Limestone II	15′
9.	*Washington Coal Bed*	Blossom.
10.	Concealed to the run	70′

The thickness of Limestone IV, as given in this section, is so great that I am inclined to believe that an error has been made in entering the figures in the note-book. Near Buffalo village, the *Washington Coal* was once opened by Mr. J. Eagleson, and was found to be four feet thick, but it is too slaty for use.

Between Buffalo village and the first fork in the road northwest, the *Washington*, *Waynesburg "a"* and *Waynesburg* are exposed, and the interval between the first and last is only ninety feet. The *Waynesburg* is well shown at the fork, but no opening has been made to determine its thickness or quality. Immediately south-east from the village Mr. S. Work formerly mined the *Washington* to supply the steam mill with fuel. The *Waynesburg "a"* is there seen at forty feet below that bed and underlying Limestone I, which is ten feet thick. The coal is at least two feet. Two-thirds of a mile west of north from Buffalo village, on land belonging to Mr. Maxwell, the *Washington* was once opened, but the opening has been abandoned. Half a mile north-west from this locality, the following section was seen near Mr. T. M'Keever's residence:

1. Sandstone	Fragments.
2. Concealed	130′
3. *Washington Coal bed*	Blossom.
4. Concealed	60
5. Sandstone	20′
6. Shale	10′
7. *Waynesburg Coal Bed*	3′ 6″

The heavy sandstone No. 1, is the one directly underlying the Upper Washington Limestone, and has been quarried in this neighborhood by Mr. Adamson for building purposes. The *Washington Coal* was once opened by Mr. M'Keever, but this, like nearly every other opening in the bed seen in this township, has been abandoned. The *Waynesburg* has been opened on the road leading from Buffalo village to Cross creek by Mr. W. Hunter, who finds it three feet six inches thick, with one inch of clay at one foot from the top. The coal is of very poor quality, but is used by some of the farmers in the vicinity. About a mile east from West Middleton on this road the massive sandstone, No. 1 of the section, is seen in loose blocks of great size near Mr. D. Brown's residence.

West Middleton is about seventy feet above the Upper Washington Limestone, which is seen just north-east from the village. The interval between it and the *Washington Coal* on this road is one hundred and fifty feet. The coal was opened on property belonging to Mr. James Thompson, but was found to be worthless. The *Jolleytown Coal* is exposed at the cross-roads near Mr. W. Craig's house, where it is forty feet below Limestone VI, and seems to be two feet thick. Its blossom is shown quite frequently on the hill-tops in this portion of the township.

48. Independence Township, in Washington County.

This lies directly west from Hopewell, and adjoins West Virginia. It is drained mostly by tributaries of Buffalo creek, though some small streams flow through its northern portion into Cross creek. The section extends from the Great Limestone to Limestone VI. A synclinal passes through its south-eastern corner.

Along Buffalo creek the *Waynesburg* and *Washington Coals* are accessible for nearly the whole distance, and in the extreme south-western corner there are a number of openings in the latter. The section obtained at Mr. T. Hagerty's bank may be taken as typical. It is,

1. *Coal*, slaty	6''
2. Clay	6''
3. *Coal*, slaty	5''
4. Clay	4''
5. *Coal*, slaty	10''
6. Clay	4''
7. *Coal*	5' 2''

Only No. 7 is mined. The quality here is the same with that at the openings belonging to Messrs. Smith, Morgan and Jones in the vicinity. Some of the coal is good, but the greater part is very bad. Still the farmers who have opened it, prefer burning it to hauling the *Pittsburg Coal* four or five miles for, as they say, it burns well enough if enough of it can be got together. This condition requires the construction of large grates. At ninety-seven feet below this opening on Mr. Hagerty's place, the *Waynesburg* is seen in the bluff and dark bituminous shale is exposed at eighteen feet lower. The *Waynesburg* is only *eight inches thick*. The rocks dip east to Bryson's store, where the direction is changed, and thence they rise. At the second fork in the road above the store the *Waynesburg* is seen associated as follows:

1. Sandstone, seen		20 0''
2. Shale		2' 0''
3. Slaty Limestone		4' 0''
4. Clay		1' 0''
5. Bituminous shale,	*Waynesburg Coal*,	8'
6. Clay		1' ''
7. *Coal*		10''
8. Shale		5' 0''
9. Shaly sandstone		5' 0''
10. Concealed to the creek		30' 0''

The thinning of this coal is remarkable, and seems to begin not far from the mouth of Brush run, for an opening made just above that point seems to have found nothing. A similar section is exposed at the store.

At the mouth of the first tributary to Brush run, the *Waynesburg* was once mined by Mr. Smith, but the working did not prove profitable, and was abandoned. On this little stream, at a mile from its mouth, Mr. T. R. Law has an opening in the *Washington*, which shows the bed in two divisions, as at Hagerty's bank. The upper contains two benches of coal, respectively six and eight inches, while the lower is an almost solid mass of coal, five feet six inches thick. This division contains about forty inches of very fair coal, which burns to a fine ash. Near the middle is a "brick" coal, similar to that observed at a few other banks. The interval to the *Waynesburg* is one hundred feet.

On Sugar Run the following section is exposed about one mile east from the West Virginia line, on a road leading south:

1.	Limestone VI	Fragments.
2.	Concealed	35′
3.	*Jolleytown Coal Bed*	Blossom.
4.	Limestone	8′
5.	Sandstone	20′
6.	Coal (?)	Blossom.
7.	Limestone IV	20′
8.	Concealed	20′
9.	Limestone III	Fragments.
10.	Sandstone	20′
11.	Concealed	35′
12.	*Washington Coal Bed*	Blossom.
13.	Concealed	160′
14.	Great Limestone in the run.	

No. 14 is the top of the Great Limestone, and the *Uniontown Coal* is seen resting on it, in a hollow entering from the north. There it has been mined by stripping, and has a thickness of two feet. At the mouth of Camp run, two-thirds of a mile farther up, the blossom of the *Waynesburg* is twenty-five feet above the run, and one-third of a mile beyond it is at the water's level, and seems to be two feet thick, with four or five feet of shaly limestone resting on it. From this point the rise of the rocks is more rapid than that of the stream, so that at Mr. M'Connell's the *Uniontown* is in the run and two feet thick.

It is an imperfect cannel, and contains fish scales throughout, while a layer at the bottom has many bivalves. Where the road from West Middleton to Independence crosses the run, the *Uniontown Coal* is again seen, and at sixty feet higher is the blossom of the *Waynesburg*, so small that the bed cannot be more than one foot thick. At the head of the run, near the residence of Mr. A. Manchester, the blossom of the *Washington Coal* covers a vertical space of nearly twenty feet, so that the bed must be at least ten feet thick. Limestone VI is exposed in the road on the township line, two thirds of a mile west from West Middleton.

In the vicinity of Independence the blossom of the *Washington Coal* is seen, and the following section was obtained in descending to M'Guire's run:

1. *Washington Coal Bed*	10′
2. Shale	25′
3. Sandstone	15
4. *Waynesburg "a" Bed*	2′
5. Concealed	30′
6. Sandstone	20′
7. Concealed, place of *Uniontown Coal*	4′
8. Great Limestone, upper division	6′
9. Shale	15
10. Great Limestone, lower division	50′

This is an exceedingly interesting section, as forming one of a series made in this portion of the county. On Buffalo creek, in this line, the interval between the *Washington Coal* and the upper division of the Great Limestone is one hundred and sixty feet; here it is ninety-six feet; at six miles farther north it is seventy, and at Eldersville it is only fifty feet. There is no satisfactory exposure of the rocks below the Great Limestone to the *Pittsburg Coal* on M'Guire's run, and the *Sewickley Coal* is not seen. The *Pittsburg* is mined by Mr. Magee, about a mile from Independence, and by Mr. Bell on the other side of the run. At these banks the coal shows in its lower division:

Coal, two feet nine inches; *coal* and partings, three inches; *coal*, one foot ten inches; total, four feet ten inches.

At Mr. Magee's, the roof division is not exposed, but at the other opening there are two benches of impure coal, ten and nine inches, above which is the bituminous shale. The Pittsburg sandstone is here only an arenaceous shale. The coal shows its usual character. The upper bench is quite pure and portions of it are excellent for smiths' use; the "brick" has the accustomed planes of cleavage and comes out in blocks, while the bottom is bad and not mined. Descending M'Guire's run to its mouth the following section below the *Pittsburg* was obtained:

1.	Shale, with iron ore	10′
2.	Limestone	8′
3.	Sandstone	12′
4.	Limestone	5′
5.	Shaly sandstone	45′
6.	*Bituminous shale*	1′
7.	Flaggy sandstone	50′
8.	Massive sandstone	25
9.	Sandstone and red shale	80′

Near Mr. Carman's residence, No. 7 shows a fossiliferous layer containing remains of plants. This is not persistent, and at the mouth of the creek the whole stratum is massive, forming, with the one below, cliffs along the creek sixty or seventy feet high.

Ascending Cross creek, the strata are seen falling eastward, so that the coal, which at the mouth of M'Guire's run, is two hundred and forty feet above the creek, is soon caught in the hills directly bordering on the creek, and is opened by almost every farmer along this line. Measurements made at a number of openings show the following to be the extent of variation:

Roof, two feet six inches to three feet four inches; clay, three inches to sixteen inches; *coal*, four feet ten inches to five feet.

The "bottom" coal, though very impure, is usually taken out and burned here.

CHAPTER XII.

49. Jefferson Township, Allegheny County.

This is the southern township on the west side of the Monongahela river, and is crossed by Peters creek from west to east. Its section extends from eighty feet above the *Pittsburg Coal* to three hundred and fifty feet below it, but above the coal the exposures are very unsatisfactory. The *Pittsburg* is accessible in by far the greater portion of the township, and aside from the Pinhook and Peters creek axes, is the only feature of interest.

On Lick run, which forms in part the western boundary of the township, separating it from Snowden, the coal remains above the stream from its mouth to beyond the Jefferson line. From the point in Snowden township where it disappears under the run, it rises south-east to within two miles of Peters creek, where the dip is reversed, and thence to Peters creek it descends in that direction. The first opening seen in going down the run is at the cross-roads near Wilson's saw-mill, where the coal is roofed by ten feet of shale and shows the following section:

Roof division, 6 feet 2 inches; clay, 13 inches; lower division, 5 feet 2 inches.

In the roof, there is at the base a bench of coal two feet six inches thick, but above that there is only clay with thin layers of coal, excepting for eight inches on top, where there is a bituminous shale. The bed is fifty-five feet above the run. Half a mile below, it is one hundred feet, and at a mile and a half below the saw-mill, it is one hundred and sixty feet, the maximum, for here the Pinhook axis crosses. The coal is worked here by Mr. Miller, at whose bank the section is,

Roof division, seen, 1 foot 11 inches; clay, 11 inches; lower division, 5 feet.

At eighty-five feet below the coal, is a limestone five feet thick, but the rest of the interval is concealed. Lick run falls rapidly, but not so fast as the rocks, for at Cochran's mill, less than half a mile from this place, an opening was seen at one hundred and fifty feet above the stream, which shows the following structure:

Roof division, 4 feet 4 inches; clay, 1 foot 2 inches; lower division, 5 feet.

In Mr. Miller's opening the roof showed two benches of coal, ten and eight inches respectively, but here the benches are seventeen and thirty inches. In this section the bituminous shale above is separated from the roof division by six inches of clay, and is seven inches thick. The limestone seen at the other locality at eighty-five feet below the coal is wanting here, as appears from this section below the bed to the run,

1. Concealed	60′
2. Limestone, seen	2′
3. Sandstone	3′
4. Red shales	15′
5. Sandstone	15′
6. Clay shales	10′
7. Impure limestone	1
8. Concealed to the run	45′

At the mouth of the run the coal is ninety feet above the stream. Descending Peters creek the first opening is found at the mouth of Rocky run, where the *Pittsburg* is mined by Mr. Castor, at one hundred feet above the creek. The bituminous shale is eight inches thick, and rests directly on the roof division. The measurement is,

Roof division, 3 feet; clay, 8 inches; lower division, 5 feet 5 inches.

Opposite the mouth of Beams run, at about one mile due east from the last, the coal is opened by Mr. Bedell at 130 feet above the run, and shows,

Roof division, 2 feet 6 inches; clay, 7 inches; lower division, 5 feet 5 inches.

The roof is single, and consists wholly of bony coal. At the mouth of Lewis run, somewhat less than a mile east north-east from the last, the coal is mined by Mr. Large, and the following section is exposed:

1. *Pittsburg Coal bed*	8′
2. Shale	20′
3. Limestone	4′
4. Shale and sandstone	60′
5. Limestone	5′
6. Red shale	50′
7. Shaly sandstone to creek	40′

In all about one hundred and eighty feet. The synclinal between the Pinhook and the Peters' Creek axis evidently crosses the creek below the mouth of Rocky Hole run.

On Lewis' run the coal is mined by Mr. Elliott near Jones' saw-mill, two miles from the creek, where it is one hundred feet above the run, and shows :—

Roof division, concealed ; clay, 6 inches ; lower division, 5 feet 8 inches.

Half a mile farther up, the stream forks. On the West fork the coal disappears near Mr. Syckman's residence, which is about one-quarter of a mile from the township line. On the other fork it is worked by Mr. M'Ilhenny, at whose opening it shows :—

Roof division, 2 feet 6 inches ; clay, 5 inches ; lower division, 5 feet 6 inches.

On the river line the openings are numerous. Those belonging to Messrs. J. Walton & Co., near West Elizabeth, are extensive. In their tunnel at West Elizabeth the coal has the following structure :

Roof division, 4 feet 4 inches ; clay, 8 inches ; lower division, 5 feet 10 inches.

Above the coal the hill reaches nearly seventy feet, but, aside from a few fragmentary exposures of shaly sandstone, everything is concealed. Immediately overlying the coal is shale, eight feet, containing much nodular iron ore of inferior quality. Below the coal the following section was obtained :

1. Imperfectly exposed	35′
2. Coal	Blossom.
3. Shaly sandstone	35′
4. Compact sandstone	8′
5. Micaceous shaly sandstone	40′
6. Compact sandstone	9′
7. Impure limestone	2′
8. Flaggy sandstone	10′
9. Sandy shale	28′
10. Concealed to track	58′

No. 6 is quarried to some extent, and is micaceous, soft and of a bluish-gray color. Its compactness is merely local, for at two hundred yards farther down the river it is shaly, and the mass, from 3 to 6, inclusive, is one. About a mile below West Elizabeth, No. 10 is exposed and consists of arenaceous shale or shaly sandstone. At the other coal works in this vicinity the structure is the same as at the Walton works.

At the mouth of Peters' creek the hills are low and show nothing, but the section is below that just given and the coal is evidently not less than three hundred feet above the railroad. From this point it falls very rapidly north-west to Coal Valley, where it is less than two hundred feet above the track, and in the works of Foster, Wood & Co. is dipping north-west. At that locality the bed is eighty feet below the top of the hill, but above it everything is concealed by from four to eight feet of debris. No connected section can be made below the coal, there being only occasional exposures of shale or shaly sandstone. The dip is very irregular, snd swamps, or "jumps," are of frequent occurrence. In the most extensive of these, the coal drops twenty-three feet six inches within forty yards, and aftewards rises six feet before resuming its regular dip. This annoying "swamp" is found north-east in Lynn, Wood & Co.'s works, and is a source of trouble as far south-west as Lewis' run. It evidently marks the synclinal between the Peters' Creek and Pinhook axis, which crosses here and strikes through the old O'Neill colliery, near the township line. This latter, now known as the Pine Run Coal Works, is at somewhat more than a mile from the river, and the pit-mouth is barely two hundred feet above low water mark, being rather lower than the level of the Blackburn colliery, which now belongs to Foster, Wood & Co. Here the synclinal crosses, and in the entries now worked the dip is south-east. From this point down the river the rocks rise to Camden.

50. Mifflin Township, in Allegheny County.

This adjoins Jefferson on the north, the Monongahela river is its northern and eastern boundary, and Street's run at the west separates it from Baldwin township. The section extends from two hundred feet above the *Pittsburg Coal* to three hundred and sixty feet below it. The township is crossed by the Pinhook axis.

Near the head of Pine run, which flows along the southern edge of the township to the river, the coal is worked by Mr. Blank, whose opening shows the lower division of the bed thus:

Coal, 2 feet 11 inches; *coal* and partings, 4 inches; "brick" *coal*, 9 inches; "lower bottom," 14 inches.

Here the dip is south-east, and so it continues to be to the river, the coal falling one hundred and ten feet in that distance, which is barely two miles. At a little way below the last is Mr. M'Gowan's opening, which shows:

Roof division, 2 feet; clay, 1 foot; lower division, 5 feet 2 inches.

The upper bench of the lower division is three feet thick. The brick and lower bottom benches are distinct. The latter is very inferior.

On Thompson's run, which flows through the eastern part of the township, and enters the river nearly three miles below the mouth of the Youghiogheny, the openings are numerous. Mr. M'Ilhenny's bank, on a little tributary coming in from the south, at three-fourths of a mile from the river, shows the bed thus:

Roof division, 2 feet 2 inches; clay, 1 foot; lower division, 5 feet 11 inches.

The lower bottom is fourteen inches thick, and, as usual, is of very poor quality. At somewhat more than a mile and a half from the river, the coal is only eighty feet above the run. At two miles and a half from the river, the coal is mined at Mr. Stone's works, where the following was seen:

Roof division, concealed; clay, 8 inches; lower division, 5 feet 10 inches.

Below the coal there is one foot of fire-clay resting on a coarse limestone, of which only two feet are exposed. The coal goes under the run just below the election house, about four miles from the river. On the ridge between this and Pine run the *Pittsburg Coal* is deeply buried, being not far from two hundred feet below the road at the Presbyterian church.

On Street's run the coal continues above the stream until the extreme south-west corner of the township, where it is mined by Mr. M'Kee, whose opening shows:

Roof division, 3 feet 4 inches; clay, 8 inches; lower division, 5 feet 1 inch.

On the roof division rests one foot of bituminous shale, above which sandstone is exposed for ten feet. The worthless lower bottom of the coal is thirteen inches thick, and the roof division shows two benches, eight inches and twenty-two inches, both of which are bony. The local dip at this bank is south-east, but the true dip is north-west, for followed south-east to the head of Pine run the coal shows a rise in that direction of sixty feet. But there it is dipping south-east, so that the Pin-hook axis must cross not far west from the head of Pine run. On Blackburn's run, which enters Street's run about two miles below the last locality, the coal is opened by Messrs. Irwin and Risher. The bank belonging to the former does not show the roof division, but gives the following as the structure of the lower division:

Coal, 3 feet; *coal* and partings, 3 inches; "*brick coal*," 8 inches; "*lower bottom*" *coal*, 14 inches.

At the mouth of this stream the coal is two hundred and forty feet above the run. Along Street's run the hills become low before reaching the river, and do not contain the coal.

On Patterson's run, which enters the river below Braddock's Station, there are a few openings, which, however, show no material differences from those already enumerated.

As far down the river as Braddock's Station the coal is caught in the hills, and is mined extensively for shipment. The Camden Coal Works are near the station of that name, about one mile from the southern boundary of the township. The entrance to the pit, which is at some distance from the river, is nearly two hundred and thirty-five feet above low water mark, and the coal is sent down an incline two thousand six hundred feet long. The works are quite extensive, there being nearly two miles of direct tunneling. The section here is:

1. Concealed		20′
2. Sandstone		2′
3. Clay shale		9′
4. Clay		3′
5. Pittsburg Coal		9′ 10″
Roof division	3′ 1″	
Clay	1′ 0″	
Lower division	5′ 9′	
6. Fire-clay		1′ 0″
7. Limestone		2-8′ 0″

Below this to the railroad track there are occasional exposures of shale and sandstone, but no connected section can be made out. In the tunnel through the first hill, which is one mile long, the coal rises north-west at the rate of fifty feet per mile, but in the tunnel through the second hill the rate is greatly diminished and is only sufficient to admit of easy drainage. Of the coal, the "lower bottom" is too poor to work and the "brick" holds a rather persistent binder of pyrites near the top. The rest of the lower division is very good, stands shipping well and is said to be an excellent gas coal. The roof division shows two benches of coal, fifteen and ten inches, separated by one foot of clay, which sometimes disappears, and the roof becomes three feet of poor coal. The main clay parting occasionally becomes compact, and then shows much iron and limestone. The shale above the coal is sometimes fifteen feet thick.

Opposite the mouth of the Youghiogheny river there is a cluster of mines. At all of these the dip is south-east, but is irregular, and "swamps" are a cause of much annoyance. The coal is clean and the slack is coked in ovens by one of the firms. The coke is compact and is said to be used at the M'Keesport Furnace. Here the river bends towards the east and the coal falls, so that at Mr. W. Neal's works, just below M'Keesport station, it is barely two hundred feet above low low water mark. There the bed shows:—

Roof division, 5 feet 1 inch; clay, 10 inches; lower division, 5 feet, 8 inches.

The overlying rock is shale twelve feet thick, above which is sandstone to the top of the hill, twenty-six feet. There is no evidence of any other coal above the *Pittsburg*. From the pit-mouth to the railroad track only sandstone and shale are exposed, and these are imperfectly shown, except in the lower one hundred feet. The roof division of the coal has two benches, fourteen and thirty-seven inches thick, separated by ten inches of clay. Unfortunately this mass of coal is worthless. The coal from the lower division bears shipping well, and the slack is made into coke, which the owner of the works says is of good quality. The dip is irregular, but the prevailing direction is south-east.

Below this to Thompson's station there are no exposures along the river. The hills are low and do not catch the *Pitts burg Coal.* Just below that station the crinoidal limestone is exposed at the level of the track, varying in thickness from eight inches to one foot. It is bluish-gray, coarse-grained, resembling sandstone, looks soft, but does not disintegrate readily, and stands out in blocks along the bluff. This rock is exceedingly fossiliferous here, ten species of mollusca, together with immense numbers of crinoidal plates and spines, having been found in one block. The coal is three hundred and sixty-three feet above the river at this place.

The limestone remains in sight to Braddock's station, where the coal is mined by Mr. Redman at three hundred and sixty-two feet above the river. In his works the dip is south-east. From this locality to the township line there are no exposures. The hills are all low for some distance back from the river, and the next coal works are at Six-Mile Ferry, just over the line in Baldwin township.

The Pinhook anticlinal crosses not far below Braddock's station. Its crest is sufficiently well defined, but its north-western slope in this region is very obscure.

51. Snowden Township, in Allegheny County.

This adjoins Jefferson on the west, along the Washington county line. Piney fork of Peters' creek flows near its southern border and Little Piney fork flows south-east through the centre. The section extends from three hundred feet above the *Pittsburg Coal* in the north-west portion to one hundred and ten feet below it in the south-east. Connected exposures above the coal seem to be entirely wanting, and those below are devoid of interest.

In the north-west part of the township, the blossom of a thin coal is occasionally seen on the highest hills. It is probably the *Waynesburg* "*a*," though possibly it may prove to be the *Washington.* Near school-house No. 7, three coals are found, the top one being that just referred to. This is associated with fragments of limestone. Fifty feet below it is another coal, and at forty feet lower is the third. It is not at all impossible that the bottom coal is the *Waynesburg*, and the top coal the *Washington.* In the extreme north-west corner of the township, at the cross-roads on the ridge near Mr. W.

Ager's residence, the upper beds are seen sixty feet apart, with a yellowish limestone associated with the lower one. The interval between the topmost bed and the *Pittsburg* is about three hundred and fifty feet, so that it is most probably the *Washington*. Near Mr. Ager's it is eighteen inches thick.

On the headwaters of M'Laughlin's run the great limestone is seen, showing a thickness in all of nearly seventy feet. The exposure is incomplete, and the mass is far from being wholly limestone. The same rocks are exposed in a similarly imperfect manner along the upper portion of Piney fork, near the village of Library. On this stream the rocks are rising southeast, so that the *Pittsburg Coal* is brought up at a short distance below that village, and at two-thirds of a mile below, it is worked by Mr. H. Potter, at whose bank it shows,

Roof division, concealed; clay, 18 inches; lower division, 5 feet 7 inches.

The "lower bottom" is nineteen inches thick, and as usual, is too impure to be mined. The rest of the bed is very good. At Mr. Siebold's opening, a little farther down the creek, the coal is seventy-five feet above the stream, and the lower division is five feet five inches thick. The roof is concealed, but is said to be only one foot. Half a mile farther down, the following section of the coal is seen:

Roof division	1′ 6″
Clay	3″
Lower division	5′ 7″
Concealed to creek	110′ 0″

The dip is reversed at the mouth of Little Piney, and thence the coal falls to the mouth of Piney Fork, where it is only seventy-five feet above Peters creek.

On Little Piney Fork the *Pittsburg* is mined by Mr. Kiddoo, near the old steam mill. The roof is concealed, and the lower division is as follows:

Coal, 3 feet 3 inches; coal and partings, 4 inches; coal, 1 foot 5 inches; total, 5 feet.

Immediately underlying the coal is a bed of nodular iron ore, which is entirely local. It is somewhat calcareous, and the nodules are in some instances as large as a half bushel.

Rather more than half a mile farther up the stream is Mr. Glenn's opening, which shows,

Roof division, 6 feet; clay, 6 inches; lower division, 4 feet, 11 inches.

The roof is a mass of shale, with many thin streaks of coal. At a short distance above this opening, the coal passes under the creek.

On Lick run the coal goes under near Mr. Wallace's house, not far from the Jefferson township line. At one time a little coal-bed was mined at half a mile farther up the stream, but its relation to the *Pittsburg* was not ascertained.

52. BALDWIN TOWNSHIP, IN ALLEGHENY COUNTY

This adjoins Mifflin on the west, and has a river front of barely two miles. Street's run separates it from Mifflin, and Saw-Mill run flows northward through it to the river. The section extends from three hundred and fifty feet above the *Pittsburg* to three hundred and fifty feet below it, but as in the other townships already described, exposures above the coal are very imperfect.

Along the river there are some important coal-works. Near the Mifflin border, just below the mouth of Street's run, are the Hays Coal works, where the coal dips south-east at the rate of fifty-six feet per mile. It is evident that the trough between the Pinhook and Washington axes crosses the river east from this point, but not far, as the altitude above the river is about three hundred and forty feet. The section here is,

1. Concealed to hill-top		80′ 0″
2. Shale		10′ 0″
3. *Bituminous shale*		6″
4. *Pittsburg Coal Bed.*		
Roof division	4′ 6″	10′ 1″
Clay	3′ to 2″	
Lower division	5′ 5″	
5. Fire-clay		3″
6. Limestone		3 0″
7. Sandstone and shale		60′ 0″
8. Concealed to track.		

No. 6 is coarse and brecciated, and contains minute univalve shells in considerable numbers. The roof division of the coal consists about equally of coal and clay, the alternating layers being about two inches thick. The main clay parting varies in thickness from two inches to three feet, and always at the expense of the coal below. These horsebacks are frequent and annoying, as they cut out nearly all the available coal, the lower bottom being fourteen inches thick and worthless. Associated with them are clay veins, which are scarcely less troublesome. The coal itself is somewhat sulphurous, apparently more so than is usually the case along this portion of the river. The pit mouth is one mile and a half from the river.

About a mile below the mouth of Street's run Mr. Walton mines the coal at three hundred and sixty feet above low water. Above the bed four feet of shale and five feet of sandstone are exposed, the former resting on one foot of bituminous shale. The coal shows:

Roof division, 4 feet 2 inches; clay, 10 inches; lower division, 5 feet 5 inches; total 10 feet 5 inches.

The lower bottom is fourteen inches, so that the available coal is only four feet three inches thick. This thickness seems to be maintained with great constancy throughout the eastern portion of Allegheny county. Three hundred feet below the coal the Crinoidal Limestone is seen near the track level at Monterey. The dip is south-east.

At the mouth of Reiley's run, which enters Street's run two miles from the river, the coal is one hundred and sixty feet above the stream. It is opened near the head of the run by Mr. Cowan, who mines it to supply the farmers in the vicinity. There the coal shows:

Roof division, 2 feet; clay, 1 foot; lower division, 5 feet 6 inches.

The lower bottom is fifteen inches. The rest of the bed yields good coal.

Along the Brownsville road, southwest from the Whitehall Hotel, the blossom of a coal is frequently seen at three hundred and twenty feet above the *Pittsburg*. In the vicinity of the hotel it was found by Mr. Nies, while he was digging a cellar. He reports it as three feet thick. At twenty-five feet above it there is a constant limestone, four feet thick, light blue, quite pure, and breaking with a flinty fracture. The interval to the *Pittsburg* is concealed, but the coal is the same with that of the highest coal in Snowden township, and is either the *Washington* or the *Waynesburg* "*a*," more probably the former.

There are a number of openings along Sawmill run, but they show no material difference from those already enumerated.

53. LOWER ST. CLAIR TOWNSHIP, IN ALLEGHENY COUNTY.

This little township adjoins Baldwin along the river, and extends to opposite Pittsburg, including Birmingham within its limits. The section extends from two hundred and twenty feet above the *Pittsburg Coal* to three hundred and sixty feet below it.

Near Ormsby Station, and at one mile back from the river, are the mines of Messrs. Jones and Laughlin, which are operated chiefly to supply fuel for the extensive iron works belonging to that firm. The exposed section is a short one, as follows:

1. Concealed from top of hill		60′ 0″
2. Shale		8′ 0″
3. *Pittsburg Coal Bed.*		
Roof division	3′ 8″	
Clay	1′ 6″	10′ 1″
Lower division	5′ 9″	
4. Fire-clay		3″
5. Limestone		2″ 6″

The dip is south-east. The roof is irregular and the distribution of the coal is variable. At the mouth of the main entry the roof is single and shows two feet nine inches of coal, but at sixty yards within, this breaks up into four benches, separated by layers of white clay. These gradually come together, and within one hundred yards there is again but one bench of coal. At one place in the mine, as stated by the pit-boss, the roof consists of twelve feet of shale, containing thin streaks of coal throughout. Where there are two or more benches of coal, the coal is bright and compact, but when there is but a single bench it is shaly. In all cases it is impure, and, owing to the presence of much sulphur, even the more compact portions soon lose their brightness upon exposure.

The lower division is much affected by clay veins. The largest of these is in the main tunnel, at five hundred yards from the mouth, and is two feet six inches wide. Its course is north-east and south-west; others, however, follow a contrary course, and there seems to be nothing definite in the direction. They can be traced in different rooms, but usually maintain their integrity only for short distances, and sub-divide, though before disappearing or thinning out they often re-unite. They extend from the under-clay through the lower division, and in some cases not only through the entire bed, but also through some of the overlying rocks, for several of them are reported to have been found in digging wells. A similar statement may be made respecting clay-veins originating in the main clay parting, which pass through the roof division into the rocks above. Where these with so great vertical extent occur, the coal is more or less distorted on each side of the clay. The main parting varies from two to eighteen inches, and occasionally develops into a horse-back, which cuts out two or three feet of the coal below. These, however, are not wide and cause but little annoyance. They are always boat-shaped, and those observed taper toward the north-east and south-west. The lower bottom averages fifteen inches, and is much broken by vertical cleavage, as well as by binders of pyrites. It is too brittle and too impure to be mined The brick and bearing in benches are somewhat pyritous, but the upper bench, which averages not far from three feet, is a very fair coal, though somewhat variable in quality.

In the same vicinity are the Ormsby Coal Works, belonging to Messrs. J. Keeling & Co., at which the coal is extensively mined for sale in Pittsburg. There the following section was obtained:

1. Shales and sandstone		140′	0″
2. Ferruginous limestone		1′	0″
3. *Coal*		Blossom.	
4. Shale and sandstone		70′	0″
5. *Pittsburg Coal Bed—*			
Roof division	7′ 0″	13′	4″
Clay	10″		
Lower division	5′ 6″		
6. Concealed		25′	0″
7. Limestone and clay shale		18′	0″
8. Limestone		1′	0″
9. *Carbonaceous shale*			6″
10. Variegated shale		13′	0″
11. Sandstone		23′	0″
12. Limestone			8″
13. Shale		30′	0″
14. Sandstone		35′	0″

The dip is south-east. In the interval No. 1, the exposures are incomplete, but they are sufficient to show that the Great Limestone has disappeared, or if present it is utterly insignificant. Not a fragment of it occurs anywhere. Between the little coal and the *Pittsburg* there is no rock except shale or shaly sandstone, and no limestone occurs above it, aside from the ferruginous limestone given in the section.

The roof of the *Pittsburg* is of extreme thickness, and contains nine distinct benches of coal varying from two to six inches in thickness, which are separated by dull clays varying in like manner. This coal is very poor in all parts, and is not mined. The lower division does not differ in character from that at the last opening referred to. The slack is coked extensively, there being twenty ovens, but the coke is not pure enough to be used in the manufacture of iron, and is employed only in the drying of brick and similar coarse work.

The limestone, No. 8, is quite coarse, and in some parts is slightly brecciated. It contains many minute univalves. The sandstone, No. 11, is bluish gray, micaceous, and some layers show fine ripple markings. It is quarried to some extent for building purposes, but the proprietor of the quarry informed me that, though it makes a handsome stone, it does not withstand the weather. The same objection applies to No. 14. No. 10 and the clayey portion of No. 7, are used in making brick, the nodular limestone of the former being first removed. They answer well, the brick being of good quality.

On the road leading up from the Birmingham station, on the Panhandle railroad, the upper portion of the lower barren series is well exposed. The following section was obtained there, but being measured down the dip, shows a total error of twenty-five feet below the coal, the true distance from the coal to the Crinoidal Limestone being three hundred feet, as ascertained by levelings both above and below this locality. The section is,

1.	Shale and sandstone, imperfect exposure	150′	0″
2.	*Pittsburg Coal bed*	——	0″
3.	Concealed	33′	0′
4.	Shale, with iron ore	28′	0′
5.	Limestone	1′	4″
6.	Shale	14′	0′
7.	Limestone	2′	5″

8.	Sandy shale	27′	0″
9.	Limestone		10″
10.	Variegated shale	23′	0″
11.	Calcareous sandstone	2′	0″
12.	Sandstone	52′	0
13.	Variegated shale	11′	0
14.	Dark shale	45′	0
15.	Dark shale, not bedded	3′	0″
16.	Limestone	3′	0″
17.	Shale	24′	0″
18.	Crinoidal limestone	1′	0″

No. 14 is a laminated shale with vertical fissures, in which the lamination is vertical. It breaks out in blocks, and seems to be almost fibrous in its structure, weathering into strings like asbestus. The vertical joints are of considerable extent, and the face of this stratum is a marked feature of the abrupt hills opposite Pittsburg. Associated with the Crinoidal Limestone is a thin and impure coal, sometimes immediately under the limestone, but at others, separated from it by several feet of clayey shale. It is from six to eighteen inches thick.

Both the coal and the limestone come up above the track at a short distance below the inclined plane. The former is found to be very compact containing much ash and a comparatively small proportion of sulphur. It resists weathering well. The limestone is quite variable, and bears little resemblance to itself, as seen at Braddocks Station and near Six-mile ferry. There it is compact, bluish gray and somewhat granular; but here it is a tough, nodular rock, rather ferruginous, and showing a reddish tint on the freshly exposed surface. It is imbedded in a dark calcareous shale, which contains great numbers of fossils. Most of these are distorted, but *chonetes granulifera* occurs in good condition and in great profusion.

54. Union Township, in Allegheny County.

This, which is of small extent, adjoins Lower St. Clair on the west, and has a similar section.

Along the Panhandle railroad the Crinoidal Limestone and its coal are seen rising westward until within two-thirds of a mile of Temperanceville, or directly back of the Sheffield Steel Works. Here seems to be the crest of the Washington Anticlinal, and the *Pittsburg Coal* is upwards of four hundred feet above the river. Between the township line and this point the hills are extremely abrupt, but they afford no satisfactory exposures, as the vertical jointing of the shales has caused great slides, which conceal everything.

Just before reaching the Sheffield Steel Works the Crinoidal Limestone and its coal are seen in the bluff rising steadily toward the west, but back of these works the limestone and its underlying rocks are suddenly cut out by the shale No. 17 of the Birmingham section. This condition continues until within sight of Temperanceville, when the Crinoidal Limestone again comes in, but is now dipping north-west. At Temperanceville the litt'e coal is about six feet above the track.

On the hill, back of Temperanceville Station, there are several deserted openings in the *Pittsburg Coal*, and the interval between it and the Crinoidal Limestone is found to be three hundred feet. In the greater part of this township the coal has been almost wholly removed, and mining is carried on with such energy that before long the little that now remains will have been removed. No exposures occur above the *Pittsburg*, and that bed shows no difference from its character in localities already described.

55. Chartiers Township, in Allegheny County.

This adjoins Union on the west, and has Chartiers creek for its western boundary. The section extends from a short distance above the *Pittsburg Coal* to nearly four hundred feet below it. The valley of Chartiers creek occupies the western portion of the township, and in it all the higher rocks have been removed by erosion. The *Pittsburg Coal* is found only in small patches in the eastern part, and these, for the most part, have been worked out.

Along the railroad from Temperanceville to Nimick's Station the Crinoidal Limestone and its coal continue in sight, and they are evidently rising toward the north-west. This condition apparently lasts to the mouth of Chartiers creek, where a massive sandstone is seen, which is finely exposed on the other side of the river, at Woods' run, on the Pittsburg and Fort Wayne railroad. Underneath it, at that place, are dark shales, containing a thin coal. The Claysville axis seems to cross at this place.

At Nimick's Station the limestone and its coal are exposed in the bluff resting on a dull, ferruginous clay, fourteen feet thick. They remain in sight to the tunnel, where the following section was obtained:

1.	Concealed	60′ 0″
2.	Limestone	Fragments.
3.	Concealed	125′
4.	Laminated shale	30′
5.	*Coal*	0′ 4″
6.	Variegated clay	12′
7.	Limestone.	1′ 5″
8.	Variegated clay	10′
9.	Sandy shale	35′
10.	Clay shale, with nodular limestone	12′
11.	Sandy shale	6′
12.	Dark shale	20′
13.	Crinoidal limestone and shale	5′ 1″
14.	Coal	1′
15.	Shale to track	5′

The *Pittsburg Coal* occurs near by, and is two hundred and sixty feet above the track at the south end of the tunnel, where No. 9 of the section is the lowest rock seen. The Crinoidal Limestone has the following structure here:

Ferruginous limestone, 17 inches; sandy shale, 15 inches; red limestone, 14 inches; black calcareous shale, 15 inches; total, 5 feet 1 inch.

It is, however, very irregular, for at one spot in the northern approach to the tunnel, it is a mass of nodular limestone nearly eight feet thick. Each of the limestone layers contains many fossils, and the black shale at the base is very rich in individuals of a few species. The underlying coal is rather more sulphurous than it is along the river. Nos. 4, 9 and 12 of the section show marked fissures, all of them nearly vertical and filled with vertically laminated shale. Nos. 4 and 12 have the fibrous structure already mentioned as characterizing a stratum in the Birmingham section, which occupies the relative position of No. 9 in this.

Of the section, all the strata below No. 9 pass under the track in the tunnel, and at Ingram station, at the end of the southern approach, only ten feet of that rock are in sight. The little coal, No. 5, goes under the track about midway between Ingram and Crafton. In the cuts beyond Crafton the following section was obtained, which occupies the lower part of No. 3 in the tunnel section:

1. Sandstone	30′	0″
2. Clay, with limestone six inches	16′	0″
3. Sandy shale	22′	0″
4. Limestone	2′	0″
5. Sandstone	2′	0″
Clay, with nodular limestone	40′	0″
7. Nodular limestone	1′	5″
8. Sandy shale	10′	0″

No. 8 is No. 4 of the tunnel section. From Idlewild station, the road descends to Mansfield and the rocks fall in that direction at nearly the same rate, so that this section, as far down as the middle of No. 6, remains in sight quite to the township line, being seen in several cuts. Near the Pennsylvania Lead Works, the *Pittsburg* is one hundred and ten feet above No. 6.

56. Scott Township, in Allegheny County.

This lies south from Union and west from Baldwin township. Chartiers creek forms its western boundary. The section extends from about three hundred and fifty feet above the *Pittsburg Coal* to one hundred and twenty feet below it, but away from Chartiers creek there are no connected exposures, and everything above the *Pittsburg Coal* is practically concealed.

The coal already referred to in Snowden township is seen here on the high ridge near Mt. Lebanon P. O., and is exposed at the cross-roads half a mile east from that village. A light-colored limestone occurs twenty-five feet above it, and at Mt. Lebanon there is a coal seventy feet below it. The upper coal is most probably the *Washington*, though it may possibly be the *Waynesburg* "*a*." About three-fourths of a mile east from Woodville station, near Mr. Ford's residence, there is seen in the cross-road a bituminous shale, which may represent the *Sewickley Coal*. It rests on a limestone in the bed of the run.

Along Chartiers creek and its immediate vicinity, the *Pittsburg Coal* and its associated rocks are well exposed. At barely half a mile south from Mansfield the coal is mined by Mr. Glenn, at whose opening it shows the following structure:

Roof division, 1 foot 3 inches; clay, 1 foot; lower division, 6 feet 4 inches; total, 8 feet 7 inches.

The thickness of the lower division is no doubt less than that given. Of the lower bottom only six inches are exposed, but I was informed that it varies from eighteen to twenty-four inches. It is probable that in this estimate the "brick" coal is included. For the most part, the coal is clean, but in the top for fifteen inches there are numerous binders of pyrites, which render mining somewhat difficult. At the railroad tunnel just beyond this the following section is exposed:

1. Sandstone, seen	2′	0″
2. Sandy shale	20′	0″
3. *Pittsburg Coal Bed*	7′	0″
4. Clay	2′	0″
5. Sandy shale	10′	0″
6. Limestone and shale	8′	0″
7. Shale and shaly sandstone	10′	0″
8. Shale and *Coal*		6″
9. Limestone	3′	0″
10. Shale	9′	0″

The *Pittsburg Coal* is single in this section, the roof and clay being absent.

For some distance above this, the railroad runs on a shelf cut out from the rocks, so that the section below the middle of No. 2 is well exposed. The Limestone No. 9 is brecciated, and contains immense numbers of minute univalves. At the Glass Works of Messrs. Lindsay & Co., the coal is forty-two feet above the track and is mined. The section is,

Roof division, 3 feet 4 inches; clay, 1 foot; lower division, 5 feet 5 inches; total, 9 feet 9 inches.

The roof shows two benches, four and eighteen inches, but the lower one sometimes becomes only six inches thick. The main clay parting contains many thin streaks of coal, most of which are connected with the roof division. The *coal* is good throughout the lower division, even the lower bottom being clean, so that the whole bed is mined. Blasting is not necessary, and the coal is brought out by wedging, after the removal of the "bearing-in" bench. An extensive clay vein four feet wide is seen at fifty yards from the mouth of the pit, and crosses the main and one lateral entry. Another was cut in the main entry at two hundred yards from the mouth. The dip is very irregular, and "swamps" are a source of constant annoyance.

At Woodville Station, barely a half a mile farther up the creek, the railroad passes through a cut which gives the following exposure:

1. Sandy shale	20′ 0″	
2. *Bituminous shale*	0″	10″
3. *Pittsburg coal*	11′ 2″	13′ 7″
4. *Bituminous shale*	4″	
5. Argillaceous shale	0-3′ 0″	
6. Limestone	4′ 0″	
7. Shale	5′ 0″	
8. Clay	2′ 0″	

The *Coal* itself shows as follows:

Roof division, 4 feet 6 inches, to 6 feet 8 inches; clay, $\frac{1}{8}$ inch to 3 inches; lower division, 6 feet 8 inches.

The main clay parting is very insignificant; the coal is much tossed and the dip is irregular. The lower division of the bed is mined here by Messrs R. Lea & Bro., who state that the whole of it is clean and fully marketable.

At Bowerhill Station, nearly half a mile farther up, the coal is mined in the Bowerhill Coal Works. The structure here is similar to that at the last locality.

57. UPPER ST. CLAIR TOWNSHIP, IN ALLEGHENY COUNTY.

This lies west from Snowden and south from Scott. Chartiers creek forms its western boundary, separating it from South Fayette. The section extends from the Great Limestone to thirty feet below the *Pittsburg Coal.* Like the other townships already described, this is so deeply covered with debris that, above the coal, everything, with rare exceptions, is concealed. Fragmentary outcrops of the Great Limestone in the eastern portion of the township, show that that mass is becoming earthy, which may in part explain the paucity of exposures. Owing to the influence of the Washington axis, which is deflected eastward in this township, the *Pittsburg Coal* is available for a considerable distance along the runs in the north-east corner. The land nowhere rises very far above the level of the creek.

At Sodom, the lower layers of the Great Limestone are exposed in M'Laughlin's run. Somewhat more than a mile below the village, the *Pittsburg Coal* appears, and at a short distance beyond, is worked by Mr. M'Millan, at whose opening it shows,

Roof division, 3 feet 6 inches; clay, 10 inches; lower division, 5 feet 6 inches; total, 9 feet 10 inches.

Above the roof is one foot of bituminous shale, so rich that it has a fracture like cannel. The lower bottom is thirteen inches thick and worthless. At another opening farther down the run the coal shows a total thickness of twelve feet seven inches, this great increase being due to the presence of a layer of clay, three feet thick, in the roof division. This contains many thin streaks of coal. Above the coal there is shale, ten feet, on which rests sandstone, exposed to the thickness of forty feet. Near Bridgeville there are two openings worked by Mr Samuel Rimmel. In both, the lower division is five feet eight inches thick, but the roof is seen only in the lower one, where it is three feet ten inches. The main clay parting varies from seven to twelve inches. At fourteen feet below the coal is a limestone, rather coarse in texture, and containing many minute univalves. The interval is occupied by clay, which is more or less micaceous.

In these openings the coal is good, except in the "lower bottom," which is too poor to pay for working. In the upper bench, the top fifteen inches contains binders of sulphur, one of which, at ten inches from the clay, is occasionally strong enough to hold up the coal above. In this bench there are several layers of cannel, and as a whole the coal is rather open-burning. In the lower pit, extensive clay-veins occur, one of which is seven feet wide. Many of the larger ones have a north-east and south-west trend, but there are many others which have a different course. Those following the north-east and south-west direction seem to affect the coal, which is somewhat distorted on each side, while the others appear to have exerted no influence upon the structure.

About one hundred yards above the station the coal comes down to the track and is finely exposed in a cut for several hundred feet. The variations in the roof division are well shown here, as appears from the following measurements :

1. *Bituminous shale*	8″	10″			
2. Clay	6″	5″			
3. *Coal*	10″	1″			
4. Clay	3″	25″	14 ″	20″	18″
5. *Coal*	10″	10″	10 ″	8″	8″
6. Clay	25″	2″	½″	9″	2″
7. *Coal*	18″	10″	10 ″	11″	9″
8. Main clay parting	12″	10″	14 ″	6″	10″
Total, less No. 8	80″	63″	34½″	48″	37″

This portion of the bed is much troubled by clay-veins, of which two are shown in the cut, both extending from the main-clay parting through the roof to the overlying rocks. The coal is digged at several holes along the creek, just above Bridgeville, and shows a total thickness of thirteen feet, the roof being six feet ten inches. At twenty-three feet above it is a coal, two inches thick, which may represent some portion of the *Redstone*. Above this there seems to be only sandy shale for seventy feet. The coal goes under the creek at a short distance above this point, and farther up the creek the Great Limestone is exposed.

In the northern part of the township the coal is mined at several banks on Panther run. The general structure is shown at Mr. Donald's opening, where the section is:—

Roof division, 1 foot 6 inches; clay, 2 to 4 inches; lower division, 5 feet 3 inches; total, 7 feet.

The roof division is worked here and is said to be a very fair fuel. As usual, it yields a bulky ash, which is white and powdery.

58. South Fayette Township, in Allegheny County.

This adjoins Scott and Upper St. Clair at the west. It has Chartiers creek for its eastern and Robeson's run for its northern boundary. Miller's run crosses it from west to east. The section in the north-western portion reaches to fully four hundred feet above the *Pittsburg Coal*, but so thick is the covering of debris that nothing can be seen, aside from fragmentary exposures of limestone. Even the upper coals are wholly concealed. In the south-eastern portion the Great Limestone is exposed along Chartiers creek.

From the Washington county line to Hastings Station, exposures are frequent and satisfactory along Chartiers creek. The following section is exhibited between those points, the dip being southward all the way:

1. Concealed	70′ 0″
2. Limestone	2′ 6″
3. Clay shale	4′ 0″
4. Sandstone	5′ 0″
5. Concealed	25′ 0″
6. Sandstone	5′ 0″
7. Concealed	10′ 0″

8.	Limestone	40′ 0″
9.	Clay shale	1′ 0″
10.	*Bituminous shale*	1′ 6″
11.	Shale and sandstone	30′ 0″
12.	*Bituminous shale*	1′ 0″
13.	Limestone and shale	7′ 0″
14.	Dark shale	2′ 0″
15.	Limestone and shale	12′ 0″
16.	Shales	50′ 0″
17.	*Pittsburg Coal Bed*	11′ 8″

The base of the section is obtained from the shaft on the coal at Hastings' Station, where the bed is thirty-eight feet below the track, or twenty-eight feet lower than at Bridgeville Station. The section of the coal is:

Roof division, 5 feet 4 inches; clay, 6 inches; lower division, 5 feet 10 inches.

The lower bottom is sixteen inches, and is not mined. The rest of the lower division is extensively mined for shipping, and the coal finds a ready market. For thirty feet above the bed the rocks are clearly exposed in the shaft, and no coal is present. For sixty-five feet above the coal the rocks appear to be shale, but there is evidently some limestone in the upper portion of this interval, where the rocks are poorly shown. Above this for nearly fifty feet everything is concealed, and the missing strata are not found until near Boyce's Station. Of No. 8 probably nine-tenths consist of solid limestone. It is somewhat earthy, fractures with a smooth, dull surface, and on exposure breaks down into many sided angular fragments. This feature characterizes the lower division of the Great Limestone almost everywhere. The concealed intervals above do not seem to contain much limestone, and No. 5 appears to be chiefly sandstone.

The rocks immediately overlying the coal at Hastings' Station are argillaceous, but at a short distance north, where the railroad crosses the creek, they are sandy, and some of the layers are sufficiently thick and compact to be employed as building stone, for which purpose they are occasionally quarried. In a cut between Bridgeville and Bowerhill, the strata underlying the coal for a short distance are exposed. At twenty feet below the coal is a sandy shale, about thirty feet thick, of which some layers contain very good specimens of the more common species of plants.

Miller's creek enters Chartiers near Bridgeville, and at its mouth the *Pittsburg Coal* is almost level with the stream. The fall of the creek is not great, and its course in this township is from between west south-west and south of west, so that at no time up to the Washington county line is the coal more than a few feet below it, while frequently in the northward curves the roof division is exposed in the banks. At the mouth of Mohawk run Mr. Dinsmore's opening shows:

Roof division, 4 feet 8 inches; clay, 3 inches; lower division, 5 feet; total, 9 feet 11 inches.

Resting on the bed is one foot of bituminous shale, which is quite compact. The roof division has two very thick benches of coal, respectively twenty-four and twenty-six inches, both of them inferior, though the upper one is quite compact. As this is on the west side of the synclinal, the rocks are rising rapidly toward the north-west and the coal remains above the run for about a mile and a half, going under near the school-house. There are several openings, but they show no material differences from the one just given.

On Jackson's run, which enters Miller's creek at a short distance below the mouth of Mohawk, the coal is under the surface for nearly three-fourths of a mile from the creek, but above that, for somewhat more than a mile, it is available. The openings on this run are very similar to each other, and the following section obtained at Mr. Jackson's bank is characteristic of them all:

Roof division, 5 feet 11 inches; clay, 1 foot; lower division, 5 feet 8 inches; total, 12 feet 7 inches.

The roof shows two coal benches separated by twenty-nine inches of clay. This parting shows great variation, being only three inches on Mohawk.

Tom's run enters Chartiers creek nearly two-thirds of a mile below the mouth of Miller's creek. The synclinal passes near its mouth, and the coal is exposed along the run to above the mouth of Leggett's run. At Mr. Wilkinson's Coal Works the bed shows,

Roof division, 5 feet 5 inches; clay, 1 foot 2 inches; lower division, 5 feet 6 inches; total, 12 feet one inch.

On it rests bituminous shale eight inches, above which clay shale is exposed for ten feet. At Mr. Hopper's bank, one-third of a mile farther up the run, the roof is five feet eight inches, and the lower division only five feet three inches. The bituminous shale is absent, and the overlying rock is a sandstone, of which ten feet are exposed. At each locality the roof division shows two benches of coal separated by a thick clay parting, but the coal is quite inferior. In the lower division the bottom is thirteen inches, and impure. On Leggett's run the coal is mined by Mr. Nesbit, whose openings resemble those just given.

Along Robeson's run, few openings were observed until the north-west corner of the township was reached. There the coal is extensively mined at Oakdale, Noblestown and Willow Grove. The structure of the bed is very similar at all these works, the roof division being concealed, and the lower division varying little from five feet six inches. The dip is very irregular, and the pit-bosses complain much of "swamps." They claim that the coal seems to dip from all directions toward the centre of the hills. The upper bench is rather free-burning, and contains many small binders of pyrites. The coal is good throughout, and the whole is mined and shipped.

At one-third of a mile west from Willow Grove an opening near the railroad shows the following:

Roof division, 5 feet 5 inches; clay, 10 inches; lower division, 5 feet; total, 11 feet 3 inches.

The roof contains three feet six inches of coal in four benches.

59. North Fayette Township, in Allegheny County.

This lies north from South Fayette, and west from Robinson. Robeson's run is its southern boundary. The section extends from nearly two hundred feet above the *Pittsburg Coal* to nearly two hundred feet below it.

Just north from the Montours Presbyterian Church, which is on the Steubenville and Pittsburg pike, at the extreme eastern line of the township, the *Pittsburg Coal* is worked by Mr. S. Stewart, at whose opening it shows,

Roof division, 5 feet 10 inches; clay, 1 foot; lower division, seen, 4 feet.

The bottom bench is not exposed, but is said to be about ten inches thick. The "brick" is shown eight inches, resting on a thin parting. The roof shows four feet ten inches of coal in three benches, the middle one being really shale and coal in equal proportion. At twenty feet from the mouth of the pit, there is an extensive clay-vein, which has distorted the coal on each side. Above the coal there is no exposure, except a few feet of shale resting on the roof division.

About two miles west from this locality, along the pike, the coal is opened by Mr. A. M'Bride, at a short distance north from the road. There the coal shows,

Roof division, 4 feet 6 inches; clay, 1 foot; lower division, 4 feet 9 inches; total, 10 feet three inches.

There are no exposures above, excepting a few feet of sandy shale. The roof has four benches of coal near the mouth, which are ten, eight, five and nine inches respectively; but Mr. M'Bride stated that at one spot, where the roof had fallen and exposed the whole, the coal is five feet thick in one mass. The upper bench below is two feet ten inches, and contains some pyrites. The proportion cannot be large, for at the time the pit was visited, there was coal on the dump, which had been exposed to the sun and air for several months and was still compact, without any tendency to slake. The bed is jointed, and the coal comes out in handsome blocks. Along the faces at the joints, films of calcite are common. No clay-veins have been met in this opening, and the main clay parting never becomes injuriously thick.

Between this opening and Montour's run there are no exposures. The same is true of the road leading from Stewart's bank to the run.

Still following the pike, one finds, about a mile west from Mr. M'Bride's, another opening just east from Fayetteville, on the north side of the road. There the section is:

Roof division, 6 feet 7 inches; clay, 1 foot; lower division, seen, 3 feet 11 inches.

Only three inches of the bottom bench can be seen, and the probable thickness of the lower division is four feet six inches. The workable portion is only three feet eight inches, and of this four inches are lost in the "bearing in." The great thickness of the roof is owing to the dispersion of the top bench through four feet of shale. Directly under this is a bench eighteen inches thick, one third of which is clay. It is by no means improbable that the five feet of coal said to occur in M'Bride's opening may be like this, for if the mass were compact the whole might easily appear to a careless observer to be coal. The lower division at this opening yields coal of fair quality.

Near Shirland Postoffice, about two-thirds of a mile west from the last locality, the coal is worked just north from the pike. The roof was not seen in detail, but seemed to be about five feet thick. The lower division is four feet ten inches thick, and shows a double "bearing-in" bench, a local feature which was not observed at any other bank in the township. There was no coal out at the time this pit was examined, but in quality it is said to resemble that obtained from the other openings in the vicinity. The bed is again seen just west from Shirland, but beyond that the road rises above it, and so continues to near the western edge of the township, where it comes down to the coal at North Star Postoffice. North from the pike there are no openings, as the surface soon falls below the horizon of the bed.

Between Shirland and North Star, the road at one point reaches one hundred and twenty feet above the coal. At several places there are exposed two thin limestones, thirty and sixty feet above the *Pittsburg*, and on the upper one is a black shale. Nothing is seen above this until at the school-house, east from North Star, the top of the lower division of the Great Limestone is shown in the road at one hundred feet above the coal. Its thickness is not far from twenty-five feet. It has the usual magnesian look, but is earthy. The upper division, if present, is concealed.

At the school-house a road turns off southward, leading to the north branch of Robeson's run. It rises at once to two hundred feet above the coal, but the *Waynesburg* is not exposed, though no doubt it occurs in this interval. About a mile from the school-house, on this road, the *Pittsburg Coal* is seen in the run, and at a short distance beyond it is worked by Mr. Wilson, at whose opening it shows:

Roof division, seen, 2 inches; clay, 1 foot 3 inches; lower division, 4 feet 10 inches.

The coal is of good quality. There are several other openings in the vicinity, but they all have the same character. At about two miles from the pike the coal was formerly mined by Mr. M'Kee. Thence to the bend in the stream there are only two openings, Farmer's and Whitmore's, which do not differ in any essential particular from that just given.

At the saw and grist mill the coal is ninety feet above the run. On the road from this point to Noblestown, the coal is seen where the old road is crossed, say half a mile from the school-house on the hill, but the exposure is imperfect. Thence to Noblestown the road passes through the lower barrens, and everything is concealed except sandstones.

On the road leading from Oakdale to Fayette Postoffice, on the Steubenville pike, a deserted opening was seen about one-third of a mile from the north branch of Robeson's run. Beyond this the road rises rapidly, and the coal is not reached again until a short distance north from Fayette Postoffice. In this interval everything is concealed.

Along Robeson's run no openings were seen in the coal, and the exposed rocks belong wholly to the lower barrens. The only points of interest have already been referred to in another connection.

60. ROBINSON TOWNSHIP, IN ALLEGHENY COUNTY.

This lies east from North Fayette and Moon, has Chartiers creek for its eastern and the Ohio river for its northern boundary. Robeson's run at the south separates it from South Fayette. The section extends from two hundred feet above the *Pittsburg Coal* to nearly four hundred and fifty feet below it. The road leading from Ewing's Mills P. O., on Montour's run, south of east to Chartiers creek, is the extreme northern limit of the coal. Between it and the river, as well as along certain portions of the road itself, the coal has been removed by erosion. South from it, the coal area is almost continous to the Panhandle railroad, which follows Chartiers creek and Robeson's run.

Along the railroad the coal is worked near the Pennsylvania Lead Works, at a short distance north from North Mansfield.

The roof division is concealed and the lower division is five feet eleven inches thick. The coal is clean, but the top bench shows some persistent binders of pyrites. Just west from Mansfield the following section of the bed is shown at the Grant Coal Works.

Roof division, 5 feet; clay, 1 foot; lower division, 5 feet to 6 feet 4 inches.

There are no exposures above or below the coal. In the roof are three benches of coal, in all twenty-two inches thick. The whole of the lower division is mined and proves marketable, though the bottom is sometimes of inferior quality. Clay-veins occur in number sufficient to cause some trouble. The bed is one hundred and ten feet above the track, and is rising north-westwardly. A cut near by shows the following section, which belongs to the lower part of this interval:

1.	Sandstone	25′ 0″
2.	Sandy shale	20′ 0″
3.	Variegated shale	4′ 0″
4.	Limestone	1′ 6″
5.	Variegated shale to track	10′ 0″

This section continues in sight in the several cuts until opposite the hamlet of Fort Pitt, but there in a cut the sandstone No. 1 suddenly displaces all the rest of the section, and evidently extends far below the track, so that at Fort Pitt the following section is obtained.

1.	Concealed		60′ 0″
2.	Sandy shale		15′ 0′
3.	*Pittsburg Coal Bed.*		
	Roof division	5′ 10″	11′ 10″
	Clay	1′ 0″	
	Lower division	5′ 0″	
4.	Limestone		2′-4′ 0″
5.	Concealed		15′ 0″
6.	Limestone		Fragments.
7.	Sandstone		15′ 0″
8.	Concealed		10′ 0″
9.	Sandstone to track		65′ 0″
	Sandstone below track		10′ 0″

The coal here is extensively mined for shipping and seems to be very good, except in the bottom, which is not taken out. The roof has at its base a bench of coal twenty-eight inches thick. Horse-backs occur occasionally and clay-veins are very troublesome. The latter vary from two to five feet in width.

No. 9 of the section is the sandstone referred to as displacing the associated rocks in the cut. It is at least seventy-five feet thick here, and may include also Nos. 7 and 8. No. 9 is massive towards the base, where it contains a layer full of small nodules of iron ore.

Near Walker's station this mass is better exposed. There and for a considerable distance westward, almost to the first cut beyond the station, it forms a cliff along the creek about fifty feet high. The upper portion is shaly to flaggy, but the lower part is massive, coarse, felspathic, micaceous, cross-bedded and soft, weathering with rounded or honey-combed surface. Some of the layers are six to ten feet thick. The massive part is a handsome bluish-gray rock, which is quarried for building stone, though it seems to be rather soft for that purpose. The layers containing the nodular ore, being unfit for building stone, are largely employed for ballasting the road.

In the cut, nearly midway between Walker's and Hays' Station, the sandstone is gone. It disappears abruptly, about two hundred yards east from this, but there being no excavation to expose the condition, the structure cannot be fully traced out. The rock evidently disappears as suddenly here as it appeared at Fort Pitt, for the bluff cliff along the stream breaks down within a distance of twenty-five feet, and becomes a gentle slope. In the cut there are exposed variegated shales and limestone underlying those seen at Fort Pitt. At Hays' Station the section is as follows:

1. Debris		6′ 0″
2. *Pittsburg Coal bed*		
Roof division	4′ 2″	11′ 1″
Clay	1′ 8″	
Lower division	5′ 3″	
3. Concealed		75′ 0″
4. Flaggy sandstone		20 0″
5. Shale		3′ 0″
6. Limestone		2′ 0″
7. Concealed		20′ 0″
8. Sandstone and shale		20′ 0″
9. Limestone		1′ 6″
10. Sandy shale to track		10′ 0″
11. Same below track		25′ 0″
12. Variegated shale		12′ 0″

No. 6 is the limestone cut out at Fort Pitt. It is fine-grained and dark gray in the upper layer, but brecciated and dark blue, with earthy light blue included fragments in the lower portion. It is fossiliferous throughout. No. 9 is seen also in the cut just east from the station. The rocks are rising north-west much faster than the track. The coal has evidently been mined quite extensively here, but the works were in bad condition when visited, and seem to have been deserted for some time. A bad clay-vein cuts the whole bed from under-clay to roof-shales at a short distance from the mouth of the pit. Its course is north-west and south-east, and the coal is much distorted on each side of it.

There are no other openings along the railroad, and the sections obtained along that line are devoid of interest.

Near the head of Scott's run, which enters Robeson's run at Walker's station, the *Pittsburg Coal* is worked by Mr. W. M'Clane, at whose bank it shows,

Roof division, 5 feet 11 inches ; clay, 1 foot ; lower division, 5 feet 4 inches ; total, 12 feet 3 inches.

On it rests one foot of bituminous shale, above which is clay shale 4 feet. The roof contains forty-four inches of slaty coal.

On the South Fork of Campbell's run the coal is worked by Mr. Smith, at a short distance south from Palmersville, and on the North Fork by Mr. Hack and Mr. Nickel, in the vicinity of Harrison's school-house. At those localities it resembles the section just given. At the junction of the two forks of Campbell's run, the massive sandstone seen at Walker's Station on the railroad, occurs in bluffs, and shows the same features as on Robeson's run. It has been quarried for building purposes.

Where the Pittsburg and Steubenville pike crosses Chartiers creek, the Crinoidal Limestone is in the stream. A direct measurement up the dip from this point to Mr. Yeager's opening in the *Pittsburg Coal* on the pike, gives the interval between the two strata at two hundred and eighty-five feet. At that opening the coal shows,

Roof division, 4 feet 9 inches ; clay, 1 foot ; lower division, 5 feet 6 inches ; total, 11 feet 3 inches.

At Remington P. O., on the pike, the coal has been worked quite extensively, as is evident from the piles of slack which were seen in the hollows near some deserted openings. Several banks are now operated in the immediate vicinity of the village, but none of them shows the roof in full. One at the roadside in the village has the following structure:

Roof division seen, 1 foot 2 inches; clay, 1 foot; lower division, 5 feet 3 inches; total.

The roof as exposed consists only of coal. The lower division is very clean and evidently somewhat free-burning.

Near Remington, as well as at several other localities along this pike, there is a bituminous shale resting on a nodular limestone, while above it there occurs some more limestone. This shale is about sixty feet above the Pittsburg Coal, and is undoubtedly the same with that observed on this pike in North Fayette, near Shirland P. O., and in Hanover township, of Washington county, east from Florence. Near Summitville the coal is worked by Mr. Edmundson. It is mined also on the road leading from the pike at Montour's church to Montour's run. Mr. Guy's opening here shows:—

Roof division, 5 feet; clay, 1 foot; lower division, 5 feet 3 inches; total, 11 feet 3 inches.

North from the pike openings are quite numerous, but the coal rises rapidly north-westward and is soon out of the hills, so that the northern boundary is approximately marked by the road leading from Ewing's Mill's P. O., on Montour's run, to Chartiers creek. Starting northward from the pike at half a mile east from Remington P. O., one sees at the junction of the road pike the massive sandstone previously observed on the railroad. Under it are reddish shales extending to one hundred and eighty feet below the Pittsburg Coal. Between the sandstone and the coal there seems to be nothing except shales and sandstones, aside from two thin limestones, respectively twelve and thirty feet below the coal. On this road, at rather more than a mile from the pike, there is an opening in the coal belonging to Mr. Hearn. The roof division is not fully exposed, but seems to be not far from three feet thick. The exposure is as follows:

Roof division seen, 4 inches; clay, 1 foot 5 inches; lower division, 4 feet 10 inches.

The coal is soft, but is said to be good fuel and clean. The bed is growing thinner and the following is the general structure in the opening here:

1. *Coal*	2′	11″
2. *Coal* and partings		6″
3. "Brick" *coal*		8″
4. "Lower Bottom" *coal*		9″
Total,	4′	10″

In all the openings in this little area, the "bottom is poor and brittle, containing much ash and pyrites. At some of the openings, however, it is mined. From this locality to the Still school-house, the coal is frequently exposed at the roadside. It rises northward, so that at the school-house it is one hundred feet higher than at Mr. Hearn's bank. This point is very near the extreme north-east outcrop of the coal in this township. About half a mile north-west from this place is Mr. Alex. Speer's opening, which is on the extreme northern line of the bed. The exposure there is:—

Roof division seen, 2 inches; clay, 1 foot; lower division, 4 feet 4 inches.

Only four inches of the bottom are exposed, so that the thickness of the lower division is evidently the same as at Mr. Hearn's. The area of coal is very limited and the roof to the bed is so thin that the coal throughout resembles "crop-coal," being stained by iron; but it is said to be very good fuel and evidently finds a ready market, as the bank is kept steadily in operation.

At little more than one-third of a mile west from this place, the road falls below the coal and soon descends to Moon run, where, at two hundred and thirty feet below Mr. Speer's opening, the Crinoidal Limestone is seen, with a little coal at fifteen feet above it. Immediately above and below the limestone are variegated shales. From this run to Ewing's Mill, on Montour's run, the road lies wholly north from the area of the coal, and at the latter place is not far from three hundred and fifty feet below the *Pittsburg*, the dip being calculated in addition to the measured interval.

At somewhat less than a mile from Ewing's mills, on this road, and near a little hamlet, a knob rises which catches the coal, and is continuous with the coal area at the south. The first opening seen south from the road is that of Mr. W. B. Stewart, on the way to Summitville, where the section is,

Roof division seen, 10 inches; clay, 10 inches; lower division, 5 feet.

The roof is said to be very nearly five feet. At a short distance further on, is the opening belonging to Mr. W. Andrews, which shows,

Roof division, concealed; clay, 1 foot 3 inches; lower division, 5 feet 2 inches.

In the lower division there is a double "bearing-in" bench, and all the partings are unusually thick. The upper bench is a very handsome, almost semi-cannel coal. It is very clean. This seems to be more or less characteristic of the bed in all the openings in this vicinity.

Between this and the pike there are no other openings. There are no exposures above the coal.

61. Moon Township, in Allegheny County.

This faces the Ohio river, and lies between Robinson and Findlay. The section extends from seventy-five feet above the *Pittsburg Coal* to about five hundred feet below it, so that the exposed rocks belong, for the most part, to the lower barren series. Above the *Pittsburg* there are no exposures, and it is impossible to determine even the character of the rock immediately overlying it. The coal is confined to narrow strips, one of which extends along the ridge road south-east from Sharon for nearly four miles, and the other is in irregular patches along the road from Sharon to Clinton. A small outlier of the coal occurs about a mile and a half from Sharon, on the road to Middletown. The numerous tributaries to Montour's run, which flow through the southern and western portions of the township, have cut their channel-ways deeply, and so have removed the coal from the hills in those parts. Exposures of the rocks below the coal are so rare that no connected section could be obtained. The ancient river-channel spread over the hills for three miles south from the present channel, and its detritus has been distributed so as to cover everything.

On the road from Sharon to Clinton an opening was seen near the township line, but it could not be examined. Between this point and Sharon the coal is not worked, though it is frequently reached by the road and its blossom exposed. At Sharon there are several openings, but none of them shows the roof. The lower division has a parting in the upper bench and the detailed section is as follows:

1. *Coal*		10″	5′ 10″
2. Parting		1″	
3. *Coal*	2′	2″	
4. *Coal* and partings		6″	
5. "*Brick*" *Coal*		9″	
6. Lower B. *Coal*	1′	6″	

It is by no means improbable that No. 1 is part of the roof division.

This thickness seems to be maintained in one of the banks for a considerable distance, but in some of the others the bed is thinner. North from this village the coal is wanting as the country falls off rapidly towards the river. On the road leading east from Sharon to Ewing's mills, the coal is caught occasionally to a little way beyond the first cross roads. There it is in the hill, but is not worked. About midway between that point and Middletown, an outlier is found on land belonging to Mr. Trunick, but is of small extent. The surface falls off from it in all directions.

About two miles south-east from Sharon, on the Ridge road, the coal is mined by Mr. J. M'Cormack, at whose bank it shows,

Roof division, 3 feet 9 inches; clay, 10 inches; lower division, 4 feet 10 inches; total, 9 feet 5 inches.

The parting in the upper bench of the lower division is present here, and divides that bench into two nearly equal portions, the total thickness being two feet eleven inches. The roof division shows three feet of coal in four benches, all of which is bony. The coal from the lower division is caking, leaves little ash and has but a small proportion of sulphur. On the same road, and south-east from this opening, is one belonging to Mr. Nolte, at which the section, as far as exposed, is as follows:

Roof division, seen, 1 foot; clay, 1 foot 6 inches; lower division, 4 feet 7 inches.

The roof division is not far from four feet thick, as appears from the blossom. The partings in the lower division are unusually thick, and the lower two benches cannot be distinguished. At a little distance farther south-east Mr. M'Cormack mines the coal, which there shows a section like the last. This is the end of the area, and the surface falls thence in all directions towards Montour's run.

Exposures of rocks below the *Pittsburg Coal* are few and unsatisfactory. Occasionally a coal blossom was seen at sixty feet below, and near Guy's mill, on Montour's, run, another blossom was seen at two hundred and twenty feet below. The Crinoidal Limestone was observed on both forks of Meek's run, where the road leading south from Middletown crosses them.

62. Findlay Township, in Allegheny County.

This is the western township, and borders on Washington and Beaver counties. The divide between the waters of Raccoon creek and those of Montour's run passes west of south along its central line. On this divide the *Pittsburg Coal* is found. A small outlying patch of the coal occurs in the south-eastern corner on the ridge, about two-thirds of a mile north of west from M'Clarn's mills. The section extends from one hundred and thirty feet above the *Pittsburg Coal* to about two hundred feet below it. Aside from the narrow limits already given, the coal is altogether wanting.

On the road leading from North Star Postoffice to Clinton, the coal is mined at somewhat less than a mile from the former place by Mr. J. Cox, at whose opening it shows:

Roof division, 5 feet 10 inches; clay, 1 foot; lower division, 4 feet 9 inches; total, 11 feet 7 inches.

The roof contains coal and clay in nearly equal proportions, each in thin layers. The coal from the lower division is a very fair fuel, but evidently has more sulphur than is ordinarily found in this bed. The pyrites is distributed throughout. About a mile and a half from this, on the same road, the coal is mined by Messrs. Hays and Auld, whose openings show a section similar to that just given. Just a little way east from this road, on the first fork south from Clinton, Mr. Lewis has an opening, at which the clay and lower division are exposed, as follows:

Clay, 1 foot 6 inches; lower division, 5 feet.

The brick and lower bottom are not separate, and the "bearing-in" bench is double, there being three middle partings. The coal is of the usual quality, and is an excellent fuel. Clay veins seem to be annoying, owing to the frequency of their occurrence. They do not cause any distortion in the coal, and in several instances have no connection with the under clay. One, which is eighteen inches wide, reaches down only to the top of the "brick."

Near Clinton the coal is mined by several persons. The area here is comparatively extensive, embracing in all not less than two square miles, and, owing to the heading of numerous streams in this vicinity, it is accessible on all sides. The ridge rises to one hundred and thirty feet above the bed, and the coal is thoroughly sound. At Mr. Irwin's opening, just west from the village, the following measurement was made:

Roof division, 4 feet 8 inches; clay 1 inch; lower division, 5 feet 3 inches; total, 10 feet.

The top bench of the lower division in this portion of the county rarely exceeds two feet nine inches, but here it is three feet eight inches. The bottom bench of the roof is mined, but is bony. Throughout, the coal is soft enough to be brought down with the pick, is free burning, and has a high reputation as fuel. No clay veins have been encountered in this mine, which is quite extensive. There are no horsebacks in the lower division.

On the road leading from Clinton to Sharon the coal is worked near the former place by Mr. Twyford and others. Mr. Twyford's coal is said to be the best obtained in this township. The lower division of the bed is somewhat thinner than at Mr. Irwin's bank, near Clinton. From this locality the road runs almost on the level of the coal, being seldom much above or below it until within half a mile of the township line, where the coal is seen in the road for the last time. This is practically the last exposure northward in this township.

On the road leading southward from Clinton to M'Clarn's Mills, by way of Groomsville, the coal remains in sight for nearly a mile from the place first named. There have been many openings here, but for the most part they have been deserted. Mr. M'Cullough works a bank at a little way off the road which yields a good coal. In the vicinity of Groomsville the hills are too low to catch the coal, and no openings are seen until, after crossing both branches of Henry's Fork of Montour's run, Mr. Tomlinson's bank is reached, on the little outlier already referred to. The exposure here is:

Roof division seen, 3 feet 2 inches; clay, 1 foot; lower division, 4 feet 6 inches.

The roof is not fully shown. The coal from the lower division is said to be rather softer than that brought from Pittsburg, but is preferred as a steam coal, having been fully tested at M'Clarn's Mills. It is free-burning, forms no cinder or clinker, and leaves a bulky, but light and powdery white ash, resembling that from hickory wood.

About two miles south from Clinton there is an opening belonging to Mr. J. M. Stewart, in which the lower division is four feet five inches thick and shows a two inch parting at eight inches from the top. This thin top bench and the bearing-in contain much pyrites. It is possible that the former belongs to the roof division. The coal is caking throughout. Clay-veins are a source of much annoyance, not only because of the loss of time in cutting them, but also because in many instances the coal is crushed and distorted for some distance on each side of them. Horsebacks occur, but they seldom cut away more than the thin top bench.

CHAPTER XIII.

BEAVER COUNTY SOUTH FROM THE OHIO RIVER.

BY I. C. WHITE.

This is the north-western county of the district. The exposed section is approximately as follows:

1.	Shale and debris	30′	0″
2.	*Coal*	0′- 5′	0″
3.	Sandstone and shale	40′	0″
4.	*Pittsburg Coal Bed*	8′	0″
5.	Shale	8′	0″
6.	Limestone	5′	0″
7.	Concealed	100′	0″
8.	Shale	10′	0″
9.	*Coal*	1′	6″
10.	Sandy shale	35′	0″
11.	Limestone	4′	0″
12.	Calcareous shale, fossiliferous	3′	6″
13.	Slaty *coal*	1′	0″
14.	Shale	10′	0″
15.	MORGANTOWN SANDSTONE	60′- 70′	0″
16.	Shaly sandstone	35′- 50′	0″
17.	*Coal*	0′- 3′	0″
18.	Flaggy sandstone	25′- 35′	0″
19.	*Crinoidal limestone*	2′- 5′	0″
20.	*Coal*	1′	4″
21.	Variegated shale	25′- 30	0″
22.	Bluish sandy shale	50′- 60′	0″
23.	Red clay shale	0′- 20′	0″
24.	*Cannel coal*, local	6′	
25.	Laminated sandstone	90′-100′	0″
26.	Limestone or calcareous shale, black	0′- 5′	0″
27.	Dark shale	0′- 15′	0″
28.	*Coal (Elk Lick?)*	0′- 2′	6″
29.	Sandy shale	25′- 35′	0″
30.	MAHONING SANDSTONE	30′- 70′	0″
31.	Shale	0′- 12′	0″
32.	*Upper Freeport Coal Bed*	0′- 4′	0″
33.	Fire-clay and shale	5′- 15′	0″
34.	Freeport limestone	2′- 4′	0″
35.	Sandy shale	55 - 65′	0″
36.	*Lower Freeport Coal Bed*		10″
37.	Shale	8′- 10′	0″
38.	FREEPORT SANDSTONE	75′- 85′	0″

39. Shale	0′- 30′	0″
40. *Coal*, "strip,"	1′- 2′	0″
41. Shale, with iron ore	20′- 45′	0″
42. *Kittanning Coal Bed*	3′- 1′	6″
43. Fire-clay	5′- 8′	0″
44. Flaggy sandstone	60′- 75′	0″
45. Iron ore		6″
46. Ferriferous limestone	1′- 15′	0″

The upper coals are found only in small outliers near the southern line of the county, and the rocks belonging to the lower coal series—those below the Mahoning sandstone—are seen in full only along the Ohio river and its immediate vicinity. The strata forming the mass of the hills in the greater part of the county, belong to the lower barren series, or that portion of the column included between the *Pittsburg Coal* and the Mahoning sandstone. The section is merely approximate, for as will be seen by reference to the details given hereafter, the variations in the character of the rocks are excessive.

63. Hanover Township, in Beaver County.

This township includes the highest measures found in the county, reaching up to the top of the general section. The *Pittsburg Coal* is seen in small outlying patches near the Washington county line, which, in all, contain not far from six hundred acres.

The Bulger axis crosses through its south-western corner, entering it near the mouth of Big Traverse creek, and passing out of it into Washington county, near Mr. Armour's residence. The synclinal, or trough, north-west from this axis, crosses Big Traverse creek at Mr. Kiefer's mill, and enters Washington county near the village of Frankfort. Though this synclinal is very obscure, and amounts to little more than a flattening of the dip, yet it is of some importance in its effects, as had it not existed, there would have been no *Pittsburg Coal* in Beaver county, for that bed would have been carried above the tops of the highest hills. The lowest stratum in the township is the black shale, usually occurring at one hundred feet above the *Upper Freeport Coal*, so that with the exception of the small area of the upper coals in the southern portion, the only rocks exposed in this township, belong to the lower barrens.

At a mile and a half east from Frankfort the *Pittsburg Coal* is mined by Mr. Cooley, at whose opening the following section is shown:

Roof division.			
Coal	2′ 6″	3′ 3″	9′ 8½″
Clay	5″		
Coal	4″		
Main clay parting		0′ 6″	
Lower division.			
Coal	2′ 1″	5′ 11½″	
Parting	3″		
Coal	9″		
Coal and partings	4½″		
Coal	2′ 6″		

The roof division is not mined, as it is full of slate. The upper bench of the lower division is a very good coal, coming out in almost rectangular blocks, and can be used for smithing. The best coal, however, is in the little bench below it, as that is quite clean and free from sulphur so as to be good for smithing purposes. Of the bottom, which includes both the brick and the lower bottom benches of the other parts of the district, only about eighteen inches are removed, as the portion below is very impure. The upper part is here known as the "brick" coal, from the shape of its blocks.

In the road above this opening is seen the coal No. 2, of the general section.

One mile east from Mr. Cooley's bank the same coal is worked by Mr. Babbett, near the county line, where it exhibits the following section, which is in striking contrast with that just given:

Roof division		3″	5′ 11½″
Coal		0′ to 0′ 3″	
Main Clay parting		1′ 4″	
Lower division—			
Coal	2′ 8″	4′ 4½″	
Coal and partings	3½″		
Coal	1′ 5″		

The roof is occasionally absent at this locality, which is odd, as the overlying rock is a shale. One mile north-west from Frankfort Mr. Leeper's bank shows as follows:

Roof division, 2 feet 3 inches; clay, 1 foot 2 inches; lower division, 4 feet 6 inches; total, 7 feet 11 inches.

Here the roof division is coal, and the upper bench of the

lower division is divided nearly midway by a three-inch clay parting. At this bank horse-backs are very annoying, sometimes cutting out almost all the coal. The largest area of the coal is in the immediate vicinity of the village of Frankfort, where it is mined by Rev. Mr. Rockwell and others.

As previously stated, No. 26 of the general section is the lowest rock exposed in the township. It is seen near the mouth of Big Traverse creek, where it is only eighteen inches thick.

No. 15, the Morgantown Sandstone, is seen all along Big Traverse creek, and is always a compact rock. Three-fourths of a mile above Kiefer's mill, on a small tributary of that creek, mineral springs issue from this rock at a place known as Frankfort Springs, where the stream has cut a miniature cañon, very narrow, and with vertical walls nearly forty feet high. This locality has been a summer retreat for fifty years, extensively patronized by persons from Pittsburg, Wheeling and other cities along the Ohio river. The water of the springs is said to have valuable medicinal qualities. This sandstone occurs in bold cliffs at the grist mill near the head of Big Traverse.

No. 19, the Crinoidal Limestone, was not found in the western portion of the township, where it should occur, on King's creek, but it is frequently exposed on both Big and Little Traverse creeks. Near the mouth of the former it is one hundred and sixty-five feet above the stream, but as the bed of the creek rises very fast and the dip flattens, the limestone goes under about two miles above Keifer's mill.

The northern part of this township is the divide between the waters flowing into Mill creek, and those flowing into Raccoon, and there being no streams to expose the rocks, everything is concealed. On this divide there are large and almost perfectly level tracts of land.

64. Independence Township, in Beaver County.

The strata exposed in this township are those embraced between Nos. 14 and 35 inclusive of the general section. The Bulger axis passes nearly through the centre, crossing Raccoon creek just below the village of Independence, where it brings up the *Upper Freeport Coal.*

No. 15, the Morgantown Sandstone, is well exposed in this township, being seen almost continuously along Raccoon creek from the Washington county line, where it is well up in the hills, to the mouth of Service creek, where it caps the highest points. It is, almost without exception, compact, and, with the shaly sandstone below it, sometimes forms perpendicular bluffs more than one hundred feet high. This is the rock of the "backbone" at the mouth of Big Traverse.

The blossom of No. 17 is seen at the "backbone," and in the north-western part of the township that bed was once worked by Mr. Olney, half a mile south from Service creek. He reports that it is an excellent coal, well fitted for smiths' use, but the bed varied so abruptly that working it was unprofitable. The extremes of thickness there are one and three feet.

No. 19, the Crinoidal Limestone, is the most widely exposed stratum, as it can be seen in almost every hill in the township. Where Raccoon enters the county it is about forty feet above the water, and as the stream flows northward the limestone gradually rises, until in the northern part of the township it is found in the tops of the highest hills. It varies from two to four feet in thicknesss, and is frequently burned by the farmers for use on their lands, though by no means so frequently as it should be, for the land needs lime as a dressing.

The shales and sandstones between Nos. 20 and 26 vary considerably in thickness. At the "backbone" they are only one hundred feet thick, whereas at Bock's mills they are one hundred and ninety feet. In the "backbone" section No. 26 is seen at ten feet above the creek, and only six inches thick, but at a short distance below the mills it becomes four feet. Sometimes it is a very hard black limestone, but at others it is only a loose mass of black calcareous shale. The blossom of the little coal under it, No. 28, is seen at numerous places along Raccoon, but it never attains workable thickness.

No. 29, the Mahoning Sandstone, is first seen on Raccoon creek, at a short distance above Independence, where it occurs in massive beds along the stream, but below this it becomes shaly, and does not regain its compactness until Service creek is reached, where it is found in huge blocks and cliffs as far up as Sterling's mill. The course of the stream while in this stratum is very tortuous, sometimes describing almost a complete circle. The fall is quite abrupt and often precipitous. At the mill the fall is twenty-two feet.

No. 32, the *Upper Freeport Coal*, comes to the surface at a little way below Independence, but being rarely more than one foot thick, it is practically of no economical importance, though a few persons get their fuel by stripping it out of the bed of Raccoon. It is never more than ten feet above the stream in this township, so that the Freeport Limestone is seen only at one locality, half a mile below the mouth of Service creek. where it is a layer of ferruginous nodules in the bed of Raccoon creek. As this township is entirely destitute of workable coals the people obtain their fuel either from the *Pittsburg* bed in Hanover and Hopewell, or from the *Upper Freeport*, farther down Raccoon creek, where it becomes more important.

65. Hopewell Township, in Beaver County.

In the south-eastern portion of this township, there are three or four small patches of the *Pittsburg Coal* near the summits of some high hills, that rise to six hundred feet above the Ohio river. At Mr. Cortney's bank the following section is exposed:

Roof division—			
Slaty *coal*	2′ 0″		
Clay	5″		
Coal	10″	4′ 9″	
Clay	8′		
Coal	10″		
Main clay parting		0′ 2″	8′ 9″
Lower division—			
Coal	1′ 9″		
Coal and partings	5½″	3′ 10″	
Coal	1′ 8″		

This bed is worked by Mr. Thompson and Mr. Wallace in the same neighborhood. It is often inferior, as it is covered by only thirty feet of rock, which does not always form a good roof. When not rotten or half slaked, it comes out in slaty looking blocks. In this vicinity, Mr. Eachel once had an opening from which the Economites obtained coal for their factories as long as fifty years ago.

This township shows a long column of rocks, beginning with the *Pittsburg Coal* and extending downwards to No. 40 of the general section. No exposures, however, occur in the lower barrens above the Crinoidal Limestone, except that the massive sandstone. No. 15, is seen near Mr. Eachel's residence.

The northern outcrop of the Crinoidal Limestone commences near the northern border of the township, on Raccoon creek, and extends in a very irregular line to the Ohio river, near the upper end of Crows' island. In the hill opposite Economy it is three hundred and fifty feet above the river and five feet thick. It also occurs near the saw-mill on Tramp Mill Run, where it is quite siliceous. The variegated shale, No. 21, which constantly underlies that limestone, is often seen in broad bands stretching across cleared fields, where the grass is not well set, or encircling high knobs. The shaly sandstones, below this to No. 27, are well exposed in the hills opposite Economy, where they rise in perpendicular cliffs one hundred and fifty feet high. They are never massive or compact, but in all cases are thinly laminated and shaly.

No. 26 is seen at numerous localities. At Mr. Figley's, near Raccoon creek, it is four feet thick and quite compact, and in the section opposite Economy it is five feet thick and rich in fossils. It has never been burned for lime, and it is doubtful whether the stuff would slake, as it contains much silica and other foreign matter, The little coal under it, which is probably the representative of the *Elk Lick Coal* of Rogers, was once opened at the head of Logstown run by Mr. Temple, but being only thirteen inches thick was abandoned. In the section opposite Economy it is twenty inches and rests immediately upon sandstone, with fifteen feet of shale between it and the limestone above. This is the same coal which Mr. Figley attempted to open on Raccoon, and reported as being two feet six inches thick.

No. 30, the Mahoning Sandstone, was found compact at only one locality within this township. On Logstown run, two miles above its mouth, it occurs in great cliffs near Mr. Spaulding's, where it rests directly upon the *Upper Freeport Coal.* At all other points it is a mass of sandy shale and sandstone. It sinks under the Ohio at a short distance above the Beaver county line.

No. 32, the *Upper Freeport Coal* has been worked by drifting at only two places. Mr. Spaulding, on Logstown run, found it twenty-two inches thick, but his opening has fallen in, and the bed cannot be examined. One mile below Mr. Spaulding's it is mined by Mr. Snider, at whose bank the exposure is:

1. Sandy shale		3′ 0″
2. *Upper Freeport Coal.*		
Coal	1′ 6″	2′ 0″
Clay	1″	
Coal	5″	
3. Fire-clay and shale		10′ 0″
4. Freeport Limestone		3′ 0″

Though very thin, the coal has been mined here for local use, but it is very poor, as it contains much slate and sulphur. The clay parting seen at Snider's bank does not usually occur where the bed is thin, but in Greene township it is characteristic. This bed has has been stripped out of the creek on Raccoon, near the mouth of Tramp Mill run. The Freeport Limestone at Snider's is in two layers, quite compact, and of a light buff color. A minute univalve occurs here in the limestone. The rock is frequently exposed along the Ohio river in this township.

The rest of the strata down to No. 40 of the general section are reached in the northern portion of the township, along the Ohio, but they are so concealed by the terraces that nothing can be said respecting them.

At a short distance above the western end of the ferry, at Economy, a shaft was sunk by Mr. M'Donald. From one of the workmen who helped sink it, I learned the following facts: The shaft was commenced at thirty feet above the Ohio river, and carried to a depth of one hundred and sixty feet. Two small beds of coal were cut, and at the bottom of the shaft one was reached by drilling, which was supposed to be five feet thick. The water, however, proved so annoying that the enterprise had to be abandoned. As the shaft begins about twenty-five feet below the *Upper Freeport Coal*, the one struck at the bottom is probably the *Kittanning.*

66. Moon Township, in Beaver County.

As we pass from Hopewell into this township, we find the strata of the former running out in the summits of the hills, while new ones are rising from the bed of the Ohio, so that in Moon township we get very nearly to the bottom of the column as given in the general section. No. 44 of that section is exposed almost wholly at some points between Phillipsburg and the mouth of Raccoon creek. The Crinoidal Limestone was observed at no place, though some of the high knobs on the southern border must catch it, or come very near doing so. The highest rock exposed is shaly sandstone to the thickness of one hundred and fifty feet above No. 26.

In the section at Moffatt's mills, No. 26 occurs as a thin arenaceous shale, quite fossiliferous, but containing very little calcareous matter.

The Mahoning Sandstone, No. 30, forms an almost continuous belt for more than half way round the township, being compact all along the Ohio and up Raccoon to Moffatt's mills. It is extensively quarried for three miles along the river above Phillipsburg, and it is a very important stratum, as it is an excellent building stone, and within easy reach of transportation. The quarries in this township furnish Pittsburg with a great proportion of the building stone used in that city. An extensive quarry was seen back of Phillipsburg. Of this rock the massive cliffs near the summits of the river hills below Phillipsburg are composed. Along Raccoon creek the Mahoning Sandstone occurs in vertical walls along the hills, and frequently huge blocks of it have broken off and fallen into the stream below.

No. 32, the *Upper Freeport Coal*, is worked at many points along Raccoon creek, in this township. Mr. M'Millan's bank, two miles above Moffatt's mills, shows,

Coal, 7 inches; clay, ¼ inch; *coal*, 2 feet.

This is the average thickness of the bed, though the clay parting is not always present. The coal is of very fair quality and is used by smiths. It burns with a rather dull flame and makes a very hot fire, but leaves a good deal of ash. It is opened near Moffatt's mills by Mr. Alison. Two miles above the mouth of Raccoon creek is Mr. Figley's opening, which the last in that direction. Following the coal up the Ohio from the mouth of Raccoon creek, we find it running out altogether below Phillipsburg, though it re-appears within a few yards with a thickness of six inches. Above Phillipsburg, along the Ohio, there are no openings, and if the coal exi ts there it is certainly very thin. On the southern branch of Elkhorn run it has been worked to a slight extent by Mr. A. Black, at whose opening the following exposure occurs:

Mahoning sandstone, seen	20′ 0″
Bluish shale	5′ 0″
Upper Freeport Coal	1′ 4″

The coal is mined by drifting, and is worked by removing some of the underlying clay. On the other branch of Elkhorn, this bed is mined by Mr. Bloom at a mile and a half above Stewart's mill, where the structure is,

Coal	4″	3′ 0½″
Shale	2′	
Coal	2′ 2″	
Slate	½″	
Coal	4″	

The coal is slaty and contains much sulphur.

The Freeport Limestone, No. 34, occurs at a few feet below the coal at both of the openings just referred to. Near the mouth of Elkhorn run it has been burned by Mr. Brooks, who says that it slaked well and made excellent lime.

The *Lower Freeport Coal* is often seen on the little run that enters the Ohio near Phillipsburg, where some one has attempted to open it, but finding only ten inches of coal, abandoned the work. Though frequently found in other parts of the township, it was at no time more than ten inches thick. Two miles up Raccoon, near Mr. Figley's residence, it is a cannel.

The Freeport Sandstone, No. 38, is well developed in this township, and forms the lower line of cliffs along the river below Phillipsburg. It is generally hard and micaceous, so that it dresses with difficulty, and is seldom used as a building stone. On Raccoon, near Moffatt's mill, the hard layer near the middle is seen in the stream, where it is polished as smooth as glass by the water. The same layer is well shown on Elkhorn run, near Stewart's mill.

The coal, No. 40, while quite thin, is very persistent in this township. It has been worked at two localities. Mr. Brooks, a short distance above the mouth of Elkhorn run, found it sixteen inches thick, but very pure and good, so that it was a fine smithing coal. At a short distance below Hog island it is worked by Mr. Black, who has it one foot ten inches thick. Though so thin, it has been mined to a considerable extent. The massive Freeport Sandstone rests directly upon it here. This condition is occasionally seen elsewhere, but usually there are from twenty to thirty feet of shale intervening.

No. 41 is commonly a dark, thinly laminated, sandy shale, and generally contains nodules of iron ore, which in some localities are quite abundant, though never in sufficient quantity to warrant mining.

The *Kittanning Coal*, No. 42, rises from the Ohio river half a mile above the mouth of Elkhorn run, and has been worked near that stream on the land of Mr. M'Cullough. There it is two feet two inches thick and yields a coal which is quite inferior, as it contains a great deal of pyritous slate. This bed is mined back of Phillipsburg by Mr. Patterson, and many years ago was worked on the river bluff below that place. At no locality in this township was it seen more than two feet six inches thick, except on the land of Mrs. Mansfield, near the mouth of Raccoon creek, where it is three feet and of much better quality than is usual. The fire-clay, No. 43, everywhere underlies the *Kittanning Coal*, and is the same with that used at the fire-brick works below Phillipsburg. Near the mouth of Raccoon, seventy feet of No. 44 are exposed, and of the same stratum, forty-five feet can be seen below Phillipsburg.

67. Raccoon Township, in Beaver County.

In this township the highest rock exposed is the Morgantown Sandstone, No. 15, of the general section, and the lowest is the Ferriferous Limestone, the latter being the lowest stratum shown in the southern portion of Beaver county. The former caps the hills in the vicinity of Mechanicsburg, where its fragments are widely scattered.

The Crinoidal Limestone, No. 19, runs out about one mile from the southern boundary of the township. It is seen in the summits of the hills along Service creek, and at the roadside, a little way south from Green Garden school-house, where it is very full of fossils and contains *producti* as large as one's fist. Between it and the Mahoning Sandstone no strata are well exposed, as they occur far up in the hills and are not uncovered by ravines.

The Mahoning sandstone is quite massive at most localities along the Ohio, and is extensively quarried on the land of Mr. Allen, two and a half miles below the mouth of Raccoon creek. The stone used in building the piers of the new bridge across the Monongahela at Pittsburg is obtained here.

The *Upper Freeport Coal*, No. 32, has a maximum thickness in this township of two feet six inches. On Service creek it comes to the surface near the United Presbyterian church, above which it was worked on the land of Mr. Campbell, but the openings have been abandoned. There it is twenty feet above the creek, but, as the stream comes from the south-west, the coal soon passes under it and is accessible altogether for a distance of little more than one-fourth of a mile. Just before passing under the creek, it is worked by Mr. Cotter, at whose opening the following exposure was seen:

Mahoning sandstone	6′	0″
Sandy shale	5′	0″
Upper Freeport Coal Bed	2′	8′

The coal is of fair quality, contains little slate and not enough sulphur to unfit it for smithing. It leaves a white ash and no cinder in burning. It is mined by Mr. Alexander at a mile and half west from Moffat's mill, and by Mr. Springer in the same neighborhood. At both of these openings the section is almost precisely the same as that just given. On Crabapple run, which enters the Ohio at the western line of the township, it has been mined somewhat extensively by Mr. Clear, at whose works it shows the following structure:

Shale, containing streaks of *coal*	1′	0″
Coal	3′	2″
Clay		2″
Coal		6″

The bottom bench is usually the purest coal. Two miles and a half below the mouth of Raccoon creek this coal is worked on the land of Mr. Allen, where it is in two benches, respectively fourteen and twenty-four inches, separated by two and a half inches of clay. The coal is slaty, and much inferior to that obtained at other localities in the township.

The coal, No. 40, was once opened by Mr. Braden, on Raccoon creek, but he found it only twenty inches thick, and abandoned it. No other opening was found. The *Kittanning*, in like manner, is opened at only one place. Mr. Manor, at a mile and a quarter above the mouth of Raccoon creek, has it two feet nine inches thick, but very impure.

Nos 45 and 46 of the general section are exposed along the river near Mr. Allen's, and at two miles and a half below the mouth of Raccoon. Where first seen the Ferriferous Limestone is twelve feet thick, but in tracing it down the river, one finds it gradually becoming shaly and arenaceous, until it becomes only one foot thick and very impure. The iron ore rests on the limestone, and seems to vary little from six inches in thickness. The sandstone overlying it forms cliffs sixty feet high, and in this neighborhood is compact. The Ferriferous Limestone, which has been quarried extensively, is full of fossils, among which the following are the most common species: *Productus longispinus*, *Spirifer opimus*, *Spirifer lineatus*, *Spirifer cameratus*, *Pleurotomaria grayvilliensis*, *Pleurotomaria turbinella*, *Polyphemopsis peracutus*, and *Euomphalus subrugosus*.

68. Greene Township, in Beaver County.

This township exhibits a long column of rocks, extending from the Morgantown Sandstone to the bottom of the general section, though the lower strata are all covered by the terrace deposits.

The Morgantown Sandstone is found along the southern line of the township, where large level tracts of land occur above it. Owing to the height of the hills, the Crinoidal Limestone is found in their summit over a large area, and reaches even to the hills skirting the river near Georgetown. The little coal, No. 17, was once opened on the road between Georgetown and Shippingport, by Mr. Wright, but being only two feet thick, was abandoned. Exposures between the Crinoidal Limestone and the Mahoning sandstone are very poor, as the interval occupied lies far up in the hills. No. 26, however, is seen with its underlying coal near the head of Little Mill creek, as well as on Big Mill creek, one mile above Hookstown.

The Mahoning Sandstone is quite massive over a large part of the township, and is seen in the river hills forming vertical cliffs near Georgetown, from which many huge blocks, usually conglomerate, have rolled down toward the river. Along Big Mill creek it occurs as massive walls on each side of the stream almost to the steam mill above Hookstown, where it is in the bed of the creek. On Little Mill creek it shows the same characteristics, and disappears under that stream at two miles and a half from the Ohio. In the north-eastern portion of the township it is often shaly, and sometimes only thirty feet thick.

The *Upper Freeport Coal* is of no little economical importance in this township, as it is never less than three feet six inches, and commonly four feet thick. Ordinarily it is an excellent fuel, and seldom contains enough sulphur to unfit it for use in smithing. On Negro run, at two miles above Shippingport, Mr. Wilson's bank shows the following structure:

Coal	3' 4''
Clay	3'' to 4''
Coal	8''

This is typical of the bed as it occurs in Greene township. The clay parting is termed by the miners the "bearing-in" place, as they mine the coal by first digging out the clay, and then knocking down the upper bench, after which the bottom is taken up, The bottom is rather softer than the top, and is thought to be somewhat purer. This bed is extensively mined below Hookstown by Messrs. Swearingen, Todd and others. In the vicinity of Shippingport it is mined to a considerable extent by Mr. Cristler, and above that village, on Negro run, the works of Messrs. Swaney, Wilson, Thorn and others supply a large territory with fuel. It is worked in the river hills above Georgetown by Mr. Poe and Mr. Peters, and on Little Mill run by Mr. Calhoun.

The Freeport Sandstone, No. 38, is very massive, and is found in solid walls on almost every stream.

The coal, No. 40, though little more than two feet thick, is so pure that it is mined along the river and Little Mill creek to supply the local demand. Mr. Diehl's bank, below Georgetown, shows it in two benches two feet and three inches, separated by a very thin clay parting. The upper bench is somewhat slaty near the out-crop, but at some distance under the hill, it is solid and comes out in handsome blocks. This coal is bright, oily and so free from sulphur that the blacksmiths prefer it to the best quality of *Pittsburg Coal.* The bench below the parting is usually sulphurous and slaty. On Little Mill creek this bed is worked by Mr. Johnson and Mr. Bryan. At the latter opening the following section was obtained:

1. Shale	10′ 0″
2. *Coal*	2′ 5″
3. Fire-clay	5′ 0″
4. Shale	12′ 0″
5. *Kittanning Coal bed*	3′ 0″
6. Fire-clay	7′ 0″

The *Kittanning Coal* was once worked here to supply fuel to the engines while the oil-borings were in progress. It is very impure, and contains so much pyritous slate that it is commonly known as the "sulphur vein." Both of the clays seem to be quite good, though no use has been made of them here.

The upper coal of this section—No. 40, of general section—is worked also on Big Mill creek by Mr. Smith, who has it only twenty inches thick. On the same creek, near the old saw-mill, the *Kittanning* was once opened, but owing to its impurity, was not much used.

At one mile above Georgetown the river hills catch the Crinoidal Limestone near their summits. In descending from the top to the river, the following section was obtained, which, though imperfect, is of value as showing the intervals between important strata:

1. Concealed to top of hill	20′
2. Crinoidal Limestone	5′
3. Concealed	100′
4. *Cannel Coal*	6′
5. Concealed.,	200′
6. *Upper Freeport Coal bed*	4′
7. Concealed	180′
8. *Coal*	2′
9. Concealed	20
10. *Kittanning Coal bed*	3′
11. Concealed to river	75′

The Crinoidal Limestone is literally crowded with fossil remains. The cannel, No. 4, is a purely local deposit, and was seen at no other locality in southern Beaver. The upper half is an impure cannel, while the lower half approaches more nearly to semi-cannel. It was opened here on the land of Mr.

Peters, and before the discovery of petroleum, oil was manufactured from it. It is said to burn well in the grate, but to leave abundance of ash.

Some oil has been procured in this township, but this matter has been referred to in another connection. Several of the borings were begun in the bed of Big Mill creek, and passed through fifty feet of gravel deposits before reaching rock.

PART IV.

CHAPTER XIV.

TABLES SHOWING DEPTHS OF COAL AT VARIOUS LOCALITIES IN GREENE AND WASHINGTON COUNTIES.

Introduction.

In the following tables the determinations are for the *Pittsburg* and *Waynesburg* only, as the other beds of the upper groups are so variable that, for the most part, they are of no value.

The intervals between the more important strata show a constant diminution northward, but the exact ratio could not be determined, owing to the great distance between points where direct measurements of the lower rocks could be obtained. In Greene county, no direct measurement between the *Waynesburg* and *Pittsburg* can be made, except in the extreme south-eastern portion of the county, so that the interval must be calculated for the rest. In fully two-thirds of the county the interval between the *Waynesburg* and *Washington Coals* is, for the most part, below the surface. A similar difficulty exists throughout southern Washington county.

The general fact that the intervals diminish northward having been ascertained, I thought it best to calculate a special section for each township, beginning at the south, and using as the basis for determining the variation the last long section obtained on that side. In all probability, therefore, the calculations will be found nearly exact along the southern border of a township, but becoming too large as the northern line is approached.

In some places a local difficulty occurs which causes great perplexity. Thus, the interval between any two strata may be abruptly and greatly diminished, as that between Limestones VI and X, on the ridge between Grays' and Brown's forks of Ten-Mile, in Greene county, or that between Limestone VI and the

Washington Coal, as in eastern Richhill township, of the same county. In such localities an accurate determination is utterly impossible, and I have given two calculations. The truth lies somewhere between them.

A more serious difficulty than either of those mentioned exists in the fact that a very considerable portion of the district is deeply covered with debris, which sometimes forms an unbroken sheet for several miles. Under such circumstances the only dependence is the barometer, but this instrument is so delicate that even slight changes in atmospheric conditions are sufficient to render the results valueless, if any considerable interval of time elapses between the observations.

It is believed, however, that the calculations are as nearly correct as it is possible to make them without the aid of lines, carefully leveled with instruments, and that the error will in no case be found considerable.

The points given in the following tables are cross-roads or forks in the roads. To mark a locality, the name of the person living near by is used. In the Washington county lists, the names are taken from the map published by Pomeroy and Treat in 1861, and for Greene county, Mr. M'Connell's excellent map was used. In the lists certain localities are marked with an asterisk. These are given on the maps accompanying this report.

1. East Bethlehem Township, in Washington County.

In this township, the interval between the *Waynesburg* and *Pittsburg Coals* is regarded as varying little from three hundred and sixty feet. To ascertain the probable depth of the *Pittsburg*, add that number to the depth given for the *Waynesburg*.

1. National road—		
*Beallsville	75	W.
J. Hough	65	W.
J. Young	45	W.
A. Horton	325	P.
*Centreville	300	P.
M. E. Church	325	P.
*Corner of Pike Run Township	380	P.
Toll-house	290	P.
2. Road leading South from Beallsville—		
First cross-road	300	P.
*Mrs. Regester	100	P.
Second Fork south	250	P.

Cross-roads north from G. Crumrine	55	W.
Cross-roads at mill south-west from Mrs. Regester	200	P.
3. Road leading south-east from Pike near Centreville—		
Fork near Linton estate	400	P.
Westland Friend's Church	400	P.
J. & C. Cope	375	P.
J. Hormell	260	P.
4. Between roads 1 and 3—		
Near Wesleyan Church	150	P.
Near J. Hamilton	100	P.
Near J. Watkins	40	P.
5. Between roads 2 and 3—		
North from N. Baker	200	P.
G. Hill	160	P.
W. Sargent	60	W.
A. Beauly	80	P.

2. West Bethlehem Township, Washington County..

Here, as in East Bethlehem, the only coals worthy of note in this connection are the *Pittsburg* and *Waynesburg*. No direct measurement can be had to determine the interval between these beds, for the former is not exposed in the township. As this interval seems to diminish westward, it is probably not more than three hundred and forty feet in this township. By adding that number to the depths given for the *Waynesburg*, the place of the *Pittsburg* will be ascertained.

1. Along Daniel's run —		
*Mouth of Little Daniel s run	340	P.
S. Hedge	80	W.
*Taylor estate	85	W.
Above D. Ulery	90	W.
West from D. Hildebrand	160	W.
S. Ross	200	W.
Mrs. Densor	90	W.
2. East from Daniel's run—		
Free Church	250	W.
South from R. Hill	50	W.
3. West from Little Daniel's run—		
Walton's store	450	W.
*T. M'Farland's store	190	W.
*North from F. Rosel	40	W.
4. Little Daniel's run—		
*D. Eller	95	W.
G. Egy	145	W.
*D. Grable	140	W.
5. National Road—		
S. Ross	200	W.

*S. Yarley	275 W.
J. Ames	300 W.
6. North from National road—	
A. Spoon	W.
School-house No. 13	W.
Cross-roads near S. F. Nichol	W.
Fork north-west from T. Gan's	220 W.
7. South from Ten-Mile creek—	
*J. Inglis, on county line	530 W.
D. Bennington, on county line	420 W.
W. Arnold	230 W.

3. AMWELL TOWNSHIP, WASHINGTON COUNTY.

In this township no direct measurement can be obtained between the *Waynesburg* and *Pittsburg*, as the latter is not reached. The interval is not more than three hundred and ten feet at the north, and is probably three hundred and thirty or three hundred and forty at the south. In the north-western portion of the township the *Waynesburg* is likely to prove thin, and shafting for it would be injudicious. As, however, that bed is quite thick in the southern portion of the township, the depths are calculated for it. The *Pittsburg* is probably nowhere more than three hundred and forty feet below it.

1. On Ten-Mile creek—	
*Ten-Mile village	140 W.
*Cross-road from Amity	140 W.
Township line	140 W.
2. Between Ten-Mile and Pleasant Valley—	
*E. Baker	250 W.
3. Between Ten-Mile and the National road—	
Amity	330 W.
Pleasant Valley	W.
*Hughes' mill	320 W.
Baptist church	175 W.
West from T. Tucker	340 W.
North-west from J. Pratt	300 W.
J. Reed's mill	150 W.
*J. Doak, on National road	390 W.
J. Weyandt	280 W.

4. MORRIS TOWNSHIP, WASHINGTON COUNTY.

In this township there are no strata exposed lower than the *Washington Coal.* The great Upper Washington Limestone is exposed over a large part of the township and may be used as a horizon. From this rock to the *Washington Coal* is one hundred and forty-five to one hundred and sixty feet here. From the *Washington* to the *Waynesburg* the assumed interval is one hundred and forty feet, according to the nearest direct measurement, and from the *Waynesburg* to the *Pittsburg* is not far from three hundred feet. The references are made out for the *Waynesburg* Coal, as it is evidently persistent all along the southern border of the county. At the same time its quality is so poor that I am not justified in advising any one to shaft for it to a distance of more than a very few feet.

1. Along Ten-Mile creek—
   School-house No. 6 .... 185 W.
   *Lindley's Mill .... 185 W.
   Mouth of Short creek .... 310 W.
   *Prosperity .... 335 W.
2. Short creek—
   *Days' saw mill .... 325 W.
   Lindley's P. O. .... 315 W.
   Parcel's store .... 420 W.
   *S. Day .... 600 W.
   J. Cooper .... 240 W.
   J. Day .... 210 W.
3. Craft's run.
   W. E. Craft .... 180 W.
   Near A. Elliot .... 250 W.
   *J. Simpson .... 230 W.
   T. M'Carroll .... 520 W.
   School-house No. 9 .... 360 W.
4. Road from Van Buren south-east—
   A. Wier .... 380 W.
   J. Sanders .... 330 W.
   T. Sanders .... 260 W.

5. East Finley Township, Washington County.

Except in the southern portion of this township, the *Waynesburg Coal* is likely to be worthless. The nearest approach to it is at the township line, on Hunter's fork, where the bed is one hundred and ninety feet below the surface. Judging from exposures of the lower rocks at the west, the *Pittsburg* is likely to be the only workable bed of our series underlying the township. The interval between it and the *Waynesburg* is about three hundred feet.

1. Gordy's Fork—
   *Mrs. M'Donald .... 305 W.
   J. Montgomery .... 2[illegible]0 W.
   J. S. Knox .... 295 W.
   *J. Rockefeller .... 290 W.

2. Templeton's fork—
   - E. Alexander .... 290 W.
   - H. Alexander .... 290 W.
   - Schoolhouse No. 5 .... 300 W.
   - A. Towne .... 260 W.
   - J. Scott .... 310 W.
   - W. Montgomery .... 530 W.
3. Robinson's fork and State road.
   - *J. Patterson .... 320 W.
   - Head of run on State road .... 570 W.
   - *Cross roads west from W. Warrell .... 320 W.
   - Blacksmith shop .... 520 W.
   - *Pleasant Grove .... 490 W.
   - First fork east from last .... 470 W.
4. Hunter's fork—
   - T. Carroll .... 400 W.
   - West from W. Cumpson .... 310 W.
   - *Second fork below last .... 250 W.
   - *Township line .... 190 W.
5. West fork of Buffalo creek—
   - *J. Oliver .... 260 W.
   - J. Woodburn .... 200 W.

6. West Finley Township, Washington County.

In the southern portion of this township the *Waynesburg* is of workable thickness, but northward becomes very thin and of no value, as shown by the exposures in West Virginia. The numbers are made out for the *Waynesburg*, as that bed is available at the south, but for the greater portion of the township the *Pittsburg* should be regarded as the only workable coa. the *Washington* and *Waynesburg* being alike worthless. The interval between the *Waynesburg* and *Pittsburg* is assumed to be three hundred feet.

1. Hunter's Fork—
   - *Mouth of Owens' run .... 170 W.
   - Mouth of Mill run .... 125 W.
   - *Mouth of Robinson's fork .... 275 P.
2. Templeton's Fork—
   - R. & C. Wallace .... 255 W.
   - W. M'Nae .... 240 W.
3. Robinson's Fork—
   - J. Whitman .... 80 W.
   - *Forks of stream .... 220 W.
   - Road leading to West Finley .... 260 W.
   - M. Gunn .... 260 W.
   - Saw-mill .... 270 W.
   - J. Chase .... 265 W.
   - H. Jordan .... 290 W.
   - *Good-Intent .... 290 W.

R. Frazier .... 340 W.
Township line .... 320 W

4. Western portion of Township—
Mouth of Croup's run .... 290 P.
S. Beymers .... 550 W.
W. M'Whorter .... 530 W.
*Fork at West Virginia line .... 560 W.
S. Powers .... 550 W.
North from School-house, No. 2 .... 540 W.
*D. Hominger .... 520 W.
Mrs. Frazier .... 550 W.
R. Burns .... 570 W.

5. Middle Wheeling Creek—
*A. M'Cleary .... 170 W.
S. Kimmins .... 205 W.
Above R. Stewart .... 245 W.

7. East Pike Run Township, in Washington County.

The *Pittsburg* is the only available coal here, the *Waynesburg* being worthless.

*Cross-roads near J. Jackman .... 240 P.
Union school-house .... 390 P.
*J. Richards .... 270 P.
H. Robinson .... 220 P.
Near Jeff. Duval .... 200 P.

8. West Pike Run Township, in Washington County.

The *Waynesburg* and *Pittsburg* are both available here. The interval is about three hundred and fifteen feet.

1. Pike run—
A. Jeffries .... 25 P.
G. Deems .... 100 P.
Distillery .... 165 P.
*R. Arnold .... 200 P.
James Hill .... 265 P.
S. Borom .... 350 P.

2. Little Pike run—
*South-east from J. Nicholson .... 170 P.
North-west from B. Cleaver .... 140 P.

3. West from Little Pike run—
J. John .... 140 W.
Baptist church .... 85 W.
J. Baker .... 280 P.

4. South from Pike run—
Near I. Grimes' .... 150 P.

9. Allen Township.

Only the *Pittsburg* is available here.

Speer's mill .... P.
*School-house No. 7 .... 250 P.

S. Jackman	300 P.
Next south-west from last	310 P.

10. Fallowfield Township, Washington County.

In the western portion of this township the *Waynesburg* is thick and workable, but in the eastern portion it becomes very thin and is worthless. The interval between it and the *Pittsburg* below is taken as three hundred and ten feet, this being the maximum measurement obtained in the township.

1. Pigeon creek—	
*Mouth of Saw-Mill run	70 P.
T. Richardson	75 P.
J. Redd	100 P.
Above H. Newkirk	135 P.
T. Johnson	125 P.
2. Saw-Mill run—	
B. Dickey	230 P.
Near Mrs. Williams	175 P.
M. E. church on West fork	260 P.
3. South from Pigeon creek—	
R. Richardson	145 P.
*J. Jones' store	220 P.
W. Beasel	200 P.
*S. Withrow	215 P.
4. Eastern part of township—	
School-house No. 5	400 P.
P. Snider	180 P.
J. Conrad	20 P.
*Cooper's grist-mill	70-P.
*M. E. Church	70 P.

11. Carroll Township, Washington County.

Here the *Redstone* Coal seems to be persistent at fifty to fifty-five feet above the *Pittsburg*. The *Waynesburg* is worthlsss.

Eastern part—	
*J. G. Maple	340 P.
*Heirs of R. Grant	150 P.
Western part—	
*Brownsville road and Monongahela pike	200 P.
*Brownsville road and Pigeon creek	50 P.

12. Somerset Township, Washington County.

The *Waynesburg* is the important coal in this township. The interval between it and the *Pittsburg* is about three hundred and ten feet.

1. South fork of Pigeon creek—	
*Junction of forks	180 P.
Bentleysville	160 P.

First fork above Bentleysvile	170 P.
*S. Imes	230 P.
School-house No. 3	310 P.
*W. M. M'Ilwain	290 P.
Vanceville	10 W.
*Above J. M'Murry	100 W.
2. South from this fork—	
H. Richardson	120 W.
School-house No. 2	295 P.
*J. Hawkins	50 W.
M. Mankey	290 W.
*M. Tombrough	300 P.
3. North fork of Pigeon creek—	
A. Hetherington	190 P.
D. Kerr	205 P.
*U. Wright	210 P.
School-house No. 7	200 P.
*T. M'Corkle	290 P.
J. Barr	150 W.
4. Western part of township—	
A. W. Smith	130 W.
Seceder church	160 W.
*South from Seceder church	165 W.
Mrs. Wier	175 W.
5. North-east part—	
Near H. Leyda	120 W.

13. South Strabane Township, Washington County.

In this township the *Washington* and *Waynesburg* are worthless, and the *Pittsburg* is the only coal which can be regarded as workable; but, as will be seen by reference to the list below, it is for the most part at a very great depth. The average intervals in this township are as follows: Upper Washington Limestone; interval, 160 feet; *Washington Coal*; interval, 125 feet; *Waynesburg Coal;* interval, 300 feet; *Pittsburg Coal.*

1. North fork of Chartiers creek—	
R. O. Henry	665 P.
Cross roads near J. Zedeker	500 P.
U. B. Church	565 P.
*Martinsburg	625 P.
*G. Black	585 P.
2. Old Pittsburg road—	
First fork from Washington	350 P.
J. Munce	350 P.
J. M'Nary	270 P.
3. East from the last road—	
Cross road near W. J. Munce	535 P.
Fork north-east from last	585 P.
4. Pittsburg road—	
Toll house	290 P.

5. In Washington—
   Old Pittsburg road .... 415 P.
   National road at east end of borough............ 485 P.

14. Franklin Township, Washington County.

The conditions here are approximately the same as in South Strabane.

1. Chartiers creek—
   *Near Wilson and Shields'...................... 425 P.
   J. V. Wilson.................................... 410 P.
   *J. Brownlee.................................... 445 P.
   *Vanburen....................................... 605 P.
2. Along road through Point Lookout—
   *Point Lookout.................................. 730 P.
   Above W. D. Andrew.............................. 615 P.
   *A. Horn.. 615 P.
   School-house at township line................... 615 P.
   Cooper's tanyard................................ 610 P.
3. South-west part of the township—
   S. Walters...................................... 470 P.
   *Near J. Cooper................................. 465 P.
   South-west from J. Patterson.................... 425 P.
   Near school-house No. 6......................... 415 P.

15. Buffalo Township, Washington County.

In this township the *Waynesburg* is so thin that with only one exception it is everywhere utterly worthless. The only coal exposed which attains workable thickness is the *Washington*, which, however, is of very poor quality.

1. Buffalo creek, East fork—
   *Mrs. Gantz..................................... 465 P.
   J. M. Carter.................................... 625 P.
   National road................................... 425 P.
   Railroad.. 410 P.
   J. W. M'Kee .... 400 P.
   Near school-house No. 2. .... 450 P.
   *National road at township line................ 675 P.
2. Buffalo creek, West fork—
   *U. P. Church................................... 450 P.
   National road................................... 450 P.
   E. Wright....................................... 390 P.
3. Buffalo creek—
   *Junction of forks.............................. 375 P.
   Taylorstown..................................... 335 P.
   *Cross-roads, near A. Henderson................. 390 P.
   First fork below Taylorstown.................... 270 P.
   Fork on last road, near T. Knox .... 390 P.
   Second fork below Taylorstown................... 270 P.
   Near W. Grimes.................................. 330 P.

*Noble's mill .... 280 P.
Atchison .... 260 P.

16. Donegal Township, Washington County.

Here the *Waynesburg* is utterly worthless. The *Washington* is exposed at many localities, with thickness of several feet, but it is so variable as to be of little economical importance. From this coal to the *Pittsburg* is not far from three hundred and ninety feet.

1. Hempfield railroad—
   - *National road east from Claysville .... 470 P.
   - Same road west from Claysville .... 420 P.
   - *Coon Island station .... 400 P.
   - *Tunnel at West Alexandria .... 520 P.
2. Dutch fork of Buffalo creek—
   - West from H. W. Rysor .... 355 P.
   - *Baptist church .... 250 P.
   - Ritchie heirs .... 270 P.
   - Cross-roads east from J. Stoolfire .... 390 P.
   - Near J. Miller .... 500 P.
3. East from Dutch Fork—
   - J. Mehaffy .... 550 P.
   - *G. Moore .... 450 P.
   - Cross-roads at A. Chapman .... 315 P.
4. West from Dutch Fork—
   - *Cross-road at B. Dickey .... 700 P.
   - *School-house, No. 5 .... 615 P.
5. South from railroad—
   - West from J. B. Moss .... 460 P.
   - South-east from S. A. Craig .... 740 P.
   - South-west from J. Ryan .... 565 P.
   - Toll-house on National road .... 690 P.
   - J. Harrison .... 710 P.
   - E. L. Post .... 510 P.

17. Union Township, in Washington County.

The only coal of any value here is the *Pittsburg*.

1. Peter's Creek—
   - Below *Pittsburg Coal* quite to township line..
2. North from the creek—
   - *J. Salisbury .... 235 P.
3. South from the creek.
   - J. Huston, Jr .... 200 P.
   - *W. Kennedy .... 150 P.
   - Presbyterian Church .... 150 P.
   - West from T. Flannigan .... 340 P.
   - W. Crites .... 220 P.
   - M. P. Patton .... 200 P.
   - J. Nicholson, on Mingo creek .... 25 P.

18. Nottingham Township, in Washington County.

In this township, both the *Pittsburg* and *Waynesburg* are available, and the interval between them is about three hundred feet. The interval between the *Washington* and *Waynesburg* varies not much from one hundred and forty feet.

1. Mingo creek—
   - *D, Longwell.................................... 45 P.
   - J. Rankin.. 250 P.
   - South from H. Devore............................ 300 P.
   - *B. J. Mosier, at Township line.................. 600 P.
2. Fork of Mingo creek—
   - M'Millan's tannery............................... 60 P.
   - *School-house, No. 3............................ 460 P.
3. Peter's creek—
   - Below *Pittsburg* to Thomas' mill.................
   - *Rainey's mill.................................. 360 P.
   - Munntown.. 500 P.
4. Other localities—
   - North from A. Darrow............................ 175 P.
   - Mrs. M'Donnald.................................. 335 P.
   - W. Miller....................................... 470 P.
   - South from A. B. Scott.......................... 360 P.

19. Peters Township, in Washington County.

Here, both the *Washington* and *Waynesburg* are so thin as to be of little value. From the *Washington* to the *Pittsburg* is four hundred and twenty feet, and from the *Waynesburg* to the same coal is two hundred and ninety feet.

1. Brush run—
   - Bower hill...................................... 400 P.
   - *M. Ross.. 275 P.
   - M'Loney heirs................................... 210 P.
   - School-house, No. 2............................. 250 P.
   - *Presbyterian Church............................ 410 P.
   - H. Bethrow...................................... 350 P.
   - M'Claim's saw-mill.............................. 170 P.
   - First fork above Thompsonville.................. 165 P.
   - *First fork below Thompsonville................. 170 P.
2. Eastern part of township—
   - J. Boyce.. 390 P.
   - W. Allison...................................... 190 P.
3. South-west part
   - S. Bell... 455 P.
   - *J. S. Kerr..................................... 310 P.
   - J. P. Skiles.................................... 270 P.
   - J. Wilson, Sr................................... 280 P.

20. North Strabane Township, Washington County.

The only coal here is the *Pittsburg*, which is reached along

Chartiers creek The other beds are too thin to be of any value. The intervals do not differ very materially from those given for the last township.

1. North fork of Chartiers creek—

J. G. Wilson, Jr.	200 P.
J. M'Millan	170 P.
*Linden	180 P.
R. M'Clelland	270 P.
*Pees' mill	280 P.
School-house No. 2	340 P.
S. Thomas	250 P.
*J. Scott	250 P.

2. West from the creek.

Near J. Christy	540 P.
Pollock's saw mill	310 P.
D. Quail	330 P.
J. P. Norris	500 P.
M. Linn	430 P.
North from W. Quail	345 P.
*Presbyterian church	450 P.
R. C. Neal	345 P.
E. M'Clelland	220 P.

21. Chartiers Township, Washington County.

The *Waynesburg* is not exposed in this township, and its horizon is reached at but few localities. The *Pittsburg Coal* is exposed at many places. The following list is chiefly of localities on the ridges:

1. Pittsburg Pike—

Mrs. Alison	P.
*Houstonville	60–P.

2. West fork Chartiers—

*M'Connellsville	P.
Hickory road crossings	20 P.

3. Hickory road—

U. P. Church	250 P.
*H. Fergus	230 P.

4. Plum run—

*School-house No. 7	40 P.

5. Ridge road at north.

J. Bryce	240 P.
Mrs. J. Watson	160 P.

22. Canton Township, Washington County.

The *Waynesburg* and *Washington* are very poor and thin in this township. The calculations have been made for the *Pittsburg* alone, as it would be be unprofitable to shaft for either of the others.

1. Northern part.

Heirs of E. White	300 P.
Near Mrs. Slemon	550 P.
South-east from Mrs. M'Kee	450 P.
J. Kelly	405 P.

2. Southern part.

Above J. Boon	330 P.
*J. Anderson	300 P.
W. H. Cook	290 P.
E. Wolfe	250 P.
S. K. Weirick	415 P.
*Railroad and National road	400 P.

23. HOPEWELL TOWNSHIP, WASHINGTON COUNTY.

In this township the *Waynesburg Coal* is occasionally workable, and a number of openings are noted in the chapter on Washington county. The *Washington Coal* also has been opened, but both of these are variable in thickness and of very poor quality, so that shafting for them would be merely waste of money. The calculations, therefore, are all made for the *Pittsburg Coal.* For the *Washington* subtract four hundred feet and for the *Waynesburg* subtract two hundred and ninety. The calculated distance is no doubt too large in every instance, but is made up in accordance with the nearest direct measurement at the south. The basis for the *Pittsburg* is the shaft at Washington.

1. Brush run—

*J. M'Clay	315 P.
J. C. Clark	280 P.
*Work and Denny	260 P.
East from J. Brownlee	240 P.
*Township line	260 P.

2. Buffalo village to West Middletown—

*Buffalo village	520 P.
W. W. Densmore	490 P.
*D. Brown	550 P.
West Middletown	650 P.

3. North from this road—

School-house No. 4	510 P.
A. Wotring	530 P.

4. South from the road—

J. M. Rush	370 P.
West from M. Templeton	340 P.
J. Dunkle	260 P.

24. INDEPENDENCE TOWNSHIP, WASHINGTON COUNTY.

In the southern part of this township the *Washington* becomes of workable thickness and of some economical importance. It diminishes in importance northward and cannot be depended on in any portion. The *Waynesburg* is so thin as to be utterly worthless. The calculations are for the *Pittsburg* only. The intervals are regarded as the same as in Hopewell.

1. Sugar run—	
S. Williamson	270 P.
*Near J. Murphy	300 P.
South from M. E. Church	260 P.
Next south from last	265 P.
D. Houston	300 P.
2. Northern part—	
A. Manchester	470 P.
Near J. Vance	460 P.
*R. Craig	570 P.
School-house No. 2	530 P.
*Near T. M'Kimm	370 P.
Independence	400 P.
*Johnson's hotel	420 P.
3. South and east from Sugar run—	
North-east from A. Buchannan	650 P.
T. R. Law	490 P.
4. Brush run—	
A. Ramsay	265 P.
*Mrs. Grimes	260 P.
Below Bryson's store	240 P.

25. Cecil Township, Washington County.

The *Pittsburg* is reached on Miller's run, just beyond the line of this township. All the coals above it have become utterly worthless.

Creek crossing below Venice	150 P.
Johnson's hotel and store	420 P.
Grier's, on railroad	210 P.
Hill's on railroad	90 P.

26. Mount Pleasant Township, in Washington County.

This township is without immediately available coal, except on a branch of the west fork of Chartiers, along which the road to Cannonsburg passes. The *Waynesburg* is very thin and poor, and the *Washington*, though quite thick, is evidently good for nothing.

1. Raccoon creek, east fork—	
*Hickory	280 P.
*Rankin's grist mill	230 P.
*Township line	60 P.
W. P. Cherry	200 P.

2. West fork—
   First fork west from Hickory.................... 310 P.
   *Presbyterian Church........................... 350 P.
   *G. Fulton...................................... 275 P.
3. Cross creek—
   Parkinson's school-house........................ 380 P.
   Tannery... 180 P.
4. Cannonsburg road—
   T. Miller.. 30 P.
   *A. Wilson...................................... P.

27. Cross Creek Township, in Washington County.

The *Pittsburg* is available in a considerable part of this township. The other beds are of very uncertain value. The intervals employed are, from *Washington* to the *Waynesburg*, ninty feet; from the *Waynesburg* to the *Pittsburg*, two hundred and fifty feet.

1. Southern portion—
   *Woodrow P. O................................. 190 P.
   *J. M'Corkle................................... 185 P.
   *J. M'Lane..................................... 240 P.
   School-house, No. 5............................ 150 P.
   *South-west from S. Neil....................... 170 P.
   T. Criswell..................................... 40 P.
   J. Ferguson.................................... 250 P.
   *Near J. Stewart's heirs....................... 190 P.
2. Eastern part—
   T. M'Corkle.................................... 190 P.
   J. M. K. Reed.................................. 265 P.
   R. Buxton...................................... 380 P.
   W. Marquis' heirs.............................. 390 P.
3. Western part—
   *West end of Cross Creek village............... 410 P.
   Next fork west................................. 180 P.
   North from I. Perrine........................... 60 P.

28. Jefferson Township, in Washington County.

Here the *Pittsburg* is the only coal. The *Waynesburg* and *Washington* are both of workable thickness, but of such poor quality that no openings have been made in them. The intervals used in the calculation are, *Washington* to *Waynesburg*, seventy feet; *Waynesburg* to *Pittsburg*, two hundred feet; but these are much too large for the northern part of the township, where the total interval from the *Washington* to the *Pittsburg* is only two hundred and fifteen feet.

1. Eldersville and Burgettstown road—

*Near J. T. Marquis	200 P.
D. Stephenson	200 P.
*Eldersville	215 P.
Fork west from Eldersville	170 P.

2. Eastern part of towship—

*Rezin Walker	210 P.
East from J. Hohman	180 P.
Bethel church	150 P.
Grist mill near Presbyterian church	100-P.
Heirs of M'Cready	P.

29. Smith Township, Washington County.

The *Pittsburg* is available along the larger streams in a considerable part of this township, while the higher coals are of insignificant thickness. The intervals adopted here are: *Washington* to *Waynesburg*, eighty feet; *Waynesburg* to *Pittsburg*, two hundred feet.

East fork of Raccoon—

*Cherry P. O.	60 P.
*J. Keyes' saw-mill	P.
Raccoon station	P.

2. West fork of Raccoon—

School-house No. 8	200 P.
J. Sturgeon	330 P.
Near W. Stephenson	390 P.
*T. Buchanan	190 P.
*Burgettstown	P.

3. West from Burgettstown—

West from School-house No. 2	190 P.
Summit at township line	250 P.

4. North from railroad—

Bavington	100-P.
*At Vaneman's school-house	70 P.
J. Stewart	50 P.
East from Mrs. Gormly	70 P.

5. Noblestown road from Raccoon creek—

J. Keys	110 P.
*Bulger	50 P.
J. M'Burney, Jr	120 P.
*Midway	40 P.

30. Hanover Township, Washington County.

Here there are only some outliers of the *Pittsburg Coal*, and none of the higher beds is exposed.

Florence	70 P.
*School-house No. 1	20 P.
First fork east from Florence	70 P.
School-house No. 3	10 P.

GREENE COUNTY.

As the *Waynesburg Coal* is always of workable thickness wherever exposed in this county, all the calculations have been made for that bed, in those portions where it is below the surface. At the same time I should repeat the statement previously made in the body of this report, that the *Waynesburg* is invariably so poor in quality that it is not worth shafting for to a depth of more than fifty or sixty feet, except where it is otherwise impossible to obtain coal.

31. Dunkard Township, Greene County.

The *Pittsburg Coal* is available along Dunkard creek to Taylortown, and above that to the township line the *Waynesburg* is above water level. The intervals between the coals are, *Waynesburg* to *Pittsburg*, 360 feet; *Sewickley* to *Pittsburg*, 110 feet.

1. Dunkard creek—		
*Paw-Paw church..	85	P.
Mouth of Meadow run.	110	P.
2. North from creek—		
Davistown	325	P.
*Ph. Linch	160	P.
M'Clure's school-house	170	W.
Garrison's school-house		W.
Near J. M. Cumpshon	225	W.
3. South from creek—		
*J. Door	50	P.
D. Lucas	90	P.

32. Perry Township, Greene County.

The *Waynesburg* is above the surface only in the south-eastern portion of this township. Elsewhere it is deeply buried. The interval between it and the *Pittsburg* is not less than three hundred and seventy feet.

1. Dunkard creek—		
*Mouth of Colvin's run	30	W.
Creek at Mount Morris	35	W.
*Mouth of Shannon's run	105	W.
2. Colvin's run—		
M. Long	150	W.
3. Shannon's run—		
Forks of stream	120	W.
S. and J. Headlee..	170	W.
Ridge at head of run	900	W. (?)

4. South fork—
   First above mouth .... 170 W.
   Second above mouth .... 270 W.
5. Elsewhere—
   *Samuel Headlee .... 350 W.
   J. Whitlach, on Rodolph's run .... 290 W.

33. Wayne Township, Greene County.

Owing to the difficulty of obtaining satisfactory exposures in this township, and the consequent inability to carry the section, some of the figures for localities in the interior may be found materially erroneous. Those in the vicinity of Dunkard creek, however, are doubtless quite accurate. The interval between the *Washington* and *Waynesburg* is two hundred feet, and that between the *Waynesburg* and *Pittsburg* is not less than three hundred and eighty feet.

1. Dunkard creek—
   J. Worley .... 220 W.
   *Blacksville .... 200 W.
   Mouth of Hoover's run .... 300 W.
   *Mouth of Tom's run .... 340 W.
2. Roberts' run—
   N. Johnson .... 230 W.
   J. Rogers .... 265 W.
   *L. Owens .... 300 W.
   A. Gump .... 320 W.
   *Spragg's school-house .... 340 W.
   J. Strawn .... 400 W. (?)
   *Nicely's school-house .... 680 W. (?)
3. Hoover's run—
   D. King .... 360 W.
   Kuhn's school-house .... 400 W.
   D. Knight .... 700 W.

34. Gilmore Township, in Greene County.

The lowest coal exposed here is the *Jolleytown*. From that to the *Washington*, is two hundred and sixty-five feet, and thence to the *Waynesburg* is not far from two hundred feet, although the measured section is only one hundred and eighty feet. The boring at Blacksville, and a boring at Fish creek, beyond the State line, both show the larger interval to be more nearly correct. The apparent contradiction between the figures at A. Garrison's, in this township, and those at White's mill, in Spring Hill, will be explained in the report of next year's work. It has been thought unnecessary to give a long list in this township, as the coals are at an enormous depth.

*Jolleytown	465	W.
*A. Garrison	609	W.
*P. Shough	1,190	W.

34. SPRINGHILL TOWNSHIP, IN GREENE COUNTY.

Away from the line of Fish creek, in this township, the coals may be regarded as inaccessible, the *Pittsburg* being at some localities fully eighteen hundred feet below the surface. A few points only are given along the creek.

New Freeport	720	W.
*White's mill	700	W.
E. Ferguson	720	W.
*Township line	660	W.

35. MONONGAHELA TOWNSHIP, IN GREENE COUNTY.

Along the river line the *Pittsburg* is within reach almost to Gray's Landing. The *Sewickly* is exposed on Whiteley creek, and the *Waynesburg* is accessible at almost every locality where the others are below the surface.

1. Little Whiteley creek—		
West from B. Evans	330	W.
J. Hewit	220	W.
Near E. Stone	130	W.
*J. Hannan	65	W.
2. Whiteley creek—		
Below Minor's distillery	96	P.
*At Minor's distillery	96	P.
Hartley's mill	96	P.
*Creek-crossing at Mapletown	100	P.
Township line	160	P.
3. South from creek—		
*Set-still Church	240	P.

36. GREENE TOWNSHIP, IN GREENE COUNTY.

The only coal of any value is the *Waynesburg*, the *Pittsburg* and *Sewickley* being below the surface. The interval between the *Waynesburg* and *Pittsburg*, is about three hundred and sixty feet.

1. Whiteley creek—		
B. Keener	265	P.
*Willow Tree Tavern	260	P.
Above J. Mires	310	P.
*Above Garard's Fort	375	P.
J. Lantz	70	W.
*Lantz's school-house, on Township line	170	W.
2. North from creek—		
Near J. Climer	190	W.
Sam'l Miner	360	P.

3. West line of township—
   - A. Morris.. 190 W.
   - *Murdock's store.............................. 240 W.

37. WHITELEY TOWNSHIP, GREENE COUNTY.

This lies altogether above any of the workable coals. The exposures are very bad, and only a few points could be satisfactorily determined.

1. Whitely creek—
   - Mouth of Lantz' run............................ 240 W.
   - Mouth of Dyer's fork............................ 250 W.
   - *Newtown.. 260 W.
   - *Mrs. Shriver.................................... 230 W.
   - Above L. Staggers.............................. 330 W.
2. Dyer's Fork—
   - Near J. Stephen's heirs........................ 270 W.
   - J. Shriver.. 220 W.
   - Above I. Shriver................................ 220 W.
   - Stone school-house............................ 270 W.

38. CUMBERLAND TOWNSHIP, GREENE COUNTY.

The *Waynesburg* is accessible along the river line and on several streams in the eastern part of the township. The *Waynesburg* "*a*" is occasionally of workable thickness, but is too variable to be depended on. In the western part of the township the *Waynesburg* is for the most part deeply buried.

Muddy creek—
   - *Below I. Biddle................................ 75 W.
   - Below Carmichael's.............................. 350 P.
   - Near Black's school-house...................... 20 W.
   - *Near Huston's school-house.................... 70 W.
   - C. Bailly.. 80 W.

2. North from Creek—
   - Near J. Kerr...................................... 220 W.
   - H. Homer.. 80 W.
   - *A. Randolph...................................... 275 P.
   - West from J. Lucas.............................. 200 W.
3. South-east part—
   - Rea's school-house.............................. 90 W.
   - J. Stevenson...................................... 95 W.
   - Presbyterian church............................ 170 W.
   - Near T. Patterson................................ 20 W.
   - A. Burwell.. 260 W.
4. South-west part—
   - *Ceylon P. O...................................... 300 P.
   - J. Reeves.. 250 W.
   - Long's school-house............................ 230 W.
   - Joseph Cregg...................................... 70 W.

39. JEFFERSON TOWNSHIP, GREENE COUNTY.

The *Waynesburg Coal* is exposed along the river front and Ten-Mile creek, as well as on the South fork of the creek. The *Pittsburg* is reached on Ten-Mile near the mouth of the stream. In the greater part of the township, all workable coal is deeply buried.

1. Eastern part—

S. Riggle	80	W.
Price's school-house	360	P.
A. Martin	60	W.
*Murdock's school-house	85	W.
Thos. Burson	140	W.

2. Western part—

*Dowling school-house	460	W.
J. Cree	160	W.
W. Gwinn	360	W.
H. Morris	475	W.
I. Waychoff	200	W.
*H Rinehart	420	W.
Pennelberry school-house	350	W.

40. MORGAN TOWNSHIP, GREENE COUNTY.

The *Waynesburg Coal* is within reach on both forks of Ten-Mile, as well as on several of the streams leading into the south fork. No other workable coal is seen in the township.

1. Along State road and Ten-Mile—

Ruff's Creek crossing		W.
Near Harry's school-house	90	W.
Fork to Bell's run	60	W.
*D. W. Rogers	30	W.
Mouth of Hew's run	235	P.
Mouth of Casteel run	170	P.

2. Ruff's creek—

*First Fork below Martinsville		W.
Martinsville	25	W.
First above Martinsville	40	W.
Second above Martinsville	150	W.

3. Between Ruff's creek and the State road—

S. Braden	70	W.
South-east from last	50	W.

3. Casteel run—

*Baptist church at head	550	W.
Casteel school-house	60	W.
*Steward's cross road	325	P.

5. Between Ruff's creek and Casteel run—

Center school-house, on Hews' run	40	W.
Head of Hews' run	300	W.
Head of Bell's run	330	W.

A. Fulton, on ridge	260 W.
Near M. Bennett, on Ridge	530 W.
S. Crayne, on Bell's run	170 W.
M. Bell, on Bell's run	20 W.
6. North from Casteel—	
Mrs. Maylon	80 W.
Bollenfield school-house	235 W.
Amos Wise	160 W.

41. Washington Township, Greene County.

This contains no workable coal which is persistent. The calculations are all for the *Waynesburg*. The *Pittsburg* is probably three hundred and thirty feet below that bed.

1. Ruff's creek—	
*First cross-road above township line	240 W.
Second cross-road above township line	320 W.
*Brick school-house	400 W.
Mouth of John's run	410 W.
Head of Boyd's run	460 W.

42. Morris Township, Greeene County.

The conditions here are the same as in Washington township, except that the coal is more deeply buried.

Brown's fork—	
*Sargent's mill	420 W.
D. Iams	420 W.
*G. Lightner	570 W.
Lightner's run	630 W.
Nineveh	750 W.
First and second forks above the village	750 W.
*Third fork above Nineveh	740 W.
2. Bates' fork—	
Hopkin's mill	420 W.
M'Cullough's school-house	570 W.

43. Franklin Township, Greene County.

In this township the *Pittsburg* is probably three hundred and fifty feet below the *Waynesburg*. The latter coal is worked on Ten-Mile creek up to within a short distance of the borough of Waynesburg.

1. Ten-Mile creek—	
*At Center township line	400 W.
Bridge above J. Hill	375 W.
Hill's school-house	400 W.
*J. Ealy	320 W.
Toll-gate	310 W.
Jackson's run	85 W.
*Morrisville	50 W.

2. Purman's run—
   - At Waynesburg.................................. 200 W.
   - At Bridge.. 160 W.
   - At first fork in road............................ 400 W.
   - Head of Jackson's run............................ 650 W.

3. Smith's creek—
   - Mouth of creek................................... 130 W.
   - First cross-road................................. 135 W.
   - *Ingram's cross-roads............................ 140 W.
   - H. Smith... 110 W.
   - Old saw-mill..................................... 130 W.
   - *G. Thomas....................................... 140 W.
   - J. Gordon.. 220 W.

4. Ridge east from Smith's creek—
   - W. Gordon.. 425 W.
   - G. Odenbau....................................... 140 W.

5. Brown's fork—
   - Rees' mill....................................... 300 W.

6. Road to almshouse—
   - M. Porter.. 80 W.
   - Cross-roads near almshouse....................... 70 W.

44. Center Township, Greene County.

In this township the workable coals are very deeply buried, and the section reaches to the top of the series. The *Pittsburg* may be taken as three hundred and thirty to three hundred and forty feet below the *Waynesburg*.

Ten Mile creek—
   - First fork above mouth of Pursley........ 380 W.
   - Clinton.................................. 410 W.
   - *Rogersville............................. 350 W.
   - Forks of creek........................... 340 W.

2. M'Courtney's fork—
   - *Mouth of Hargus' creek.................. 340 W.
   - First above last......................... 390 W.
   - Woods' school-house...................... 415 W.
   - Woods' saw-mill.......................... 470 W.
   - *Township line........................... 520 W.

3. Hargus' creek—
   - First fork in road....................... 375 W.
   - School-house............................. 400 W.
   - First fork above last.................... 420 W.
   - Second fork above school-house........... 420 W.
   - J. Orndorff.............................. 450 W.
   - Head of stream at C. Lough,.............. 690 W.

4. Gray's fork—
   - Mouth of Clover run...................... 460 W.
   - *Mouth of Scott's run.................... 530 W.
   - Rutan post-office........................ 580 W.
   - Baptist Church........................... 675 W. (?)

5. Ridge road, north from Gray's fork—
   - Hopewell Church .... 950 W., or 850 W.
   - Head of Lightner's run .... 950 W., or 850 W.
   - Head of Brush run .... 890 W., or 780 W.
6. Pursley creek—
   - Fork below oak forest .... 280 W.
   - Fork above oak forest .... 360 W.
   - Fork below J. Hoge .... 420 W.

45. Rich Hill Township, Greene County.

The *Waynsburg* is worked here on South Wheeling creek, below the mouth of Crabapple creek. There is no other coal of any value. The interval between the *Waynesburg* and *Pittsburg*, is probably not far from three hundred feet.

1. Wheeling creek—
   - First below line of Jackson township...... 700 W.
   - Adams' mill .... 640 W.
   - *Opposite mouth of Long run .... 670 W.
   - Below Slatt's mill .... 480 W.
   - Vannatta's mill .... 120 W.
   - Ryerson's Station .... 70 W.
   - S. W. Baptist Church .... 70 W.
   - Near mouth of Crabapple .... 25 W.
   - Mouth of Crabapple .... W.
   - Near J. Bratten .... 10-W.
   - Above Crow's Hill .... 10 W.
2. South fork—
   - First above Ryerson's Station .... 70 W.
   - Third above Ryerson's Station .... 220 W.
   - *Mouth of Herod's fork .... 325 W.
3. Crabapple creek—
   - Second above mouth .... 30 W.
   - Below M'Nary's saw-mill .... 120 W.
   - Jacksonville .... 570 W. (?)
4. Gray's fork
   - Graysville .... 290 W., or 350 W.
   - Cross-roads above Graysville .... 290 W., or 350 W.
   - *Unity Presbyterian Church .... 270 W., or 330 W.
5. Owens' run—
   - S. A. Houston .... 180 W.
   - G. W. Dille .... 210 W.
   - H. Day .... 280 W.

46. Jackson Township, in Greene County.

Everything is deeply buried here, and the workable coals are inaccessible.

- *White Cottage .... 840 W.
- *J. Staggers .... 700 W.
- *T. Eisiminger .... 1,190 W.

47. Aleppo Township, Greene County.

Like Jackson, this contains the highest rocks of the series. No ravine cuts deeply enough to reach any of the workable coals, and even the *Washington* is barely touched in the northwestern part. The interval between the *Pittsburg* and *Waynesburg* coals in the southern portion cannot be far from three hundred and seventy-five or four hundred feet.

1. South fork of Wheeling creek—	
Mouth of Wisecarver's run	380 W.
*Mouth of Walnut run	685 W.
Near Caleb Evans	700 W.
*Near Windy Gap church	1,200 W.
*Near P. Ullom	1,050 W.
Near P. Lyons	1,000 W.
2. Herod's fork—	
First from township line	400 W.
At head of stream	950 W.
3. On ridge in south of township—	
Head of Walnut run	1,200 W.
Third, west from Windy Gap church	575 W.
Fourth, west from Windy Gap church	700 W.

PART V.

Economic Geology.

CHAPTER XV.

Summary of Resources.

Coal.—Of the numerous beds of coal occurring within this district, each is worked over a greater or less area, but only one maintains a workable thickness at all localities where it has been reached or exposed. A great part of the district, including central and southern Greene and south and south-west Washington, is without available coal, other than thin beds, so thin that under ordinary circumstances they would be regarded as utterly worthless. In south-western Greene the *Dunkard Bed*, which seldom yields eight inches of coal, is the main stay, while in the north-central portion of the county and the adjacent portion of Washington, the equally thin *Nineveh Coal* is mined to supply the vicinity. Of these insignificant beds no analyses have been made, as they never can become of any commercial value.

The *Washington Coal* attains workable thickness in Greene county only near Waynesburg, and there it is looked upon as of little value, owing to the proximity of the thicker *Waynesburg Bed.* Its coal, however, is not inferior to that from the *Waynesburg.* In the western portion of Washington county, where the *Waynesburg Bed* is usually quite thin, this bed is thick and is an important source of supply for local demand. The following analyses of a sample from Mr. Henderson's bank, near Taylorstown, in Buffalo township, of Washington county, gives a good idea of the *best* of this coal:

Water	1.695
Volatile combustible matter	39.150
Fixed carbon	46.658
Sulphur	1.972
Ash	10.525
	100.000
Coke	59.155
Color of ash	Gray.

(A. S. M'C.)

Usually, however, the coal is inferior to this sample, in respect both of ash and sulphur. It is made up of alternating layers of coal and clay, which cannot easily be kept separate.

The *Waynesburg Coal* is of workable thickness at all exposures in Greene county, but in Washington its variations are extreme and it is rarely of any importance, except in the eastern portion of the county. In Greene it is available along Dunkard creek to the eastern border of Perry township; along Whiteley creek almost to the western line of Greene township; along Muddy creek to Carmichaels, and along South Ten-Mile to within a short distance from Waynesburg. In the northwest corner of the county the bed is brought up along Dunkard, Wheeling creek and Hunter's fork by the Washington axis, and is available along those streams for a considerable distance above the State line; disappearing on the former near the mouth of Crabapple creek and on the latter about two miles from the State line. In the central portion of the county it is deeply buried, being at some localities fully 1,200 feet below the surface. In the eastern and south-eastern parts of Washington county it is frequently quite thick and of much local importance, but in the western portion it becomes so thin as to be traceable only with difficulty.

A large number of analyses of this coal have been made with a view to ascertain, if possible, whether or not it ever becomes good. From these I have selected eight, four of them from Greene county and four from Washington, which represent very fairly the extremes of character. The localities are as follows:

I. G. C. Sayres, below Waynesburg, Franklin township, Greene county. (A. S. M'C.)

II. L. L. Minor, near Jefferson, Jefferson township, Greene county. (A. S. M'C)

III. A. Groom, near Carmichaels, Cumberland township, Greene county. (S. A. Ford.)

IV. U. Lippincott, on Ruff's creek, Morgan township, Greene county. (A. S. M'C.)

V. Mr. Rogers, near Beallsville, Pike Run township, Washington county. (D. M'C.)

VI. J. Moninger, Pinhook, Amwell township, Washington county. (D. M'C.)

VII. Mr. Dennings, Brush Run, Buffalo township, Washington county. (A. S. M'C.)

VIII. Mr. James, near West Middletown, Hopewell township, Washingtonton county. (A. S. M'C.)

	I.	II.	III.	IV.	V.	VI.	VII.	VIII.
Water	2.265	1.175	1.180	1.235	.740	1.190	1.810	1.385
Volatile combustible matter	33.685	35.615	32.344	36.185	36.040	36.585	38.520	37.210
Fixed carbon	49.590	49.725	51.582	46.723	46 890	43.489	51.181	42.335
Sulphur	1.270	2.280	1.306	2.972	2.375	2.806	1.179	3 710
Ash	13.190	11.205	13.588	12.885	13.955	15.930	7.310	15.360
	100.	100.	100.	100.	100.	100.	100.	
Percentage of coke	66.315	63.210	66.476	62.580	63.220.	63.225	59.670	61.405
Color of ash		Pink.	Cream.	R. Gray.			Cream.	Gray.

These analyses show the coal burdened with ash, averaging about twelve per cent of the mass. At a few localities the percentage is less, but nowhere smaller than in No. VII, which is the best specimen obtained in the whole district. The percentage of sulphur is unexpectedly small in all the analyses, and in this respect, the *Waynesburg* is not far inferior to the *Pittsburg*. The amount of ash present is so great as to unfit the coal for use in the manufacture of iron, and no determination of the amount of sulphur remaining in the coke was made. It is probably very small.

In some localities the percentage of volatile matter is very large, sufficient indeed to make the coal good for manufacture of illuminating gas. No tests have been made to determine the quality of this gas, as the coal cannot be used profitably in its manufacture, for the proportion of ash in the coke, from twenty to twenty-five per cent, would render the coke unsalable.

It is evident, then, that this great bed can never be of more than local importance. Though an inconvenient fuel, owing not only to the ash in the coal, but also to the not inconsiderable proportion of clay and slate mixed with it, which is not noted in the analyses, it is of no slight value, where better coal is for the present beyond reach. It is the only fuel attainable in the greater part of Greene county. It is not worth shafting for to any considerable depth.

The *Sewickly Coal* is valuable, and is of workable thickness only in south-eastern Greene, near the Monongahela river. It occurs on Dunkard and Whitely creeks, but northward becomes very thin. When worked it is in high repute, and for domestic purposes is preferred to the *Pittsburg*. Analyses were made as follows:

I. T. Lucas, Dunkard township, Greene county. (D. M'C.)

II. —— ——, Upper Bench, near Mapleton, Monongahela township, Greene county. (S. A. Ford.)

III. —— ——, Lower Bench, near Mapleton, Monongahela township, Greene county. (S. A. Ford.)

	I.	II.	III.
Water	1.790	1.500	1.088
Volatile combustible matter	35.400	30.428	34.012
Fixed Carbon	56.818	55.038	51.783
Sulphur	1.152	1.406	2.261
Ash	4.840	11.628	10.856
	100.	100.	100.
Percentage of coke	62.810	68.072	64.900
Color of ash	Gray.	Red gray.	Red gray.

The ash is unexpectedly large, as, though bulky, it is powdery and free from cinders. The coal burns very freely, and will bear shipping well. No. 1 shows its character as found on Dunkard creek, and in the adjoining portion of West Virginia. There it is a valuable coal.

The *Redstone Coal* is insignificant, except in the north-east corner of Washington county, where in a small area it is of workable thickness, and is mined to supply the necessities of farmers when the *Pittsburg* is below the surface. Like the *Sewickly*, it burns freely, and is reduced to a fine powdery ash. Only one analyses of this coal was made, and the specimen was obtained from I. Teeple's bank, near Monongahela city. It gave as follows:

Water	1.060
Volatile combustible matter	33.590
Fixed carbon	48.688
Sulphur	2.367
Ash	14.295
	100.000

Percentage of coke	65.350
Color of ash	Gray.

(A. S. M'C.)

The *Pittsburg Coal* is the important bed of the district. It is exposed in Dunkard, Monongahela and Jefferson townships, of Greene county, along the whole river line of Washington and Allegheny counties, as well as over a large area in these two counties. Its distribution has been given in detail in another portion of this report.

In thickness the lower division varies from about four feet to nine feet, the diminution being quite gradual from the south-east corner of the district to the north-west out-crop of the bed. Occasionally, as at Brownsville and at some points along the Panhandle railroad, the coal is quite open-burning, and approaches block coal, but for the most part it is a very fat caking coal, of extreme purity and admirably adapted for the manufacture of gas or iron. A large number of analyses have been made in order to determine the degree of variation in quality. These show a narrow range of variation, and prove the bed to be of remarkable excellence throughout the whole district. As in the case of the *Waynesburg*, only a portion of the total analyses is given here. Others will be given in Mr. M'Creath's report.

1. Maple Farm, Dunkard township, Greene county. (A. S. M'C.)
2. Daniel Miller, Dunkard township, Greene county. (A. S. M'C.)
3. L. Vernon, Jefferson township, Greene county. (A. S. M'C.)
4. Mr. Liddell, Centerville, Washington county. (A. S. M'C.)
5. Mr. West, Greenfield, Washington county. (A. S. M'C.)
6. T. Redd, Fallowfield township, Washington county. (D. M'C.)
7. New Eagle Works, Monongahela city, Washington county. (A. S. M'C.)
8. Mr. Thomas, Peters township, Washington county. (A. S. M'C.)
9. Harding & Warrick, Washington, Washington county. (A. S. M'C.)

10. P. Ashurst, Chartiers twp., Washington county. (D. M'C.)

11. T. Thompson, Chartiers township, Washington county. (D. M'C.)

12. Mr. Magee, Independence township, Washington county. (D. M'C.)

13. Mrs. Bushfield, Cross Creek township, Washington county. (D. M'C.)

14. Mr. Patterson, (roof coal,) Centreville, Washington county. (D. M'C.)

	I.	II.	III.	IV.	V.	VI.	VII.
Water	1.030	0.900	0.850	1.425	1.220	0.680	1.140
Vol. comb. mat.	36.490	38.390	38.580	36.880	35.420	38.525	35.275
Fixed carbon	59.051	52.649	54.185	56.829	60.537	55.920	58.167
Sulphur	0.819	1.941	1.290	0.796	0.658	0.855	0.758
Ash	2.610	6.120	5.095	4.070	2.165	4.020	4.660
	100.000	100.000	100.000	100.000	100.000	100.000	100.000
Per ct. of coke	62.480	60.710	60.570	61.695	63.360	60.795	63.585
Color of ash	Cream.	Redd'h gray	Gray.	Cream.	Red.	Red.	Gray.

	VIII.	IX.	X.	XI.	XII.	XIII.	XIV.
Water	1.080	1.540	1.010	1.095	1.130	1.730	0.775
Vol. comb. mat.	40.350	37.825	40.995	39.790	38.720	37.735	36.770
Fixed carbon	50.311	57.063	48.769	55.033	40.253	54.561	51.467
Sulphur	2.594	0.762	2.206	1.172	3.722	1.499	2.098
Ash	5.665	2.810	7.020	2.910	16.175	4.475	8.890
	100.000	100.000	100.000	100.000	100.000	100.000	100.000
Per cent of coke	58.570	60.365	57.995	59.115	60.150	60.535	62.455
Color of ash	Red.	Cream.	Red.		Pink.	Gray.	

For comparison with these I give seven analyses of this coal as given in Volume II of the Geology of Ohio. These had been selected from a great number as representing very fairly the better coal from this bed. They show the extremes of variation, for in few localities is the percentage of sulphur much greater. The localities of the specimens analyzed are:

I. Jefferson county, Ohio. (Wormley.)
II. Belmont county, Ohio. (Wormley.)
III. Belmont county, Ohio. (Wormley.)
IV. Harrison county, Ohio. (Wormley.)
V. Harrison county, Ohio. (Wormley.)
VI. Athens county, Ohio. (Wormley.)
VII. Pomeroy, Ohio. (Wormley.)

	I.	II.	III.	IV.	V.	VI.	VII.
Water.............	1.45	1.00	1.10	2.80	2.44	2.70	4.10
Vol. comb. matter..	36.35	34.20	32.50	34.20	32.36	35.30	33.90
Fixed carbon	57.95	59.40	63.50	59.40	59.92	55.05	56.10
Ash................	4.25	5.40	2.90	3.60	5.28	6.95	5.90
	100.00	100.00	100.00	100.00	109.00	100.00	100.00
Sulphur...........	2.72	2.63	0.68	1.80	2.62	5.24	0.46

These analyses show that the *Pittsburg bed*, as found in the Greene and Washington District, yields a remarkably fine gas-coal, the average of the fourteen giving 37.974 per cent of volatile combustible matter, while in the seven Ohio analyses it is but 34.11. In only four of the results does the sulphur exceed two per cent. One of these (XIV) is the roof coal, which is never mined; another (XII) exhibits a remarkable local deterioration of the bed, while each of the others contains no more sulphur than does the celebrated Murphy Run Gas-Coal of West Virginia, while they both excel that coal in volatile combustible matter. The percentage of gas in every coal in the list is greater than that in the Penn Gas-Coal, which is the standard, while Nos. 1, 4, 5, 6, 7 and 9 are superior to it in respect of sulphur.

Mr. M'Creath states that the sulphur per cent of the coke is a little less than the sulphur per cent of the coal, and that this rule holds good in all the analyses made from this district. As the coke is compact, it is evident that a fine article should be made from Nos. 1, 4, 5, 6, 7 and nine.

The analyses show also that the prejudice existing against the coals of the Upper Monongahela river is utterly without foundation. Nos. 1 to 6, inclusive, are from the upper three pools and are certainly not inferior in quality to the others. The coal of these pools bears shipping quite as well as that from the lower pools, and one can see no reason why this portion of the district should not become valuable when the coal trade revives from its present depression.

The Upper Freeport Coal.—This coal is found only in Beaver county and is very irregular in its variations. It is of local importance near Hookstown and Shippingport, and is mined at other localities for domestic use. The analyses show a very high percentage of volatile combustible matter, but in respect of ash and sulphur the variations are so sudden that the value of the coal for gas or coke is very uncertain.

I. Cotter's bank, Raccoon township, Beaver county. (A. S. M'C.)

II. Swearingen's bank, Hookstown, Beaver county. (A. S. M'C.)

III. Todd's bank, Hookstown, Beaver county. (A. S. M'C.)

IV. Todd's bank, Hookstown, Beaver county. (D. M'C.)

V. Wilson's bank, Shippingport, Beaver county. (A. S. M'C.)

	I.	II.	III.	IV.	V.
Water	1.660	2.080	1.500	1.370	1.530
Volatile comb. mat.	37.065	39.520	39.870	37.800	41.380
Fixed carbon	51.351	54.691	46.960	54.463	49.798
Sulphur	2.709	1.249	4.595	1.587	2.467
Ash	7.215	2.460	7.075	4.780	4.825
	100.000	100.000	100.000	100.000	100.000
Coke per cent	61.275	58.400	58.630	60.830	57.090
Color of ash	Gray.	Yellow.	Red.	Gray.	Yellow.

The "Strip" Coal.—This variable coal is mined at Georgetown, in Beaver county, and is only of local value. Two specimens were analyzed, which show the coal to be of very superior quality, quite as good as on Yellow creek, Ohio, where, though little more than two feet thick, it is extensively mined for coking.

I. Diehl's bank, Georgetown, Beaver county. (A. S. M'C.)

II. Bryan's bank, Georgetown, Beaver county. (A. S. M'C.)

	I.	II.
Water	1.770	2.090
Volatile combustible	38.628	35.700
Fixed carbon	56.333	59.685
Sulphur	.717	.580
Ash	2.560	1.945
	100.000	100.000
Per cent of coke	59.610	62.210
	Reddish gray.	Cream.

IRON ORE.—In southern Fayette county, an important deposit of iron ore occurs at from 4′ to 6′ below the Pittsburg coal, which was formerly supposed to be confined to the trough lying between the Laurel (Chestnut) hill and the Fayette county axis. But recently it has been discovered on the west slope of the latter axis, and it proves persistent along the river to where the coal goes under near Gray's distillery. Numerous tests have been made on the east side of the river to deter mine its value. On the Greene county side, the rocks immediately below the *Pittsburg*, are rarely exposed, and in most cases they are covered by a thick mass of debris, so that although the horizon was always examined, the ore was observed at but two or three localities. It would be well for persons owning coal banks near the river below the mouth of Dunkard's creek to sink experimental shafts to the depth of eight or ten feet, in order to ascertain definitely the presence or absence of this ore. On the eastern slope of the Waynesburg axis this ore is wanting, and in Washington and Allegheny counties, no evidence of its presence were observed.

In the Lower Barrens there are some low grade ores within 200 feet of the Pittsburg coal, but they seem to be hardly persistent. A very fair ore occurs on the east side of the river below the mouth of Cheat, at about 325 feeet under the coal, but its horizon is concealed in Greene county. Fragments obtained on the Fayette county side contain small quantities of pyrites, and the ore is probably somewhat red-short. Should this prove to be the case, the ore, if persistent, will be valuable to the furnaces now working the ores of Fayette county, some of which are quite cold-short.

Nodular ore occurs in moderate quantity in the shales underlying the Waynesburg coal, in Greene county. At one time this was digged in several localities in Morgan township, and taken to the old furnace at Clarksville. But the undertaking was not profitable, and it was abandoned many years ago. The ore is in small quantity, and is distributed throughout a considerable mass of argillaceous shale. It seems to be quite persistent.

Ore occurs in small quantity in the black shale representing the Little Washington coal, in Greene county. An analyses is given in the general report of the Laboratory work of Mr. M'Creath. On Smith's creek, near Waynesburg, an ore is found in moderate quantity immediately above the *Washington Coal*, of which a specimen yields as follows, on analysis:

Metallic iron	37.400
Sulphur	.278
Phosphorus	.285
Insoluble residue	9.950
	(D. M'C.)

This is a very fair ore, and the quality is sufficiently good to justify careful exploration of the horizon in the vicinity. This ore is absent in Washington county.

At many localities in Greene county ore is found in considerable quantity below Limestone V. A specimen from Mr. J. Knight's farm, near Rogersville, yields as follows:

Metallic iron	30.400
Sulphur	.281
Phosphorus	1.405
Insoluble residue	12.111
	(D. M'C.)

The percentage of phosphorus by this analysis is so large as to render the ore worthless.

In Centre township, of Greene county, near the head waters of Pursley creek, there is a local deposit of ore which has been referred to in the description of that township. As nearly as could be determined, it is embraced within an area of nearly two miles, although explorations on Hargus creek may prove its presence there. Specimens of this ore were analyzed with the following result:

Metallic iron	36.000
Sulphur	.047
Phosphorus	.606
Insoluble residue	5.520
	(D. M'C).

It is unfortunate that this analysis exhibits so large a proportion of phosphorus, as otherwise it is a very fine ore. If the analysis properly represents the quality of the whole deposit the pig would contain somewhat more than 1.5 per cent of phosphorus.

Within the area referred to the quantity of ore is very great, and it lies in shape such that a large amount can be obtained at extremely low cost. Its horizon is in the shales immediately above the black shale, resting on the Upper Washington Limestone. This rock was diligently examined at all other localities in Greene and Washington counties where it is exposed, but no ore was found. The shales are always ferruginous, occasionally they contain a few nodules, but ordinarily the ore is discriminated, and the mass is worthless.

There is scarcely a farm in the whole district on which nodular iron ore does not occur in apparently considerable quantity, and many persons, judging from this circumstance, have concluded that valuable ore beds must be numerous. This conclusion, however, is by no means a necessary one. The soils owe their origin to disintegration of the rocks in the vicinity, and there is hardly a shale or sandstone in the whole section which does not contain more or less nodular ore. When these break up the nodules are set free and are found loose in the soil, when they are frequently turned up by the plough. Their presence is no evidence whatever that workable beds of ore occur, but, on the contrary, is very probable evidence that there are no such beds in the vicinity.

Limestone V, a coarse, brecciated, dull-colored rock, has been mistaken for iron ore. It is utterly worthless as ore or as limestone. Its position has been so carefully pointed out in various parts of this report, that this error need not be repeated.

OTHER METALS.—In a large portion of Greene county the belief prevails that lead ore is present in large quantities within the limits of that county. This belief has taken but slight hold in Washington county, and in Allegheny it was heard of but once.

The story of its occurrence is traditionary, and differs in no wise from that given everywhere in West Virginia and Eastern Ohio. It dates back to the early settlement of the country, and its originator was some pioneer who had been taken prisoner by the Indians. In nearly every instance the prisoner saw the Indians go off some distance and soon return with lead, which they melted down into bullets. Occasionally the tale varies, and lumps of ore were seen. In one locality I was informed that an individual was in the habit of procuring specimens of ore for a small compensation, and persons have been seen who averred that their parents knew of localities where the ore occurs in great quantities.

The story itself contains full refutation of the conclusions drawn from it. The fact that the Indians procured metallic lead, which they melted into bullets, is proof that the metal did not belong to that vicinity. A bar of iron picked up in a field would hardly be thought evidence of a deposit of iron ore. Lead occurs native no more than iron does. It is well known that the Indians, ignorant of methods for the reduction of lead ore, carefully hoarded all the lead which they could find or steal, and concealed it in magazines. Galena, or lead ore, was a medium of exchange among the Indians, as gold and silver are among civilized nations. It too was hoarded, and the deposits from which the ore has been obtained are simply the secreted treasures of the savages.

Farther than this, it is utterly incredible that any person possessed of information deemed so valuable as this, could resist the pecuniary temptation to reveal the secret. That this temptation has been successfully resisted in so many instances can be explained only by supposing that no such secret existed to be revealed.

In view of all the knowledge acquired by geologists in various portions of our country, one may very safely assert that lead ore does not exist in paying quantsties within the Greene and Washington district. If fragments be found, the finder will be justified in concluding that they have been imported. Any time or money expended in search for lead ore will be utterly wasted.

A similar remark may be made respecting zinc, tin, copper, silver and gold. The first of these is seen occasionally in minute quantity, forming the nucleus of ferruginous concretions, but in all probability the others are wholly absent.

LIMESTONE.—Everywhere throughout the district, except in south-western Greene county and in Beaver county, limestone occurs in prodigious quantity. The Fishpot, Waynesburg and Upper Washington Limestones are all of excellent quality. The first is quarried extensively at several localities along the Monongahela river, and is used in the manufacture of iron and glass. The Waynesburg makes a strong but rather dark lime, which is much used in the interior of the district. The Upper Washington is good throughout, but the dark or middle portion gives a fine white lime, which is clear enough for inside work. An analysis of this portion is given below. For agricultural purposes limestone is sufficiently abundant everywhere. Limestones I "b," IV, V and VII are too poor for this purpose or any other, but all the others yield stony lime.

For the manufacture of cement ample material is found in a large portion of the district. The upper division of the Great Limestone, that immediately underlying the *Uniontown Coal*, is the one quarried at Uniontown, Fayette county, where a cement is made which is employed at the Government Works on the Monongahela river. It is probable, however, that a better cement can be manufactured from layers at the base of the Great Limestone. These have a peculiar greasy look when freshly fractured, and on exposure readily fall to pieces. The character is the same everywhere in the district. Specimens from two localities were analyzed.

I. Great Limestone, lower division, Cannonsburg, Washington county. (A. S. M'C.)

II. Great Limestone, lower division, Cannonsburg, Washington county. (D. M'C.)

III. Great Limestone, lower division, Cannonsburg, Washington county. (D. M'C.)

IV. Great Limestone, lower division, Somerset township, Washington county. (D. M'C.)

V. Upper Washington Limestone, Washington, Washington county. (D. M'C.)

	I.	II.	III.	IV.	V.
Carbonate of lime	48.823	47.080	68.837	47.750	72.866
Carbonate of magnesia	20.621	28.528	14.649	30.943	3.813
Carbonate of iron, Alumina	7.148	7.511	3.306	5.618	2.929
Sulphur	.203	.069	.097	.126	.155
Phosphorus	.051	.127	.049	.015	.061
Insoluble residue	22.520	15.750	13.300	14.920	17.380
	99.365	99.065	100.238	99.372	97.204

STONE FOR BUILDING.—Some portions of the district are well supplied with admirable sandstone for building purposes. The great Gilmore and Fish Creek sandstone, though a little soft, have a fine color and for the most part are durable. The latter occurs in Springhill, Gilmore, Jackson, Centre and Morris townships, of Greene, as well as in Morris, cf Washington. The former is confined to the high land of south-western Greene. If an outlet existed these sandstones could be shipped with profit to a long distance. The Waynesburg Sandstone, where compact, is employed as a building stone, but it is of variable quality. At some localities it is quite durable, while at others its friability seems to increase on exposure. The compact portion of this stratum always has a handsome color and dresses easily. A good sandstone occurs under the Upper Washington Limestone at many places in Washington county. The Mahoning and Freeport Sandstones are used along the Ohio river, and the former is shipped to Pittsburg and other places in Allegheny county, where good stone is scarce. A very fine sandstone occurs locally on the Panhandle railroad, at Walker's Station, and again at a short distance east from Oakdale. It is of extremely good quality, though a little coarse in grain. It is shipped to various points along the railroad.

Good flagging stone is obtained below the Sewickley Coal, in the upper portion of the Waynesburg Sandstone, at almost forty feet above the Washington Coal and just above the Middle Washington Limestone.

CLAYS.—Fire or potter's clay occurs directly underneath all the coal seams, though the layers are sometimes very thin. In almost every instance iron is present in sufficient quantity to render the clay worthless. The only bed of persistent value is that underlying the Kittanning Coal, which is of considerable thickness and geographical extent. It is used for brick and earthenware at numerous localities along the Ohio river from Beaver county down to about midway between Steubenville and the mouth of Yellow creek. Within this district but little use has been made of this clay, and it has been thought best to defer the close investigation of it until the northern portion of Beaver county could be examined. A very good bed of clay is found under the Ohio "Strip-vein" in Beaver county, but it is not persistent. Fire-clays are of common occurrence in the Terrace deposits, but within this district their value has not been determined by any practical tests.

Good clay for brick making is found generally distributed in the subsoil. Near Pittsburg, the shales about 100 feet below Pittsburg coal, are employed for this purpose, and make a very good brick.

SAND.—Sand for the manufacture of glass is obtained opposite Belvernon, on the Monongahela river, and, no doubt, if systematic search were made at the same altitude above the river, other localities of equal value would be found. This deposit has been fully described in Chapter I. It is 180 feet above the river, and is a detrital deposit on one of the old terrace shelves. A similar sand occurs on the other side of the river at Belvernon, but is said to be hardly so well fitted for the manufacture of fine glass. That obtained on this side is used at the Belvernon Glass Works, and is shipped in large quantities to Pittsburg. The mirror glass made at Belvernon is said to be the finest made in Ameriea.

The sand of the terrace at 275 to 300 feet above the river may eventually prove valuable. No openings have been made in it, and on the terrace wall the clay and sand are indiscriminately mingled. In the vicinity of Greensboro' it might be well to make trial excavations to determine the character of the sand. The terrace is well defined there.

PETROLEUM.—Within the limits of the district petroleum has been obtained in considerable quantity on Dunkard creek, in Greene county, and on the Ohio river opposite Smith's ferry, in Beaver county. Respecting the latter locality, our imformation at present is somewhat meagre, it being thought best to defer the detailed investigation until the examination of northern Beaver is complete. The productive territory is principally on the opposite side of the river, and the rocks yielding oil do not reach the surface anywhere within the district. The relations cannot be discussed without entering into a discussion also of some points of statigraphy now in dispute, which can be settled only after a careful study of the northern region.

On Dunkard creek the oil area is extremely limited, as already shown in the description of Dunkard township, Greene county. Evidences of the presence of oil are found on Whitely and Little Whitely creeks, in Greene county, and small quantities were obtained many years ago in Fayette county, on Dunlap's creek, just above Brownsville. It is unfortunate that in all these localities the records of borings are very imperfect. Few were kept, and of those, none are now accessible. No definite information respecting the amount produced can be obtained, and most of the facts concerning the oil horizons and the productions were given by Mr. Cephas Wyley, of Dunkard, to whom I am indebted for many personal favors.

The quantity of oil in the Dunkard area is undoubtedly very great. The sandstones at the more important horizons seem to be saturated, as they undoubtedly yielded more or less in each boring, and some of the wells produced very largely for a time. Most of these promising wells suddenly ceased flowing, and gave up nothing afterward when pumped. For this reason the region lost its reputation, and the quantity of oil was regarded as inconsiderable. But the cessation of production was not owing to lack of oil. The wells were bored without casing, through the shales of the Lower Barren Series, which are cohesive enough to stand when dry. But when the oil and water poured through the hole these shales were converted into mud, and the wells "caved," thus effectually stopping the flow. Attempts to clean them out only increased the difficulty, by filling the seams and crevices of the sandstone with clay, and thoroughly sealing them. Had the wells been properly cased, the history of the locality would have been different, for those which were cased fell off in yield gradually, and quite a number of them are still productive. The character of the oil has been referred to in another portion of this report.

Three persistent sandstones have been found in the wells at 165, 425 and 560 feet below the Pittsburg coal, and their thickness are 66, 50 and 400 feet respectively.

The first or upper sandstone shows some variation, being sometimes only ten feet thick, and occasionally it is replaced by shale. It is an important oil rock throughout, as appears from the fact that in Wylie well, No. 2, oil was struck at little more than 160 feet, while in No. 1, barely 200 yards away, the strike was at 206 feet below the Pittsburg Coal. It is worthy of note, that where this sandstone is replaced by shale, the shale contains no oil.

The second, or middle sandstone, is quite constant in thickness, and is the chief repository of oil in this region. The top of the rock is usually 425 feet below the *coal*, though in two wells it seems to be barely 400. The strikes of oil in this rock are at variable distances below the *coal*, being 400 feet on the Ross farm; 425 feet in the Butler well; 443 feet in Wyley, No. 2, and 463 feet in Wyley, No. 3.

The third sandstone is double, the divisions being separated by some shale and coal, a mass of undetermined thickness, though probably not more than thirty or forty feet. It has been reached only in Wiley No. 1, and in one of the Bailey wells. In the former it yielded a little oil, but in the latter it is barren. The Bailey well is down 400 feet in this rock, and and has not yet passed through it.

These sandstones are identifiable without difficulty. The first is exposed in the river bluffs above Greensboro', and is a marked feature to a considerable distance above Morgantown, in West Virginia. This I have called the *Morgantown Sandstone*. The second is the *Mahoning*. The interval between it and the Pittsburg Coal is small compared with that at many localities farther north and east, but the structure of the rock and its relations to others underlying it leave no room to doubt the identification. It contains conglomerate layers, and rarely some carbonaceous matter. Its base is only eighty-five feet above the top of the *third* sandstone, which, as has been stated, is double. The upper portion belongs to the Lower Productive Coal series, and the lower to the Great Conglomerate. The latter is described as containing pebbles in great quantity. In the Wiley well oil was obtained at 578 and 612 feet, so that the upper division may be regarded as the oil-bearing rock. This is clearly the sandstone which has usually been termed "Tionesta," but that name has been erased from the list as a synonym of the Conglomerate, and Prof. Lesley has identified the stratum with the Piedmont Sandstone of Maryland. The horizons, then, are as follow:

1. Morgantown Sandstone, Dunkard.
2. Mahoning Sandstone, Dunkard, Whiteley, Dunlap's creek.
3. Piedmont Sandstone, Dunkard.

The rocks along Dunkard creek show no signs of disturbance worthy of comparison with that found in the interesting oil-break of West Virginia. There, as I have shown elsewhere,* the strata are so crushed and dovetailed, that no two borings show even approximately the same section; but here there is no break deserving the name. About half a mile north from the Bailey wells and near the Twin wells, a slight fold crosses the creek, having a direction of N. 20°, W. Mag. The sandstone underlying the *Pittsburg Coal* is seen dipping eastwardly and forming a dam in the stream, while, seventy yards away, the crest of the fold appears on the right-hand side of the creek, exposing the shale below for a horizontal distance of forty feet. Beyond that the sandstone dips quickly to the creek. The disturbance is slight, and cannot be traced away from this exposure, but near Taylortown, two miles farther up, it is distinctly shown in the road, where on its western slope the *Pittsburg Coal* has a dip of nearly 6°. On Big and Little Whiteley nothing of the kind was observed, but the localities on those streams, where wells were bored, are almost on the line of this axis, correction being made for magnetic variation. The well on Dunlap's creek, in which oil was found, was bored near the crest of a little fold there, which keeps the *Pittsburg Coal* above the stream for nearly two miles from the river.

The borings tell the same story with the surface. They give no evidence of faults or crushes. The intervals between the sandstones vary little in the different wells; no extensive crevices occur, and in boring the tools seldom dropped more than two or three feet.† At the same time crevices of small size are of frequent occurrence, and oil was obtained only when a crevice was struck. That these cavities communicate was handsomely demonstrated on the Ross farm, where a very fine well was tapped and drawn off by another, bored only a few yards away. The great frequency of these fissures can hardly

* Notes on Geology of West Virginia, Proc. Amer. Phil. Soc., 1875.

† "Dropping of the tools" is due in many cases to "scaffolding" in the well, and not to hollows or fissures in the rock. [J. P. L.]

be explained by supposing them to be the result of shrinkage; and it seems necessary, as well as in accordance with observed facts, to regard them as due to the action of some disturbing force exerted along the line, but without sufficient energy to produce permanent effects visible on the surface.

This seems especially probable when we consider the limited area in which the oil can be procured. The great oil-break of West Virginia is merely an anticlinal reaching certainly from the Great Kanawha river northward into Ohio, which exhibits violent faulting and crushing within the fold, in the vicinity of the Baltimore and Ohio railroad. East and west from this violently disturbed region, which is very narrow, the rocks are almost horizontal for a considerable distance, until on each side they are abruptly cut off by a fault that is on the east side at Ellenboro', being one of several hundred feet. Outside of the narrow break the rocks are regular and in no wise distorted. Within the break the borings become productive wells, whereas outside no productive wells have been obtained, though many borings have been carried down to the oil-bearing strata. There are no crevices, but the rock contains petroleum, just as within the "break. The same is true in the Dunkard region. There the productive area is less than two-thirds of a mile wide from east to west. Outside of this no wells have yielded anything, although the number of borings is very large. How far southward the area extends has never been ascertained, as the creek bed coincides with the anticlinal only in the vicinity of Bobtown. Southward from this no borings have been made along the anticlinal. The prejudice was in favor of "bottom" lands, not only because they are at a lower level than the upland, but also because it was believed that oil has a tendency to flow into hollows. It was held, also, that the extreme abruptness of the hills along the creek proved that there the disturbance was most violent. For these reasons those boring followed the creek and occupied barren ground, and no wells were sunk along the line of disturbance. On Dunlap's creek the boring made near the crest of the little axis there obtained oil. Borings made away from it, though piercing the same rock, found none.

Experience having shown in West Virginia and south-west Pennsylvania that oil could be procured in economical quantities only where there seems to have been some decided disturbance of the strata, those engaged in procuring this oil came to regard this disturbance as necessary to the original production of the oil itself. But a few considerations, I think, will suffice to show that the oil in no wise owes its origin to disturbance of strata, and that the only effect of the disturbance has been to provide reservoirs for the oil in the rock already oil-bearing.

The similarity between the products obtained by distillation of petroleum and those obtained by the destructive distillation of wood, peat, lignite, cannel coal and carbonaceous shale has led to the belief that petroleum results from the distillation of carbonaceous shales, and that the sandstones or limestones containing the oil are only receivers for the distillate. But this hypothesis is unable to account for many facts, by which, indeed, it is wholly contradicted. Thus the Trenton Limestone, in Ontario, contains petroleum in localities where it rests directly on the metamorphic rocks, whose metamorphism must have been complete before the limestone was formed. It will hardly be suggested that the petroleum came from an overlying rock, for the heat invoked to perform the distillation is from the interior, and it could not fail to affect the lower rock even more than the upper. The oil must be indigenous to the limestone in this case. Prof. T. S. Hunt adduces equally good proof that the oil is indigenous to the Corniferous Limestone of Canada.

In the Dunkard region, the Morgantown and Mahoning Sandstones are saturated with oil, for no well was bored without finding greasy, strongly odorous rock at each horizon, though, as previously stated, no supply of oil was obtained unless a cavity was struck. The quantity of oil present, therefore, must be enormous.

From the scanty information which can be procured respecting the borings, I learn that *black* shales occur between the first and second sandstones, as well as between the second and third, while between the third and the conglomerate there is a thin bed of coal, which is soft and altogether unchanged. It is asking too much of one's credulity to suppose that the few feet of carbonaceous shale in these intervals would have yielded sufficient oil to saturate two enormous beds of sandstone and yet have retained so much carbonaceous matter as to be black shale still. It is equally improbable that the shales could have yielded this matter, while the coal below remained wholly unchanged.

Underlying the oil-bearing rocks in the West Virginia oil-break is a great mass of red shale, several hundred feet thick. Its place is below the Umbral Limestone and the Waverly Conglomerate. One hastily considering the question might conclude that this shale yielded the oil during the convulsion which produced the anticlinal. But the means would not be competent to the end. The convulsion was explosive along the line of the break, and especially in the richly productive area. The whole top of the fold was blown out and the fragments falling back seemed to have braced up the sides. So violent was the movement that the force must have been dissipated in the atmosphere, very little of it being converted into heat. If it had been accompanied with sufficient heat to volatilize oils, the gases would inevitably have escaped, owing to the remarkable shattering of the rocks.

But it seems to me quite clear that the force, even if competent, was not exerted in this direction. The thin coals involved in the sides and interior of the break have been subjected to terrible treatment. They are tossed, broken and dovetailed between other strata. I have already stated that the confusion produced by this crushing and faulting is so great that the records of borings made barely thirty yards apart show little resemblance. Yet the coal is wholly unchanged in character, being the same without as it is within the break.

The question naturally arises, "Is not the assumption that the oil arises from distillation of the shales, unnecessary?" The origin of the oil can be readily explained in a simple way—in a way fully according with well-known facts and without resort to any hypothesis of violence. It is well known that petroleum is one extreme of change in organic tissue. In vegetable matter the process may take one course and give us coal, but that it is carried in the other direction is amply proved by the Trinidad deposit, which in composition is allied to the Grahamite of the West Virginia oil region. Animal matter must have produced largely the oil of the Trenton and Corniferous, in Canada, and of the Niagara, in Illinois; but in the Mahoning and Morgantown Sandstones of Dunkard creek, the oil has come from vegetable matter. Those rocks are characterized by a vast number of impressions of stems, of which nothing but impressions remain. Casts are of rare occurrence. Occasionally a thin scale of coal marks the place of the bark and then a cast of the stem is found. But an insignificant proportion only of the stems took this shape.

Here, then, was sufficient material for the production of the oil. We find the oil in the rock; we find proofs of the presence at one time of a vast amount of vegetable matter, which was capable of changing into oil. The vegetable matter has wholly disappeared; it has left no coal, but the coal is saturated with oil. It seems to me, therefore, in the highest degree probable, that this vegetable matter, instead of being converted into coal, was converted into petroleum, and that the oil is indigenous to these sandstones. (See the discussion of this argument by the fossil botanist of the Survey, Mr. Leo Lesquereux, in Report of Progress, 1874, J, pages 104–107.)

It must be understood that the above statement is confined in its application to the Southern Oil District. It may not at all apply to the loose-grained and gravel oil belts of the Allegheny River region, which can hold large quantities of petroleum without the aid of cavities or fissures produced by structural disturbance of the strata. In the southern area the rocks are fine-grained and compact, and seem to require lines of disturbance to enable them to hold stores of oil.

INDEX

To Prof. J. J. Stevenson's Report on the Geology of Greene, Washington, part of Allegheny and part of Beaver Counties, Pa. 1874.

NOTE.—In all cases (except of proper names) C. means coal; C. M. coal mine, whether drift, shaft or stripping; U. C. M. Upper Coal Measures; L. C. M. Lower Coal Measures; U. B. M. and L. B. M. Upper and Lower Barren Measures.

T. means township; W. Co. Washington County; G. Co. Greene County

L. means Limestone; U. Wash. L. Upper Washington Limestone, &c.

SS. means Sandstone; Mah. SS., Mahoning Sandstone.

	Page.
Adams (Mr.)	155
Adams' Mill	173
Adamson's Quarry	291
Ager's (W.)	304
Agriculture of District	7
Aleppo section	5, 161, 163
Alleppo T., G. Co., geology	163
" " depths of coal beds	375
Alethopteris	59
Alexander's C. M:	345
Allen T., W. Co., geology	207
" " depths of coal beds,	356
Allen's limestone	88
Allen's section (Bradford)	210
Allen's (J.) sect. Cross Cr.	288
Allen's quarries	345
Allen's C. Mine	346
Allenport	208
Allegheny Co. described	2
Allegheny Oil Well	101
Allison's (J.) C. M.	233
Ally & Sickel's C. M.	197
Ames L. of Ohio	83
Amity	28, 188
Amwell T., W. Co., geology	185
" " depths of coal beds,	353
Analyses of Waynesb'g C.	376
" of Sewickley C.	379
" of Pittsburg C.	380
" of U. Freeport coals	382
" of Strip Coal bed	383
" of Waynesburg Iron Ore,	384, 385
" of Limestones	388
Ancient river beds	19
Anderson's C. M.	113
Andrew's (W. D.) C. M.	352
Andrew's (W.) C. M	329
Andrews, Prof., Geologist	83
Annularia	59
Anticlinal axes of the district	25
Anticlinal axes of coal era age	89
Ant. axis striking N. 20° W.	393
Armour's (Mr.)	335
Arnold (R.)	
Arthur's (Jas.) C. M.	232
Ashurst's (P.) C. M.	232
Ashurst's Coal analyzed	381
Atkinson's C. M	226
Atcheson P. O. Sandstone	259
Athyris subtilita	80
Auld's limestone	160
Auld's C. Mine	331
Avenues to market	8
Babbitt's blacksmith shop	159
Babbitt's C. Mine	336
"Backbone" sandstone, Traverse Cr.	338
Bacon run	104
Bacon Street run	141
Bailey oil farms; wells	101, 393
Bailey's C. Mine	131
Baird, (Mr.)	178
Baird's C. M.	134
Baker's (N.) C. M.	180

	Page.
Baldwin's Mill	171
Baldwin T. All'y co. geology,	
Balt. and Ohio R. R. line	9
Bane's (David) C. M	183
Banfield's C. Mine	234, 236
Baptist Church; L. X	39, 41
" Centre T. Section	158
" Amwell T	189
" Donegal T. Section	267
Barnhart's (S. H.) C. M	131, 170
Barnard's (N.) C. M	176
Barometric measurements	351
Barr's (J) and (R.) C. M	220, 260
Bartleson's C. M	289
Bates fork of 10 Mile run	146, 161
" Sandstone	192
Bavington (Mr.)	271
Bayard's C. Mine	163
Beallsville	216
Beam's run	297
"Bearing-in" bench, Pitts. C., 71, 314, 322, 329.	
Beaver City terrace	16
Beaver Co. described; geology	3, 334
Beaver Co. long section, by I. C. White	334, 335
Beaver Co. terraces	16
Bedell's C. Mine	297
Bell's (Jesse) C.M	100, 144
Bell's (J. M.) C. Mine	151
Bell's C. Mine, Peters T., 223, 227, 238, 294.	
Bell's Run coals	143
Bell's Run Section, Morgan T.,	143
Belleair river-bed	19
Bellerophoon in L. IV	49, 242
Belton	5
Belvernon terraces	13
" anticlinal axis	26
" height of Pitts C	208, 213
" sandstone	390
"Bench" C. of Pitts. C. cat-a-cornered	276
Bentleyville; synclinal	27, 216
Berkheimer's (J.) C. M	177
Bethel Church, Jeff. T	284
Bigger's C. Mine	272
Bigger's run coal, L. B. M	79
Big Mill Cr.; oil wells	346, 349
Big Traverse Cr	32, 335

	Page.
Birdseye in Great Lime	230
Birmingham; Shale L. B. M.	79, 309
Blackburn colliery	299
Black Diamond run	210
Black fossil of L. B. M	76, 79
Black gravels	12
Black's C. Mine	238
Black's (A.) C. M	343, 344
Black shale; over L. X	41, 66
" " fossiliferous	111
" " Richhill T	177
" " 60, ft. x Pitts. C.	327
" " oil wells	395
Blacksville, section	26, 107, 108
Blank's C. Mine	299
Block coal, Robeson T	271
Block coal, Pitts. C	276
Blocks of S. S. on hilltops	184
" " " on Cross Cr	291
" of Mahoning S. S	342
Bloom's C. Mine	343
Bluffs of Waynesburg S. S	58
" of S. S. in Morris T	192
" " " on Cross Cr	287
" " " in Robinson T.	326
" of Morgantown S. S. Beaver county	337
" of Mahoning S. S	340
Boat-shaped horsebacks, Pittsburg Coal	308
Bobtail Oil well	101
Bobtown; axis	95, 394
Bock's mills	338
Bollenfield Sch. H. coals	141
Boon's (J.) C. Mine	247
Boon's (T.) C. Mine	236
Boring records lost	391
Borings in river beds	19
Bosworth's limestone	164
"Bottom" bench, Pitts. C	211, 276
" Independence T., 295, 304, 328	
Boulder gravels	13
Bowen's Fork coals	100
Bowerhill Station	30, 315
Bowerhill C. Works	227, 315, 318
Bowlsby's (J.) C. M	93
Bowser's (A.) C. M	176
Bowser's (D.) C. M	137
Bowser's (Jac.) C. M	136
Boyd's run	146
Brachiopods in L. IV	242

	Page.
Braddock's Fields; Station,	17, 27, 301
Braden's C. Mine	345
Brecciated limestone	325, 326
Briar Hill C. Works	271
"Brick" bench in Pittsburg Coal, 71, 211, 212, 214, 225, 229, 233, 277, 295, 308, 309, 314, 321, 332, 336.	
"Brick" bench in Wash. Co.	286
"Brick" bench in Wayne Co.	207, 293
Bridgeville; terrace...	15, 316, 317, 319.
Brister's (H.) C. Mine	175
Brook's C. Mine	343
Brown (Mr.)	107
Brown's (S.) C. Mine	92, 107
Brown's Fork rocks	4, 38, 159, 192
Brown's Island	33
Brown's Mill coals	106
Brown's (D.) sandstone	291
Brownlee's (J.) coal	251
Brownsville	20
Brownsville road	306
Brownsville dam coal	179
" Wash. Co., described,	51
Brush run,	31, 183, 231, 246, 247, 259, 293
" " section	226
" " Hopewell T., section,	289
Brushy Fork section	159, 160
Bryan's C. M.—coal analyzed,	348, 383
Bryozoans in U. B. M.	49, 231
Bryson's Store section	292
Buckingham's (A.) C. M.	229
Buckingham's (E.) C. M	176
Buffalo Creek	2, 5
" Independence T.	292
" section	292
Buffalo T. Wash. Co. geology,	253
" " depths of coal beds,	359
Buffalo village	290, 291
Buffington's C. Mine	216
Building stone, Wayne. S. S.	143
Building stone on Raccoon Cr. and Ohio	342
Building stone, Economic geology	388
Bulger	32
Bulger anticlinal traced	31, 32
" " in Beaver Co.,	275, 335, 337
Bulger tunnel section, Smith	279
Burgen's C. Mine	217
Burgetts' C. Mine	113
Burgettstown road	275, 269
Burgettstown road section	280
" station	278
" synclinal traced	32
Burson's school house	137
Burwell's (A.) C. M.	130
Bushfield's C. Mine	285
Bushfield's (Mrs.) cannel	381
Bushrock S. H. section	285
Bush run coals	136
Butler oil well	101
Buxton's section, Cross Cr.	288
Calcite films on coal faces	271, 321
Caldwell (Mr.)	289
Calhoun's C. Mine	347
Calamodendra, Wayne. C. roof	131
Camp run	293
Campbell's run	326
Campbell's C. Mine	345
Camden	299
Camden C. Works section	301
Candor	280
Cannel C. bed, L. B. M	76
" (semi) in Nineveh C.	110
" (semi) 146; Aleppo T.	165
" (semi) in Pitts. C.	233, 271
" (semi) Pitts. C., Hanover township	274, 316
" (semi) Kittanning C.	349
Cannonsburg	4, 15, 30, 230, 234, 236
Cannonsburg gravels	15
" limestone analyzed,	388
Cannon at Frankfort Springs,	337
Canton T. Wash. Co. geology,	244
" " depths of coal bed,	362
Carman's fossils	295
Carmichael's	130, 131, 133
Carmichael's terrace	11
Carroll T. Wash. Co. geology,	209
" " depths of coal bed,	357
Carson's (J.)	256
Casber's C. Mine	225
Casteele run	141
Castor's Coal Mine	297
Caves in Filmore S. S.	39
Cavities in Waynesburg S. S.	58, 133
Cecil T. Wash. Co. geology	229
" " depths of coal bed	364
Cement; in Great L.	221; 64
Cemetery Hill, Frank T.	250
" Limestone 46; section,	247

	Page.
Centreville	26, 180
Centre school house	96, 142
Centre T. Greene Co. geology,	152
" " depths of coal bed,	373
Ceylon Post Office	126
Chamberlain's (C.) C. M.	185
Champlain clay formation	21
Chartiers T., Washington Co. geology	232
" " depth of coal bed,	362
Chartiers T., Allegheny Co. geology	311
Chartiers creek	4, 186, 229, 251, 326
" " mouth rocks	311
" " section in S. Fayette T.	317, 318
" " terraces	15
Chase's (J.) lime: IV	199
Cherry Valley P. O.	270, 280, 281
Cheat river, source of drift	18
" " mouth rocks	91
Chonetes granulifera abundant	80, 310
Cincinnati C. Works	320
Clark's (S.) C. Mine	208
Clarksville; gravel	3, 12
" section	176
" coals	137, 141
Classification of the rocks	23
Clay under Strip Vein	389
Clays; economic geology	389
Clay veins in coal,	228, 276, 279, 308, 324
" " in Wash. C	53
" " in Pitts. C.,	70, 307, 316, 324, 326
Claysville	51, 262
Claysville axis traced	31
" in Canton T.	246
" on Buffalo Cr	253
" in Donegal T.	261
" in Mt. Pleasant T.	268
" crosses Ohio river	311
Clear's C. Mine	345
Cliffs of Freeport S. S	86
Clinton	28, 329, 331, 332, 333
" gravels	12
Clokeyville	241
Clouse's Mill 166, section	196
Clover run, limestone	43, 156
Clovis' (J.) C. Mine	111
Coals of L. B. M. described	78, 79
Coal IV of Ohio (Kitt. C.)	87
" VII *b* in L. B. M	80

	Page.
Coal VIII of Ohio	65, 66
" VIII A (Redstone C.)	68
" IX of Ohio	63
" X (Uniontown C.)	63
" XI (Waynesburg C.)	59
" XIII (Washington C.)	51
Coal strings descend into the clay	142
Coal between L. floor and roof,	228
Coal fragments in clay	14
Coal boulders in sandstone	285
Coal beds consolidated before others were deposited	137
Coal consumption of Waynesburg	148
Coal resources of the district,	376
Coal trade of the Ohio river	10
Coal Bluff	220, 221
Coal Lick, Franklin T.	140
" section	147
Coal Valley	299
Cochran's Mill	296
Coen's (J) C. Mine	107
College grounds	138
Coke made from slack	309, 147
" Pitts. C. slack	277
Columbia	210
Colvin's run	104
Colvin's (V.) C. Mine	214
Conglomerate in L. B. M.	81
Connellsville (M'Con.)	31, 235
Conner's)G. W. C. M.	126
Cook's Station terrace	15
Cooley's C. Mine	336
Coon's Island synclinal	261
" Station, long section	262
Coon's Island	264
Cooper's (Mrs.) C. M.	185
Cooper's (J.) L. I	252
Copper	387
Coprolites in shale, 300′ above Pittsburg C.	238
Cortney's C. Mine	339
Cost of sinking shaft	247
Cotter's C. M.—coal analysed,	345, 383
Couch's C. Mine	219
Cowan's C. Mine	306
Cox's C. Mine	141, 142
Cox's (J.) C. Mine	107, 331
Crabapple coals	167
Crabapple run	345

	Page.
Craft's (W. E.) L. IV........	191
Craft's run....................	191
Crafton section...............	312
Craig's run	146
Craig's (B.) section	252
Craig's (W.) C. Mine	291
Crayne's (Mr.) run	144, 146
Creek Vein of Yellow Cr	87
Creigh's (Dr.) section........	247
Cregg's (J.) C. Mine..........	133
Crinoidal L. (L. B. M.) described......................	79
Crinoidal L....	76, 272, 303, 309, 312, 326
" Baldwin T.......	306
" Sheffield S. W...	311
" Robinson T......	328
" Meek's run	331
" Hanover T. Beav.	337
" Raccoon Cr.......	338
" Hopewell T. Ohio	340
" Moon T..........	342, 344
" Georgetown......	346, 348
Cristler's C. Mine	347
Cross Creek..............	5, 32, 269, 295
Cross Cr. T., W. Co., geology,	284
" " depths of coal beds	365
Cross Cr. road, village........	284, 287
Cross-roads used for mapping,	351
Cronch's (Elias) C. M........	180
Crook's run section...........	90
Croup's run section	197, 198
Crow's Mill...................	167
Crow's Island	340
Crumpshon's (J. M.) C. M...	99
Crustaceans in L. VI.........	47, 51
" in black shale.........	111
" in black shale, Centre,	152
" over L. III, Nott. T....	225
Ctenacanthus Marshii........	49
Cumberland T., G. Co., geol.	125
" " depths of coal beds,	370
Curryville....................	221, 223
Dage's (J.) C. Mine..........	185
Daniel's run..................	181
Date of Erosion of District...	20
Davidson's (J.) C. M.........	134
Davidson's ferry.............	134
Davistown in Dunkard..	97, 99
Day (S.)..................	192
Day's (J.) Lime. VI.........	191
Debolt's (J.) C. Mine........	97, 121
Denning's, (C.) C. M.........	260
" coal analysed......	378
Densor, (Mrs.)...............	185
Dent Post Office.............	27
Devore's C. Mine.............	224
Dexter Coal works	147, 204
Diehl's C. Mine	347
" coal analysed.........	383
Dillner's (A.) C. M...........	91
Dinsmore's C. M..............	319
Dip in Pittsburg coal bed.....	96, 305
Diplodus (?) tooth, U. B. M.,	49
Distillation theory rejected ..	395
Distortion of fossils	310
Disturbance in oil region.....	393
Divergence of coal beds......	63
Dodysburg coals..............	26, 148
Dog Hollow..................	175
Donahy's (W.) coal..........	258
Donald's C. Mine.............	317
Donaldson's (S.) C. M........	260
Donegal, T. W. Co. geology..	261
" depths of coal beds,	360
Donley's C. Mine.............	103
Dorr's C. Mine..	92
Doulin, (Mr.)................	133
Dowling school house section,	139
Drift covering of district......	11
Drywell oil well..............	102
Dunkard town................	2
Dunkard T. G. Co. geology...	90
" depths of coal beds,	367
Dunkard Cr. waters..........	2, 3
" terraces;—coals.....	11, 35
Dunkard Sandstone..........	157
Dunkard Coal described......	41
" " Springfield T.....	113
" " Centre T	157, 162
" " Aleppo T.........	164, 165
" " economic geology,	376
Dunkard Cr. oil region.......	391
" section in Wayne T.,	107
Dunkard fork of Wheeling...	5, 165
Dunlap's Creek oil............	391, 394
Dunning's (Dr.) C. Mine.....	278
Dutch Fk. of Buffalo Cr......	261, 263
" rocks..................	267
Duvall's C. Mine.............	204
Dyer's Run coal	124
Eachel's C. Mine.............	339
Earnest's Lime. VI...........	194

	Page.
E. Bethlehem T., geology....	175
" depths of coal beds..	351
E. Finley T. geology.........	192
" depths of coal beds..	354
E. Fk. of Buffalo Cr..........	253
E. Fk. of Raccoon section....	270
" in Smith T...........	280
E. Pike run T. geology.......	201
" " depths of beds...	356
E. Richhill T. intervals vary .	351
Economy, —— ferry..........	340, 341
Economites' C. Mine.........	339
Economic Geology of District,	376
Eddy's C. Mine	107
Edmunson's C. Mine.........	327
Eagleson's (J.) C. Mine......	290
Eldersville................	48, 282, 294
" limestone	64
" section in Smith T.	283
Election House..............	300
Elkhorn run	343, 344
Elk Lick (?) coal bed........	81, 83
" " in Beaver Co...	340
Ellenborough fault in W. Va.	82
Elliott's C. Mine	298
Elliott Oil Farm.............	101
Emery's (Dr. B.) C. M.......	219
Euomphalus (U. B. M.)......	49
" in L. IV...............	242
" *subrugosus*, in Ferr. L.,	346
Enterprise shaft.............	247
Erosion of the district; date..	2, 20
" of the river beds.....	119
" in Coal Era... 89, 132, 137,	285
" of coal benches......	180
" of coal under SS	217
" of Waynesburg S. S.	58
" of S. S. Peters T.....	228
Evans (G.) C. Mine..........	119
Everly's (Steinrod) C. M	97
Ewing's (J. H.) C. M.........	233
Ewing's Mills Post Office.....	323, 327
Ewing's Mills................	330
Ewing's Station section......	240
Ewingsville..................	30
Excelsior Works C. M	147
Fairview	97
Fallowfield T. geology.......	207, 213
" depths of Coal bed,	357
Farmer's C. Mine	323
Farrar's (A.) C. M..........	289
Fayette Post Office	323
Fayetteville..................	321
Fayette Co. Axis located.....	25
Fee's (J.) C. Mine	235
Fergus's C. Mine.............	235
Ferns near Washington......	242
Ferree's distillery...........	213
Ferriferous Limestone.......	84, 88
" Raccoon T., Beaver Co.,	344, 346
Ferriferous shale	143, 144
" under W'bg. C....	144
" blackshale, Centre,	155
Ferril's (Fred.) C. M.........	185
Field's C. Mine..............	19
Fifth Terrace described	17
Figley's (J.) section..........	222
Figley's fossil limestone......	340
Figley's C. Mine.............	342
Findlay T., All. Co., geology,	331
Findlay terrace..............	15
Finley's C. Mine.............	223
Finleyville..................	223
Fire-brick clay of Phillipsburg.......................	88
First Oil Sand, Dunkard.....	79, 90
" Morgantown SS.	391, 392
Fish teeth and scales in black shale......................	111, 149
" " in U. Barren Measures,	38
" " over L. II.............	50
" " in L. III, U. B. M.....	50, 225
" " in L. IV, Donegal T....	261, 262
" " over White L. VI......	47
" " in L. X, Aleppo T.....	164
" " in Uniontown C........	63
" " in Redstone C..........	236
" " near Cannonsburg.....	238
" " in Ohio................	83
Fish Creek, section..........	5, 36
" " coals, limestone,	112, 41
" " sandstone......	36, 42, 389
Fishpot Cr. coal; limestone..	66, 67
Fishpot run section	177, 178
Fixed horizon in L. B. M	80
Flag-stones under Sewick. C.,	389
" above Mid. Wash. L.,	389
Flattening of anticlinals......	28
Flenniken's (W.) C. M	126
Flenniken's (E.) C. M........	129
Florence.....................	273, 283
Folds or anticlinals..........	25

	Page.
Foley's (Mrs.) C. M..........	233
Ford's........................	313
Fort Pitt Section.............	324
Fossils distorted in L........	310
Fossiliferous L. (L. B. M.)	80, 236, 340
Fossiliferous L. (U. B. M.) II,	225
III, IV, V..................	242, 49
Fossils in Freeport L.........	341
" Crinoidal L..............	303
" Great L................	64, 230, 231
" U. St. Clair T............	316
" Quarternary..............	22
" numerous at Glass Works,	31
Fossil shale in Moon T.......	342
Foster, Wood & Co.'s C. M...	299
Fourth Terrace described....	17
Frankfort.............	32, 273, 335, 336
Frankfort Springs...........	337
Franklin T., G. Co., geology..	146
" " depths of coal beds,	372
Franklin T., W. Co., geology,	247
" " depths of coal beds,	35
Frazier's (J.) C. Mine........	273
Frazier's (J. C.) C. M........	183
Frazier's (T.) section.........	200
Frazier's (Mrs.) limestone...	198
Frederick; Fredericktown...	175, 177
Fredericktown coal..........	68
Frederick terrace...........	13
Freeport L. described........	86
" " in Moon T.......	339, 343
Freeport S. S...........	84, 86, 344, 389
" Greene T.; Moon T.,	347, 343
Freshets......................	5
Fry's (S) section.............	153
Fuller's (J. S.) C. M.........	129
Furguson's....................	114
Gamble's (S.) C. Mine.......	219
Gantz's (Mrs.) L. II..........	253
Garard's Fort road; coal......	100, 123
Garlow (Mr.).................	93
Garrison's (A.) C. mine......	42, 98, 99
Garrison's (Lee) Dunk. C....	110
Garrison Oil Farm............	101
Gas in Pittsburg C...........	382
Gashed sandstone............	54
Gayman's (S. G.) section.....	175
Geology of the districts by towns......................	90
Georgetown.............	346, 348, 383
Georgetown section..........	348
Georgetown terrace...........	16
Gillespie's Mill...............	284
Gilmer's C. Mine.............	210
Gilmore Limestone X........	41, 161
Gilmore S. S........	36, 38, 112, 162, 389
Gilmore T., G. Co., geology..	109
" depths of coal beds..	368
Glacial drift..................	11
Glade run—coals..........	26, 130, 100
Glass flux.....................	203
Glass sand..............	13, 14, 208, 390
Glass works..................	314
Glonn's C. Mine..............	305, 313
Gold..........................	387
Goniopteris (?)...............	59
Good Intent; L. VI........	31, 194, 200
Gordy's fork..................	194
Gorby's run..................	104
Grable's C. Mine.............	215
Grafton Station terrace........	15
Graham's (R.) C. Mine.......	219
Grain crops of the district....	6
Grant Coal Works............	324
Granville.....................	204
Gray Limestone, Franklin T.,	247
Gray s (Minor) C. M.........	119
Gray's Fork; rocks..	3, 38, 156, 165, 193
Gray's Landing coals.........	115, 136
" section................	118
Graysville limestone..........	39, 167
Gravel banks.................	11
Great Limestone described...	6, 57, 63
" " under Uniontown C..	388
" " sub-divided..........	141, 176
" " in Cumberland T....	135
" " in Jefferson T........	136
" " in E. Bethlehem.....	180
" " in W. Pike Run.....	204
" " in Carroll T.........	211
" " in Union T..........	221
" " fragments, Nott T....	225
" " section, Peters T.....	226
" " Cecil T...............	229, 230
" " in Chartiers T.......	235
" " fragments, S. Strabane	240
" " at Ewingsville......	241
" " in Mt. Pleasant.......	268
" " in Smith T...........	281
" " at Eldersville........	284
" " blocks on Cross Cr...	287
" " in Snowden T.......	304

	Page.
Great Limestone in Lower St. Clair, ends	309
" " in U. St. Clair T	315
" " in S. Fayette T	318
" " in N. Fayette T	323
Great Sandstone described	112
" " in Aleppo T.,	164
Greene Co. Group, (U. B. M.)	35
Greene T. Greene Co. geology,	122
" " depths of coal bed,	369
Greene T., Beaver county geology	346
Green's (U. L.) C. Mine	147
Greensboro' section, L. B. M.	115
" terraces	18
Greenfield	202
" ravine section	207
Green Garden School House	344
Greenlee's (J.) coal	142
Gregg's run	205
Grier's old opening	231
Grist mill, Prosperity	191
Groom's (A.) coal analyzed	377
Groomsville	332
Grover's (A.) C. Mine	130
Guseman's (J.) C. M.	130
Guy's Mill	331
" C. Mine	327
Hack's C. Mine	326
Hagerty's (D.) C. M.	254
" (T.) C. M.	293
Haine's (J.) coals	237
Hamilton's C. M	234
Hamlin's Station	284
Hanover T., W. Co. geology	273
" depths of coal beds,	366
Hanover T., Beaver Co. geology	835
Harding & Warwick's C. M.	240
" " coal analyzed,	380
Hargus' Cr., Centre T.	155
Harman's creek	5
Harrison's school house	326
Harry's school house	144
Harry's (C. C.) C. M.	144
Hart's run	164
Hart's (Judge) C. Mine	185
Hartley's Mill Section	121
Hasting's Station	317
Hatfield's ferry	68, 119
Hawes (W.)	227

	Page.
Hawkins' C. Mines	140
Hawkins' Mill	176
Hays Coal Works	305
Hays Station Section	325
Hays & Auld's C. M.	331
Headlee's run section	104
Hemipronites crassus	80
Hearn's C. Mine	327, 328
Heaton's school house	145
Hempfield R. R. cut	224
" " section	255
" " extension	241
Hemphill's C. Mine	290
Henderson's (A.) C. M	256
" coal analysed	376
Henry's (R. D.) L. VI	243
Henry's Fk. of Montour	333
Herod's run, Springhill	113
Hess' (J.) C. Mine	215
Hetherington's (A.) C. M	218
Hewitt's (G.) C. Mine	135
Hews' run coal	141
Hewston's Section	131
Hewston's (J.) Section	127
Hickory	239, 268, 269
Hickory road	235
Highest point of the district	181
Highest rocks of Wash. Co.	192
High knob in Chartiers	234
Hildebrand (D.)	185
Hildreth (Dr.)	83
Hill's (J.) C. Mine	206
Hill's (J. J.) C. M	217
Hill's school house	151
Hillsborough	28
" sandstone	181, 228
" " and section,	184
" Knob	42
Hinersman's C. Mine	164
Hinerman's (Jesse) C. M	165
Hixon's C. Mines	227
Hog Island in Beav. Co	32 343
Hogbacks on the divide	109
Hoge's (J.) Section	152
Hommeger's (D.) L.	198
Honeycombed S. S	325
Hook's Mill, Franklin	147
Hook's Dam coals	149
Hookstown	85, 346
" coals analysed	383
Hoover's run, Wayne	107

Page.
Hopewell T., W. Co., geology, 288
" " depths of coal beds, 363
Hopewell Church ridge...... 38
" " limestone.. 38, 159
Hopewell T., Beav. Co., geol., 339
Hopkin's Mill................ 192
Hopper's C. Mine............ 320
Horizon in Bar. Measures.... 80
Horner's dam................ 131
Horsebacks in coal........... 276, 279
" boat shaped 308
" in Pitts. C.... 306, 308, 333
" in Redstone C.... 228
" in Waynesburg C. 148
"Horseback" C. — Waynesburg C..................... 119
Horsebacks in Washington C. 53
Horton's (J.) C. Mine........ 181
Houlsworth's Mill coals...... 139
Houston's (S. A.) C. M...... 166
Houston's (D. C.) C. M...... 233
Houstonville................. 233, 334
Howards (H.) C. M.......... 180
Huffman's (S.) C. M......... 146
Huffmann's (J.) C. M 217
Hufty's Coal Mine 95
Hughes' (S. L.) C. M........ 185
Hughes' Mill................. 186
Hunt's Mill.................. 194
Hunt's (A. D.) lime 198
Hunter's Fork; and cave .. 5, 192, 159
Hunter's (W.) C. Mine...... 291
Huston's Station Synclinal .. 70, 220
Huston's run................. 220
Huston (J.).................. 58
Hutchinson's Mill L.......... 114, 115
Hutchinson (T. S.) 213
Hydraulic L., see Cement ... 22
Hymenophyllites 59
Ice in the river.............. 9
Idlewild Station 313
Imlay's (J.) C. Mine 202
Independence; section.... 85, 338, 294
Independence T., W. Co., geology........................ 292
" " depths of coal beds, 363
Independence T., Beaver Co. geology...................... 337
Ingram Station 312
Intervals variable........... 122
Interval of 265' 266
Interval of 285' between Pitts. C. and Crin. L 326
Interval of 216' between Pitts. C. and Wash. C 283
Interval of 160' between Wash. C. and Great L 294
Interval of 225' between Pitts. C. and Wayne. C........... 287
Interval of 225' diminishes northward.................. 350
Interval between L. VI and L. X varies 350
Iron ore.................. 116, 153, 155
" " nodules............. 144
" " economic geology... 383
" " on Ferrif. L 88
" " in Centre T.......... 152
" " in Morris T.......... 159
" " under Pitts. C., Snowden......................... 304
" " under Pitts. C., Greene County............. 384
" " in S. Beaver County, 344, 346
" " in L. Bar. Meas 384
" " under Waynesburg C. 384
" " under Lime. V....... 385
" " Pursley Cr., analysed 385
Iron shales, S. E. limit of.... 132
Iron trade of Ohio river 10
Irwin & Rishter's C. M....... 301
Irwin's C. Mine.............. 332
Jackson's C. Mine............ 204
Jackson's Mill................ 274
Jackson's run C. M 319
Jackson T., G. Co., geology... 161
" depths of coal beds.. 374
Jacksonville.................. 169
James' (J.) C. Mine.......... 290
James' (C.) coal analysed.... 378
Jeffries' (A.) C. Mine........ 205
Jefferson coals................ 138
Jefferson terrace.............. 12
Jefferson T., All. Co., geology, 296
Jefferson T., W. Co., geology, 282
" depths of coal beds.. 365
Jefferson T., G. Co., geology.. 135
" depths of coal beds.. 371
Jeffry's C. Mine.............. 288
Johnson's C. Mine............ 348
" Morris T 190
" Jackson T......... 161

Page.
Johnson's Hotel.............. 230
Johnson's (M.) lime.......... 156
John's run.................. 146
John's (Scott) C. Mine....... 93
John's (A.) mill section...... 105
Joint's vertical............... 310
Jolleytown coal; described... 28, 48
Jolleytown coals... 44, 104, 107, 237, 253
" at Jolleytown...... 111
" at Cross Creek.... 291
" in Gilmore T...... 109
" in Centre T........ 152, 154
" in Franklin T..... 251, 252
" in Richhill T...... 171
" in Nottingham.... 224
" in Morgan T....... 145
" in Smith T........ 281
" in W. Finley T.... 199
" in Wayne T....... 108
Jones' saw mill.............. 298
Jones' Coal Mine......216, 285, 292, 307
Jones & Laughlin's Iron Wks. 307
Jones (Prof.) fossils.......... 238
Jumps in Pitts. C........... 299
Kammerer's C. M........... 224
Karr's run.................. 271
Keeling (J) & Co.'s C. M. sect., 308
Keener's C. Mine............ 123
Keifer's Mill................ 32
Kelly's (J.) coal............. 246
Kennedy (S.)................ 222
Kent's (E.) C. M............. 107, 150
Kerl's Mill................... 132
Key's C. Mine............... 141, 168
Kiddoo's C. Mine............ 304
Kiefer's Mill.............. 335, 337
King's creek................ 337
Kirby's school house......... 124
Kittanning C. described...... 84, 87
" " Ohio river..... 344
Kittanning C. in Raccoon T., Beaver county............. 346
Kittanning C. in Greene T., Beaver county............. 346
Knott's C. Mine............. 99, 123
Knight's ore analyzed........ 385
" (Josh.) lime........ 153
Krepp's C. Mine............ 202
Krepp's Knob; erosion....... 20
" " section....... 201
Page.
Lamellibranchs in Great L... 64
" over Redstone C.... 236
" over Limestone III, 225
Lantz (H.).................. 111, 124
Lantz's steam mill........... 107
Large's C. Mine.............. 297
Laurel run section........... 113
" Coal Mine.......... 271
Laughlin's C. M............. 307
Lawton's C. Mine............ 286
Law's (T. R.) C. M.......... 293
Lea, (R.) & Bros., C M....... 315
Leaf impressions............. 54, 59
Leadbeater & Co.'s C. M...... 203
Lead ore..................... 386
Leeper's C. Mine............ 336
Leggitt's run............ 319
Letherman's (D. M.) C. M... 184
Lewis' run section........... 297, 299
Lewis' C. Mine.............. 321
Leyda, (Mr.)................ 28
Lick run.................. 28, 296, 305
Liddell's C. analyzed........ 380
Lightner's (S.) coal.......... 160
Lgnite in sand.............. 22
" in terrace gravel...... 13
Lilly's run C. M............. 202
Limetown.................... 220
Limestone soil............... 6
Limestone pebbles; fragments, 13, 251, 263, 171, 172.
Limestone vanishes.......... 115
" of L. B. M., variable.... 78
" local at Mansfield........ 78
" crinoidal; in W. Va.... 79, 272, 82
" under Pitts. C. described, 77
" over Pitts. C.; soda...... 203
" Waynesburg........... 145
" series above Wash. Co.. 174
" economic geology....... 387
Limestone I *a*............... 124, 55
" I *b*, II, III, 55, 166, 188, 237
Limestone II, 143, 185, 230, 240, 246, 250, 250, 251, 263.
Limestone II, III........ 141, 182, 225
" II, III IV, V, VI, 189, 190, 250
Limestone III, 50, 124, 125, 133, 150, 151, 192, 196, 264.
Limestone IV....167, 221, 241, 256, 269
" very thick in Cross T.... 290
Limestone IV and VI........ 200, 237

Page.
Limestone V described, 48, 124, 125, 140, 145, 172.
" mistaken for iron ore... 386
Limestone V, VI......... 146, 150, 161
Limestone VI described, 45, 151, 153, 166, 186, 192, 228, 243, 246, 282, 286, 294
Limestone VII described, 43, 154, 156
Limestone VIII described.... 43, 151
" VIII, IX, X...... 158
Limestone IX *a* described, 43, 110, 158, 163, 192.
Limestone IX *b* described.... 42
Limestone X described, 41, 110, 112, 114, 135, 145, 146, 162, 170, 173, 188, 193
Limestone XI, *a*, *b*....... 40, 155, 160
Limestone XII described 29
Limestone XIII, XIV, Centre T........................ 159
Lindsay & Co's Glass works.. 314
Lindley's mills lime......... 190
Lippincott's coal analysed.... 377
Lisbon (Perryville) synclinal, 26
Little's (Moses) L. IV, VI.... 250
Little run.................. 153
Little creek; rocks........ 182, 185, 187
Little Daniel's run........... 185
Little Mill creek............. 346
Little Piney creek............ 304
Little Traverse creek......... 337
Little Whitely Cr. coals...... 120, 127
Little Coal Gilmore T........ 111
Little Pittsburg Coal, 77, 116, 178, 206, 208, 209.
Little Washington Coal, 44, 54, 188, 190, 239, 245, 254, 256, 262.
Little L. above Wash. C...... 190
Local L. under Wayne. C.... 62
Lock No. 4 section............ 209, 210
Logston's (J.) C. Mine....... 167
Logstown run.............. 32, 340, 341
Lone Star oil well........... 101
Long run.................... 53
Long's (G. T.) C. Mine....... 103
Lophophyllum proliferum.... 80
Lough's C. Mine............ 165
L. Bar. Meas. section......... 272, 335
Lower Prod. C. Measures.... 23, 84
Lower Prod. C. M. section.... 116
Lower Prod. C. M. in well... 179
Lower Freeport C.......... 84, 86, 343

Page.
Lower Barren Measures, 23, 75, 82, 83, 116.
"Lower Bottom" Pitts. C 71
Lower Pittsburg S. S......... 393
Lower Washington L. (II)... 44, 50
Lower St. Clair T. geology... 307
Lucas' (A.) C. Mine......... 92
Lucas' (I. T.) C. analysed.... 379
Lucas' Fork coals............ 100
Lynn, Wood & Co.'s C. M.... 299
Lyon's C. Mine.............. 168, 238
Lytle's (J.) C. Mine......... 221
Magee's coal analysed........ 381
Magee's C. Mine.............. 294
Mahoning S. S. described.... 84
Mahoning S. S...... 76, 81, 179, 335, 389
" " on Logstown run, 340
" " in Moon T....... 342
" " on Raccon Cr..... 338
" " in Raccoon T..... 348
Mahoning S. S. sinks under Ohio river................. 340
Mahoning SS. oil=2d Oil Sand, 102, 391
Main clay No. 4.............. 100
Manchester's C. Mine........ 294
Manor's C. Mine.............. 346
Mansfield.................... 313
Mansfield coal; local L.; synclinal...................... 77, 78, 31
Mansfield (Mrs.)............ 344
Map of Greene County...... 351
Map of Washington Co...... 351
Maple Farm coal analysed... 380
Maple (Oil) Farm............ 101
Maple's Woolen Mill......... 94
Mapletown; coals............ 120, 121
Mapleton coal analysed...... 379
Maretta's (J.) C. M.......... 268
Martinsburg........... 26, 29, 144, 186
Mastodon remains........ .. 22
Maxwell's C. Mine........... 291
M'Bride's (A.).............. 321
McCairn's Mills.............. 331
McCarrihan's C. M........... 197
McClain's C. M.............. 224
McClane's C. M.............. 326
McClain's Mills.............. 332
McClary's (J. C.) C. M....... 129
McClay's (D.) C. M.......... 247
McClelland's (R.) C. M...... 238
" lime;—section.. 238, 195

	Page.
McClure's (L.) C. M.........	100
McClure's (Wm.) C. M......	103
McClure's C. Mine...........	134
McComb's (R.) C. M.........	228
" (T.) C. M.........	228
McConnell's C. Mine.........	293
" Greene Co. Map,	351
McCormack's (J.) C. M......	330
McCortney's Fork; section..	3, 156, 37
McCree's C. Mine............	133
" mill................	140
McCrury's (H. & A.) C. M...	185
McCullough's C. M...........	332, 344
McDonald's station...........	230, 271
McDonald's shaft down to Kittanning Coal............	341
McGinnis' (T.) section.......	182
McGowan's C. Mine..........	300
McGuire's run section........	294, 295
McIlhenny's C. M............	298, 301
McCann's Ferry section......	126
McKeag's mill................	199
McKee (Mrs.)................	246
McKee's C. Mine.............	300, 323
McKee's rocks (S. S.)........	81
McKeesport Furnace.........	302
McKeever's C. M.; section...	291
McKinley (A.)..............	24
McLaughlin's run............	304
McLoney.....................	227
McMahon's (H.) C. M........	213
McMillan's C. M.............	215, 342
" (I.) L...........	252
" tannery section ..	223
McNary's coal...............	241
McWhorter's (W.) L........	198
Meadow run.................	97
Mechanicsburg...............	344
Meek's run..................	331
Metamorphic rocks absent from the southern drift.....	18
Methodist Churches......	124, 161, 225
Middle Wash. L. IV, 44, 48, 247, 261, 281	
Middle Wheeling creek......	198
Middletown; terraces......	330, 15, 17
Midway; Block Coal Works,	281, 276
Mifflin T., All. Co., geology ..	399
Mill run; section.............	54, 166
Mill creek....................	337 346
Millers (D.) C. M...........	92
" " coal analysed ...	330
Miller's (J. D.) C. M.........	207
Miller's (J. H.) C. M.........	270
Miller's C. Mine..............	296, 297
Miller's creek	319
Miller's run..................	229
Millsborough; section.....	12, 176, 177
Millsburg	139
Mingo creek.......	28, 213, 220, 222, 223
Mingo Colliery shaft.........	212, 213
Mineral waters, Beav. Co	337
Minor's distillery	120
Minor's (L. L.) C. Mine......	139
Minor's (S.) C. Mine.........	123
Minor's (J. J.) coal analysed,	377
Mixture of N. and S. drift ...	19
Moffatt's Mill.............	85, 342, 345
Moffatt's C. Mine............	205
Moffatt's (Lee) C. M.........	164
Mohawk run.................	319
Moninger's (J. A.) C. M	185
" coal analysed	378
Monongahela waters.........	3
Monongahela avenue to market	9
Monongahela City terraces...	14
" section; coal,	210, 211, 212
Monongahela T. geology	115
" depths of coal beds,	369
Monterey	306
Montour's run............	321, 323, 329
" " terrace........	15
" " and church...	327
" " Presb'n church	320
Moon run	328
Moon T., Beav. Co., geology..	341
Moon T., All. Co., geology...	329
Morgan T., G. Co., geology...	140
" depths of coal beds,	371
Morgan's C. Mine............	292
Morgantown anticlinal	25
Morgantown S. S...........	76, 78, 337
" Greene T., Beav. Co.,	346
" Raccoon Creek......	338
" Raccoon T., B. Co...	344
Morgantown S. S.=First Oil	
Morris run coals..............	103
Morris T., G. Co., geology....	159
" depths of coal beds..	372
Morris T., W. Co., geology...	189
" depths of coal beds..	353
Morrison's C. Mine...........	221

	Page.
Mosier (J. B.)	224
Mt. Lebanon P. O	313
Mt. Morris	103
Mt. Pleasant T., geology	267
' depths of coal beds.	364
Mountz' (J.) L. VI	253
Mowl's (S.) C. Mine	185
Mud Lick	165
Muddy Creek terraces	12, 18
Muddy Creek section; coals, 125,129,134.	
Munntown	226
Murdock's Store	124
Murphy Run gas-coal, W. Va.	382
Murray's (Levi) C. M	165
Myers' (H.) C. M	218
Myers' (J.) C. M	219
Myers' C. M.; section	236
Narrow gauge R. R	9
National Pike; section	180, 181, 187
National road ballast	46, 262
Nautilus occidentalis	80
Neal's (W.) C. Mine	302
Neel's (J. S.) C. Mine	202
Negro run; section	347, 110
Nesbit's C. Mine	320
Neuropteris	47, 59, 106, 164, 205, 242
New bridge at Pittsburg	345
New Eagle Colliery	212
' " Coal analysed	380
New Freeport	41, 112
New Sheffield terrace	16
Newtown in Whitely	124
Nichols' C. Mine	184, 326
Nies's C. Mine	306
Nimick's Station; section	311, 312
Nineveh Sandstone	81, 159
Nineveh	29, 40, 161
" coals	35
Nineveh C. in Gilmore	110
" " in Springhill	112
" " in Centre	155
" " in Jackson	162
" " in Aleppo	164
" " in Norris	192
" " in E. Finley	193
" " economic geology	376
Noble's (A.) C. Mine	134
Noble's mill	259
Noblestown	320, 323
Nodular iron ore	14, 386

	Page.
Nolte's C. Mine	330
N. Fayette, T. All. Co., geol.	320
North Fork Dunkard	4
N. Fork Campbell's run	326
N. Fork Chartiers Cr	238
N. Fork Ten-Mile run	181
North Fork terraces	12
North Mansfield	323
North Star P. O	271, 331, 322
N. Strabane T. geology	235
" " depths of coal beds,	361
Nottingham T. geology	223
" " depths of coal beds,	361
Number XII Conglomerate	23
Oakdale	320, 323
O'Connor's (J.) C. Mine	263, 264
Ohio River waters; bed	4, 19
Oil borings, Dunkard Cr	23, 125
" " in Greene T. G., Co.,	122
" " on Pumpkin run	135
" " at Forks of Pigeon,	218
Oil Centre, Dunkard	94, 101
Oil from Shale, Beaver Co	349
Oil field, W. Virginia	393, 394
Oil Break	393
Oil not from coal	395
Oil area limited to disturbance	394
Oil sands; 1st O. S.	391, 79, 90, 178
Oil well	122
Oil wells on Dunkard Cr	100
" " on Big Mill Cr	349
Oil at Blackville	108
" in Cumberland T.	127
Old Glass Works ravine sect.,	116
Oliphant-blue-lump ore	116
Olney's C. Mine	338
O'Neill's Colliery	299
Ormsby Coal Works	308
Ormsby Station	307
Orndorf (Jesse)	155
Orr's (S. C.) C. M	142, 143
Owen's run; section	30, 166
Owen's C. Mine	113
Panhandle country.	3
Panhandle R. R. coals, 8, 264, 271, 276, 323; section	309
Panthar run	317
Paris	273
Parker's (D.) C. Mine	127
Parkersburg, W. Va	45
Partings in coal variable	132, 147, 180

Page.
Patterson's coal analysed..... 344, 381
Patterson's (J.) C. Mine...... 226
Patterson's (W.) C. Mine.... 287
Patterson's (L.) Lime........ 287
Patterson's Mill section...... 32, 286
Patterson's run.............. 301
Patterson's (J.) section...... 200
Pecopteris.................. 59, 164
Penn gas coal................ 382
Pennsylvania Lead Works... 313, 323
Pepper and Salt rock (Mah. SS.)...................... 85
Perry T., G. Co., geology...... 102
" " depths of coal beds, 367
Perryville synclinal.......... 26
Pestle's (J.) knob........... 109
Peters' cannel oil............ 348, 349
Peters' C. Mine.............. 347
Peters creek.............3, 221, 225, 299
" " anticlinal......... 29, 298, 299
" " terrace.............. 14
" " in Union T.......... 220
" " in Jefferson T........ 296
Peters T., W. Co., geology.... 226
" " depths of coal beds.. 361
Petroleum, economic geology, 390
Pettits' (J.) oil boring........ 156
Phillips' (L.) C. M........... 225
Phillipsburg........... 86, 87, 343 344
" C. Mine........... 342
" terrace........... 16
Physical features of district.. 1
Piedmont SS.=3d oil S....... 179, 391
Pigeon creek........... 3, 27, 181. 211
Pigeon Creek mines......... 217
Pike run.................... 204
Pine run.................... 28
Pine run, All, Co., C. M..... 299, 306
Piney Fork.................. 28, 304
Pinhook Village, coals....... 60, 182
Pinhook anticlinal... 146, 181, 185, 299
" " in Amwell T.......... 188
" " in Morris T.......... 189
" " in Nottingham........ 223
" " in Peters T........... 226
" " in Franklin........... 247
" " in Alleghany Co...... 296
" " in Mifflin T.......... 301
" " at Braddock's Station, 303
" " traced on the map..... 27
Pinhook C.=Wayne. C....... 191

Page.
Pinnularia.................. 59
Pittsburg Coal bed described, 69
" " economic geology, 380
" " roof variable..... 208, 309
Pittsburg Coal bed........... 57, 206
" " at various heights... 26
" " Braddock's Sta. (235′) 301
" " Sheffield S. W. (400′) 310
" " Peters' Creek (300′), 299
" " dips 6°.............. 393
" " depths beneath surface............... 350
" " remarkable section (Russell's)......... 277
" " reached once to Lake Erie............... 272
" " in Ohio; coal analysed 208, 381
" " in West Virginia.... 72
Pittsburg Coal-marsh........ 137
Pitts. C. on Dunkard Cr...... 93, 94
" " at Millsburg........... 139
" " in E. Bethlehem....... 175, 179
" " in E. Pike Run........ 202
" " at Belvernon.......... 208
" " in Union T............ 220
" " on Pigeon Cr. (shaft).. 218
" " on Peters' Cr.......... 228
" " in Chartier T.......... 232
" " S. Strabane............ 240
" " in Franklin T......... 247
" " in Mt. Pleasant........ 268, 270
" " in Robison T........... 271
" " at Eldersville.......... 283
" " in Cross Cr. T.......... 285
" " in Jefferson, All. Co... 296
" " at McKeesport......... 302
" " at Ormsby Stat........ 307
" " in Scott T............. 313
" " in U. St. Clair......... 316
" " in S. Fayette.......... 318
" " in N. Fayette.......... 320
" " in Moon T............. 329
" " in Findlay, All. Co.... 313
" " in Beaver Co.......... 335
" " in Hopewell, Bea. Co.. 339
" " 200′ below Pres. Ch., Mifflin............... 300
" " North outcrop, Buffalo, 272
" " N. W. " Hanover, 273
" " shafted for by mistake at Prosperity........ 191

Page.
Pittsburg-Waynesburg interval falls from 400′ to 225′.... 228
Pittsburg Upper SS........... 57
Pitts. and Fort Wayne R. R.. 311
Plant-bearing shales....... 39, 110, 152
" " of Washington Co... 129
" " in Sewickley C...... 176
" under Waynesburg SS.. 58
" in Lime. X.............. 164
" in Independence T...... 295
Plastic clay (see clay)........ 14
Pleasant Grove sect. of L..... 194
Pleasant Valley............ 28, 182, 185
Pleurotomaria grayvilliensis, 346
Pleurotomaria turbinella..... 346
Plum run.................... 176, 234
Plummer's C. Mine.......... 285
Plumsock.................... 286
Poe's C. Mine........... 347
Point Lookout............ 250, 251, 252
Pollock's (S.) coal............ 238
Polyphemopsis peracutus..... 346
Pomeroy & Treat's Wash. Co. map....................... 351
Poorhouse terrace........... 14
Possum run.................. 8, 218
Potters' clays under Kitt. C.. 389
Potter's (H.) C. Mine........ 304
Powers' (S.) lime............ 198
Prejudice against U. Mon. C.. 382
Presbyterian churches 228, 237, 267, 300
Price's (J.) C. Mine.......... 137
Primrose Station............... 270
Pringle's (J. S.) C. Mine..... 202
Productions of the district.... 6
Producti very large, Raccoon, 344
Productus Nebrascensis....... 80
P. Pratteniauus.............. 80
P. longispinus................ 346, 80
P. semireticulatus............ 80
Prosperity...... .... 161, 191
Pumpkin run; coals; terrace, 12, 137, 135.
Purman's run; coals.......... 51, 149
Pursley Cr. section and ore... 152, 385
Pyrites in Pitts. Coal......... 72, 132
Quail's (D.) section.......... 237
Quarries; Mah. SS........... 342, 345
Quarternary *fossils*........... 22
Raccoon Cr............ 2, 4, 88, 272, 275
" terraces............... 16
" Morgantown SS....... 338
Raccoon Station.............. 278
Raccoon T., Beaver Co., geol. 344
Rainey (J).................. 225
Rainfall of the district........ 8
Ralston (Mr.)................ 267
Rankin's School H. Sect..... 269
Rea's (W.) C, Mine.......... 286
Redd's (T.) C. Mine.......... 214
Redd's (T.) coal analysed.... 380
Redman's C. Mine........... 303
Red Shales of U. B. M....... 159
Redstone C. bed, 57, 92, 136, 179, 203, 206, 209, 211, 213, 214, 236.
" " described............ 67
" " economic geology.... 379
" " in Dunkard T......... 98
" " in Monongahela..... 117
" " in E. Bethlehem.... 177
" " in Union T.......... 220
" " in Nottingham....... 223, 225
" " in Peters T........... 228
" " in Chartiers T........ 235
" " in N. Strabane....... 236
" " in S. Strabane....... 240
" " in Mt. Pleasant...... 268
" " in U. St. Clair....... 317
Reed's (J. M. K.) C. M...... 288
Reed & Co.'s (J.) C. M....... 202
Rees' (W. & T.) C. M........ 238
Rees' mill..................... 28
Register & Bair's Salt W..... 178
Riley's run.................. 306
Remington; black shale.. 32, 326, 327
Resources of the District..... 376
Rex's (J.) C. Mines.......... 138
Rhizodus, in L. III........... 50, 225
Rice's Landing............... 134
Richardson's C. M............ 180, 216
Richhill T., G. Co., geology.. 165
" " depths of coal beds, 374
Ridge road................... 133, 234
Rimmel's (S.) C. M.......... 316
Ripple marks................. 200
Risher....................... 301
River beds ancient........... 19
Robbin's Block C. Co......... 270
Roberts' run, Wayne........ 108
Robeson T., W. Co., geology, 271
Robeson's run...... 4, 271, 320, 322, 326
Robinson's run.............. 197
Robinson T., All. Co., geology, 323
Robinson's Fork............. 194

	Page.
Robinson Bros.' C. Wks	210
Rockwell's C. Mine	337
Rocky Hole run	298
Rocky Ridge SS. blocks	228
Rocky run	297
Rogers (H. D.)	86
Rogers' (D. W.) C. M	143
Rogers' (S. W.) C. M	207
" coal analysed	377
Rogersville; L. V	3, 152; 48
Rolled stones	11
Roney's Point mills	61, 62
Roof of Pitts. C. variable	208
" " " well shown,	316
Ross (B.)	92, 146
Ross Farm Oil wells	102, 393
Ross Oil well, No. 1	102
R. R. tunnel section	314
Rudolph's run	108
Ruff's creek; coals	1, 166, 144
Russell's C. Mine	277
Russell's Pitts. C. section	277
Rutan P. O., sandstone	42, 158
Ryerson's Station	165, 170
Salt	127
" in Waynesburg SS	58
" in E. Bethlehem	178
" in Franklin T	248
Sand Run, Franklin T	151
Sand deposits	390
SS. under U. Wash. L	291, 289
SS. massive (U. B. M.)	107
SS. bluffs in Morris	192
SS. bluffs in Robinson	325
SS. blocks on hill-tops	184
" Rocky Ridge, Peters T.,	228
SS. disappears	325
Sanders' (J.) Lime	189
Sargent's mill	159, 192
Saw mill run section	215, 305
" " in Cecil T	230
Sayer's (T.) C. Mine	147
" " analyses	377
School House Prosperity	191
" " No. 2, W. Finley,	198
" " " Somerset	217
" " " Buffalo T	253
" " No. 3	217
" " No. 5, Cross Cr. T,	290
" " No. 7	219
Schriver (J)	125
Scott T., All. Co. geology	313
Scott's hollow	211
Scott's run; coal	326, 61
Scott's (J.) coal	238
Scott's (J.) C. M	163
Scott's (Jos.) C. M	158, 217
Second Terrace described	18
Second Oil Sand=Mah. SS	392
Section at Belvernon	13
Sect. of L. Prod. C. Meas	84
Sect, of L. Bar. Measures	75
Sect. of U. Prod. C. Meas	57
Sect. of Wash. Co. Group	44
" " " in W. Va	45
Sect. of Greene Co. Group	35
Sect. of Wash. C. bed	52
Seller's (C.) C. M	207
Service Cr	5, 338, 344
Sewickley Cr., 57, 92, 134, 179, 204, 206, 209, 212, 268.	
Sewickley C. absent	208
" " in oil borings	218
" " exceptionally high	122
" " economic geology	379
" " bed described	65
" " bed good	117
" " bed sub-divided	119, 121
" " in Dunkard T	93, 96, 98
" " in Cumberland	130
" " on S. Fork 10 M	140
" " in E. Bethlehem	176
" " in Cecil T	231
" " in S. Strabane	240
" " in Franklin T	247
" " in Mt. Pleasant	270
" " in Hanover T	274
" " in Smith T	279
" " in Independence	294
" " in Scott T	313
Shale Veins vertical	54
Shaner's (Adam) C. M	205
Shaner's (D.) C. M	217
Shannon's run	104
Shape's C. Mine	137
Sharon	329, 330, 332
Sharon Furnace	276
Sheep raising	6
Sheffield Steel Works	310
Shells in Limestone	116
Shippingport	347
Shirland P. office	322
Short Creek, Morris	191
Shough's (P.) Knob	109

	Page.
Shriver's (J.) Lime..........	125
Sickel	197
Siebold's C. Mine	304
Sigillaria Menardi	47,152,242
Silver ore	337
Simpson's (Dr.) L. VI.......	193
Simpson's C. Mine	288
Six Mile ferry C. M	29,303
Skile (J. P.)	227
Skile's (S.) C. M.............	234
Slack of the mine coked	147
Slackwater Improvement....	9
Slatt's Mill..................	43
Slickensides in Wayne. C....	131
Slusher's (S. L.) L. II.......	186
Smith T., W. C., geology	275
" " depths of coal beds,	366
Smith (M)	204
Smith's C. Mine ..147, 164 292, 293,	326
Smith's Cr.; coals; ore....4; 150;	384
" " Franklin T........	151
Smith's run	93
Snider's C. Mine..............	341
Snowdon T., All. Co., geology	303
Sparta SS. bluffs	192
Spaulding's (U. F.) C. M.....	340,341
Speer's (Noah) grist mill	24
" " C. Mine......	208
Speer's (Alex.) C. M.........	328
Spenophyllum	59
" in Pitts Co. roof,	205
Sphenopteris................	59
Spines and plates of crinoids,	303
Spirifer cameratus...........	80,346
Spirifer planoconvexus.......	80
Spirifer opinus..............	346
Spirifer lineatus.............	346
Spirophyton	95
Spragg (C.) & (W.)..........	108
Spragg's (D.) L.......	108
Springer's C. Mine...........	345
Springhill T., G. Co., geol....	111
" " depths of coal beds,	369
Springs at Frankfort.........	337
Soda stone, limestone	203
Sodom.......................	315
Soil of the district...........	6
" " whence derived,	17
Somerset T., W. Co., geology,	216
" " depths of coal beds,	357
S. Fayette T., All. Co. geol...	317

	Page.
South Fork: coals............	3,111
" " of Ten Mile run.....	4,140
" " " " section...	140
" " of Wheeling Cr	167
" " of Cambell's run....	326
" " of Cross Cr.........	186
" " of Fish Cr..........	114
" " of King's Cr........	274
South & Keener's C. M.......	123
S. Strabane T., W. Co., geol...	239
" " depths of coal beds,	358
South's (J.) C. M	98
Stacher's C. Mine............	215
Stall's mill L................	172
Steubenville..................	32
" pike; section, 320, 326;	274
" river bed..........	19
Stevenson's (D. C.) C. M.....	126,131
Stevenson's (D. C.) Oil bore..	127
Steward's (W.) C. M	141,143
Stewart's (S.) C. M	320
Stewart's (J. M.) C. M	333
Stewart's (W. B.) C. M	328
Still school house	328
Stockton's (J.) C. M	287
Stone's C. Mine	300
Stratified rocks..............	23
Street's run	301,306
" " mouth section	305
" " in Mifflin.........	299
Strip-Vein C. described	86
" " —Middle Freeport C.	84,87
" " economic geology....	383
Strosnider's (Dr.) Well section	108
Sub-soil inexhaustible	6
Sugar run section	293
Sulphur in Pitts. C...........	282
" vein in Kitt. C	348
Summary of resources.......	376
Surface topography..........	1
Surface geology.............	11
Surface debris...............	351
Swamps in Pitts. C... 299, 302, 314,	320
Swaney's C. Mine............	347
Swearingen's C. Works	85
Swearingen's C. Mine........	347
" " coal analysed..	383
Syck's Knob on Dunkard....	94
Syckman's	298
Synclinal	124

	Page.
Synclinal at Bentleyville	216
" at Coon's Island	261
" at Kiefer's Mill	335
" on Peter's run	299
Tables of depth of coal beneath the surface of the district	350
Taggert's C. Mine	246
Taylor (John)	41
Taylor's (J. E.) C. M	93
Taylor's (A.) C. M	111
Taylor's (J.) C. M	110
Taylor's (A.) Nod. L	110
Taylor's (J. L.) C. M	205
Taylor's saw mill	205
Taylortown; long section, 254,256,258, 259.	
Teeple's (Isaac) C. M	211
" coal analysed	379
Temperanceville	30, 310
Temple's C. M	340
Templeton's run	196
Ten-Mile village	53
Ten-Mile creek	1, 2, 3
" " E. Bethlehem	175
" " terraces	12, 18
" " run coals	135
" " Franklin T	252
" " section, Centre T	154
Terraces of the district	11
" marked by drift	17, 21
Terrace epoch	21
Tharp's (S.) C. M	187
Thickness of strata	24
Thinning of U. B. M. northward	37
Third Oil S.—Piedmont SS	392
Third Terrace described	18
Thomas' saw mill	28, 225, 228
Thomas' coal analysed	380
Thomas (Mr.)	238
Thompson's (Mrs.) C. M	206
Thompson's (T.) C. M	234
" coal analysed	381
Thompson's (J.) C. M	285
Thompson's (Jas.) C. M	291
Thompson's C. M	339
Thompson's run terrace	14
" " in All. Co	300
Thompson's saw mill	205
Thompson's station	302
Thompsonville	227
Thorn's C. Mine	347
Tice's run coals	113
Tin	387
Tionesta SS	18
Titus' (L.) C. M	82
Todd's C. M	347
" coal analysed	383
Tombaugh's (S.) C. M	184
Tom's run, Wayne	107
Tomlinson's C. M	333
Tools drop in boring	393
Top of U. C. Measures	181
Top SS. of C. Measures	184
Townsend's (E. R.) C. M	228
Townsend's (Jos.) C. M	228
Township geology	90
Tramp run	32
Trampmill run	340, 341
Traverse creek	5
Trees in gravel bed	13
Trunick's C. Mine	330
Tunnel C. at W. Eliz	298
Tunnel in Mifflin T	302
Tunnel Section, Hempfield	46
Tunnel Section, Buffalo	255
Tunnel Section, Donegal	261, 265
Tunnel Section, Canton	244
Turkey Knob erosion	20
Twin oil wells	393
Twyford's C. Mine	332
Ulery's (W. H.) C. M	181
Union T., W. Co., geology	220
" depths of coal beds,	360
Union T., All. Co., geology	310
Uniontown C. described, 62, 57, 126, 134, 203, 206, 294, 388.	
" " good	127
" " on Dunkard Cr	94, 97
" " in Monongahela	120
" " in Greene T	123
" " in E. Pike Run	202
" " in Nottingham	225
" " in Union T	222
" " in Peters T	227
" " in Chartiers	234
" " in N. Strabane	238
" " at Atchison	259
" " in Smith T	281
" " in Cross Cr. T	285
" " in Hopewell	289
" " in Independence	293

	Page.
Union Pres. Ch. at S. Strabane	242
" " Buffalo T.	255
" " Raccoon T.	345
Upper Freeport C.	340
" " No. VII in Ohio	83
" " described	85, 84
" " economic geology	382
" " in Beaver Co.	335
" " on Raccoon	339
" " in Moon T.	342
" " in Greene T.	347
" " Race T.	345
Upper bench Pitts. C.	71
Upper Pitts. SS. described	69
" " turns to shale	117
Upper Prod. C. Series range	23
" " measures described,	57
" " section, Jefferson	136
Upper Barren series	34
" " range	24
" " in Washington Co.	198
" " thin northward	37
Upper sandstones	2
Upper L. thin out in Jeff.	283
Upper Wash. (White) L. 36, 44, 45, 110, 150, 156, 239.	
" " boulders	142
" " in Centre T.	155
" " in Morris T.	159
" " in Aleppo T.	165
" " in West Bethlehem,	183
" " in Amwell	186
" " in E. Finley T.	194
" " in Nottingham	224
" " in N. Strabane	236
" " in S. Strabane	240
" " in Franklin T	251, 252
" " in Buffalo T	255
" " in Donegal T.	265
" " in Smith T.	275
" " in Cross Cr. T	291
" " and SS. Cross Cr. T.	291
Upper St. Clair T. geology	315
Valleys of the district	2
Valley school house	120
Valley Grove Station sect	266
Van Buren	189
Van Buren L. VI	251
Vanceville	28, 218
Vance's Mill	122
Vance's (S.) Oil Well	122
Vaneman's Station	231
Van Varis' (J.) C. M.	92
Van Voorhees' (J.) C. M.	217
Variable limestone of L. B. M.	78
Venice	229
Vernon's (J.) coal analysed	380
Vernon's (L.) C. M	136
Vertical joints	310
Vesuvius Man. Co. C.	135
Wagon road run section	113
Walker's (S.) C. M	205
Walker's Station; SS	325, 326, 389
Wallace's C. Mine	339
Wallace's Mill	196, 305
Walnut Fork	164
Walnut Hill C. Works	276
Walter's (S.) L. II	252
Walton's C. Mine	306
Walton & Co.'s (J.) C. M	298
Warne's (J.) C. Mine	215
Warne's (H.) C. Mine	224
Warwick (Mr.)	240
Washington anticlinals traced	30
" " in E. Finley T.	194
" " in Cecil T.	229
" " in Chartiers	232
" " in Franklin	253
" " at Sheffield S.W.	310
" " deflected in U. St. Clair,	315
Washington Co. described	2
" " geology in detail	175
Wash. Co. Group U. B. M.	44
Washington C.	44, 90, 140, 218
" " =Brownsville C	51
" " mistaken for Wayne. Coal	265
" " economic geology	376
" " in Cumberland	128
" " in Jefferson	139
" " in Morgan T.	145
" " on Rupp's Cr.	146
" " in Franklin	149, 150
" " in Centre T.	156
" " in Richhill	166, 170
" " in W. Bethlehem	181
" " in Amwell	186
" " in Morris T.	190
" " in E. Pike Run	202
" " in Somerset	220
" " in Cecil T.	229
" " in Canton T.	244, 246

Page.
Washington Coal, on Hempfield R. R 249, 250
" " in Franklin.......... 252
" " in Buffalo............ 256, 260
" " in Donegal 262
" " in Jefferson.......... 282
" " in last outcrop N 283
" " in Cross Cr........... 286
" " in Independence..... 292
W. C. 10′ thick in Indep. T... 294
W. C. *a* in Perry T........... 106
" " in Wayne............ 108
Washington SS. described ... 53
Washington L. (upper) anal., 388
Washington Section.......... 242
Washington T., G. Co., geology, 155
" " depths of coal beds, 372
Water sheds of district....... 3
Wayne T., G. Co., geology.... 106
" " depths of coal beds, 368
Waynesburg anticlinal traced, 25, 176
" " at Lock 4, 210
Waynesburg................ 206
Waynesburg C. described, 99, 119, 120, 137, 147, 180.
Waynesburg Main C......... 57
" described—XI of Ohio.. 59
" economic geology....... 376
" depths beneath surface.. 350
" 1200′ deep............... 376
" variations in bed........ 186
" only 8″ in Independence, 292
" on Dunkard............ 93, 98
" in Perry T 103
" in Jefferson............. 135
" in Richhill.............. 165
" in E. Bethlehem........ 175
" in W. Bethlehem....... 181, 183
" in Amwell............. 185
" in W. Finley........ 196, 197, 199
" in Fallowfield.......... 215
" in Somerset............. 216, 218
" in Union T.............. 222
" in Nottingham.......... 224, 226
" in Peters T............. 228
" in Cecil T............... 229, 231
" in N. Strabane.......... 238
" at Taylortown........... 256
" in Buffalo............... 260
" in Donegal.............. 266
" in Hanover............. 274, 275
" in Smith................ 281

Page.
Waynesburg in Cross Creek.. 286, 291
Waynesburg C. *a* described.. 56
" " 44, 99, 123, 126, 133, 149, 180, 185, 227, 239, 290, 307, 313.
" " at Clarksville........ 141
" " at Morgantown...... 143
" " at Martinsburg...... 145
" " in Fallowfield.. 214
" " in Somerset.......... 219
" " in Snowdon.. 303
Waynesburg C. *b*........44, 55, 169, 226
" " in shaft at Prosperity, 191
Waynesburg SS. described, 57, 129, 44, 389.
" " fragments on hill tops, 286
" " in Carroll T........... 210, 212
" " in Perry T............ 103
" " in Wayne T........... 109
" " in Morgan T.......... 143
" " in Franklin........... 147
" " in Richhill............ 168
" " W. Bethlehem........ 185
" " in Amwell........... 186
" " in Somerset........... 216, 219
" " in Nottingham........ 224
Waynesburg synclinal traced, 97
Waynesburg terrace conglom. 12
Weaver (Mrs.) section....... 182, 183
Webster..................... 210
Weirick's (S. K.) Coal 246
Welsh's (J. B.) C. M......... 181
Well......................... 178
Wells on Dunkard not cased.. 391
West Alexander section...... 264, 266
W. Bethlehem T., W. Co., geol. 181
" " depths of coal beds, 352
W. Brownsville section 201
" " coal................ 202
W. Elizabeth................ 298
W. Finley T. W. Co. geol.... 196
" " depths of coal beds, 354
W. Finley Village Section L. 196
W. Fork of Chartiers 233
W. Fork of Peter's Cr........ 298
W. Middletown............. 31, 291
W. Pike Run T., W. Co., geology........................ 205
" " depths of coal beds.. 256
W. Virginia State Line 267
" " Line section... 293
" " Oil field 394
West's C. Mine.............. 204

	Page.
West's Coal analysed.........	380
Wheat crop	7
Wheeling....................	19
Wheeling Creek............	5,161
Wheeling C=Wayne. C......	197
White Cottage	41,42,43,161
White L. VI=U. Wash......	45,150
Whitehall hotel..............	306
Whitely T., G. Co., geology..	124
" " depths of coal beds..	370
Whiteley Cr. terrace's........	11
" " section above Pitts. C.	118
" " oil..................	391
White's (W. T.) C. M........	111
White's (S.) C. Mine	113
White's (Perry) C. M...	164
White's (J.) C. Mine	204
White's (W. S.) C. M........	268
Whitfield's C. M.............	184
Whitlace's (J.) C. M.........	112
Whitmore's C. M.............	323
Whittaker's (heirs) C. M....	278
Wildman's (M.) C. M	100
Wiley's Mill	222,223
Wiley Well (oil) No. 1......	101
Wilkinson's C. Works	319
Williamsport Pike...........	50
Williamsburg Synclinal	29,30
Willow Grove................	320
Willow tree tavern	123
Wilson's C. Mine	271,322,347
" Coal analysed.......	388
Wilson's (S.) L. II	250

	Page
Wilson's saw mill	285,296
Windy Gap Church..........	164
Wise's (Amos) C. M.........	185
Withrow's blacksmith shop..	214
Wolf (E.)....................	246
Wolf's (Mich.) C. M.........	208
Wood in Sandstone...........	190
Woodrow P. Office coal......	288
Woodruff's (G) C. M.........	170
Woodruff's mill..............	173
Wood's C. Mine..............	162
Wood's mill L..............	156
Wood's run..................	31,311
" " conglomerate.....	81
" " section, Jackson...	162
Woodville Stat. Sect..........	313,314
Wool raising..................	6, 7
Work's (S.) C. Mine.........	289,291
Wright's (J.) salt well.......	178
Wykoff's C. Mine	131
Wylie's (Cephas) Oil records,	391
Wylie Well No. 2............	391
Yeador's (J.) section.........	151
Yeager (Mr.)................	108
Yellow Creek; Coal..........	383; 87
Yough. River mouth coals....	300,302
Young's (Dan.) C. Mine......	168
Zeacrinus mucrospinus.......	80
Zedeker's (J.) L.............	243
Zimmerman's L..............	125
Zinc.........................	387
Zion Church.................	92
Zollersville..................	27,181

ERRATA.

Page 18, 22d line, omit "many."
" 19, 31st line, omit "of Dunkard."
" 20, 23d line, for "covreed" read "covered."
" 23, 28th line, transpose "Coal" and "Productive."
" 25, 13th line, for "cenvenience" read "convenience."
" 25, 25th line, for "Kiskiminetas" read "Youghiogheny."
" 29, 23d line, for "Ninevah" read "Nineveh."
" 38, 10th line, insert a period after "Washington County."
" 38, 11th line, insert a comma after "Center Township."
" 41, 21st line, for "overlylng" read "overlying."
" 51, 18th line, for "Washing" read "Washington."
" 55, 18th line, for "Hemfield" read "Hempfield."
" 71, 36th line, for "midway" read "Midway."
" 72, 18th line, for "usual" read "unusual."☞
" 76, 28th line, omit "south" and after "south-east" insert "corner."
" 84, 9th line, for "Measure" read "Measures."
" 84, 20th line, omit "?" after "strip vein."
" 87, foot-note, omit the third word, "with."
" 100, 31st line, for "Dunkark" read "Dunkard."
" 125, 23d line, for "betwen" read "between."
" 135, 4th line, for "70" read "7."
" 125, 36th line, for "north" read "mouth."
" 136, 4th line, for "seen" read "seem."
" 148, 16th line, for "eleven" read "four."☞
" 151, 38th line, after "limestones" insert "VI and VII."
" 152, 27th line, for "menard" read "menardi."
" 158, 4th line, for "IX 'a'" read "IX 'b'."
" 160, 34th line, after "carbonate" insert "of iron."
" 161, 11th line, before "resemble" insert "which."
" 162, 33d line, for "IX 'a'" read "IX 'b.'"
" 162, 36th line, for "IX" read "IX 'a.'"
" 162, 37th line, for "Dunkavd" read "Dunkard."
" 214, 10th line, for "divsion" read "division."
" 223, 21st line, for "cotnty" read "county."
" 228, 30th line, insert (?) after "Redstone."
" 235, 9th and 23d lines, for "Connellsville" read "M'Connellsville."
" 237, 42d line, for limestone" read "limestones."

NOTE.—In order to conform to the system of nomenclature employed in the reports generally, Limestones I and I "a" were changed to I "a" and I "b," and Limestones IX and IX "a" to IX "a" and IX "b." The manuscript was not fully corrected to accord, and the gentleman who revised most of the proof was unaware of the change. If the reader find any cases where the figures I or IX occur, he will please insert "a" after them.

J. J. S.

www.ingramcontent.com/pod-product-compliance
Lightning Source LLC
LaVergne TN
LVHW021139110826
845150LV00005B/1065

* 9 7 8 1 4 2 5 5 4 8 9 3 3 *